AF167474

Communications
in Computer and Information Science   2033

Editorial Board Members

Joaquim Filipe , *Polytechnic Institute of Setúbal, Setúbal, Portugal*
Ashish Ghosh, *Indian Statistical Institute, Kolkata, India*
Lizhu Zhou, *Tsinghua University, Beijing, China*

Rationale

The CCIS series is devoted to the publication of proceedings of computer science conferences. Its aim is to efficiently disseminate original research results in informatics in printed and electronic form. While the focus is on publication of peer-reviewed full papers presenting mature work, inclusion of reviewed short papers reporting on work in progress is welcome, too. Besides globally relevant meetings with internationally representative program committees guaranteeing a strict peer-reviewing and paper selection process, conferences run by societies or of high regional or national relevance are also considered for publication.

Topics

The topical scope of CCIS spans the entire spectrum of informatics ranging from foundational topics in the theory of computing to information and communications science and technology and a broad variety of interdisciplinary application fields.

Information for Volume Editors and Authors

Publication in CCIS is free of charge. No royalties are paid, however, we offer registered conference participants temporary free access to the online version of the conference proceedings on SpringerLink (http://link.springer.com) by means of an http referrer from the conference website and/or a number of complimentary printed copies, as specified in the official acceptance email of the event.

CCIS proceedings can be published in time for distribution at conferences or as post-proceedings, and delivered in the form of printed books and/or electronically as USBs and/or e-content licenses for accessing proceedings at SpringerLink. Furthermore, CCIS proceedings are included in the CCIS electronic book series hosted in the SpringerLink digital library at http://link.springer.com/bookseries/7899. Conferences publishing in CCIS are allowed to use Online Conference Service (OCS) for managing the whole proceedings lifecycle (from submission and reviewing to preparing for publication) free of charge.

Publication process

The language of publication is exclusively English. Authors publishing in CCIS have to sign the Springer CCIS copyright transfer form, however, they are free to use their material published in CCIS for substantially changed, more elaborate subsequent publications elsewhere. For the preparation of the camera-ready papers/files, authors have to strictly adhere to the Springer CCIS Authors' Instructions and are strongly encouraged to use the CCIS LaTeX style files or templates.

Abstracting/Indexing

CCIS is abstracted/indexed in DBLP, Google Scholar, EI-Compendex, Mathematical Reviews, SCImago, Scopus. CCIS volumes are also submitted for the inclusion in ISI Proceedings.

How to start

To start the evaluation of your proposal for inclusion in the CCIS series, please send an e-mail to ccis@springer.com.

Miguel Mujica Mota · Paolo Scala
Editors

Simulation for a Sustainable Future

11th Congress, EUROSIM 2023
Amsterdam, The Netherlands, July 3–5, 2023
Proceedings, Part II

 Springer

Editors
Miguel Mujica Mota 🅳
Amsterdam University of Applied Sciences
Amsterdam, Noord-Holland, The Netherlands

Paolo Scala 🅳
KLM
Hoofddorp, The Netherlands

ISSN 1865-0929   ISSN 1865-0937 (electronic)
Communications in Computer and Information Science
ISBN 978-3-031-68437-1   ISBN 978-3-031-68438-8 (eBook)
https://doi.org/10.1007/978-3-031-68438-8

© The Editor(s) (if applicable) and The Author(s), under exclusive license
to Springer Nature Switzerland AG 2024

This work is subject to copyright. All rights are solely and exclusively licensed by the Publisher, whether the whole or part of the material is concerned, specifically the rights of translation, reprinting, reuse of illustrations, recitation, broadcasting, reproduction on microfilms or in any other physical way, and transmission or information storage and retrieval, electronic adaptation, computer software, or by similar or dissimilar methodology now known or hereafter developed.
The use of general descriptive names, registered names, trademarks, service marks, etc. in this publication does not imply, even in the absence of a specific statement, that such names are exempt from the relevant protective laws and regulations and therefore free for general use.
The publisher, the authors and the editors are safe to assume that the advice and information in this book are believed to be true and accurate at the date of publication. Neither the publisher nor the authors or the editors give a warranty, expressed or implied, with respect to the material contained herein or for any errors or omissions that may have been made. The publisher remains neutral with regard to jurisdictional claims in published maps and institutional affiliations.

This Springer imprint is published by the registered company Springer Nature Switzerland AG
The registered company address is: Gewerbestrasse 11, 6330 Cham, Switzerland

If disposing of this product, please recycle the paper.

Preface

The EUROSIM Congress 2023, the 11th EUROSIM Congress, was organized by the Dutch Benelux Simulation Society, in Amsterdam (DBSS) on July 3–5, 2023. EUROSIM (www.eurosim2023.eu) is the Federation of European Simulation Societies; it provides a European forum for regional and national simulation societies to promote the advancement of modeling and simulation in industry, research, and development. Under the EUROSIM umbrella, EUROSIM Member Societies and cooperating societies and groups organize conferences, produce publications on modeling and simulation, and work in standardizing or technical committees, among other activities.

This volume contains a selected number of full articles of papers presented at the congress. Each submission was reviewed by at least by 3 scientific committee members of Ph.D. level. Each paper included in this volume is at the forefront of development in its scientific area, and each was presented during the conference where there were discussions for improvement. The topics are diverse, but in all of them simulation is a key component of the solution presented.

During the conference we could witness the evolution of simulation as a tool that can be complemented with novel data science techniques like ML and optimization to develop future tools that will influence our world. Some of the applications aim at solving problems that will make the future more sustainable. The readers will find in this volume very interesting applications, solutions, and developments where simulation, or different techniques together with simulation, are applied for diverse topics ranging from medicine to transportation.

We thank all the authors for participating in the conference and the effort to shape their work to fit the requirements of this volume. Special thanks go to the International Scientific Committee which allowed us to put together all the material on time and provided valuable feedback to the authors. We also thank our sponsors Sentient Hubs, Wolfram Research, IGAMT, Incontrol, and UNAM. Finally, we thank Amsterdam University of Applied Sciences for providing the location for the congress.

July 2023

Miguel Mujica Mota
Paolo Scala

Organization

General Chair

Miguel Mujica Mota · Amsterdam University of Applied Sciences, The Netherlands

Program Committee Chairs

Arnold Heemink · Delft University of Technology, The Netherlands
Miguel Mujica Mota · Amsterdam University of Applied Sciences, The Netherlands
Paolo Scala · KLM, The Netherlands

Steering Committee

Margarita Bagamanova · Amsterdam University of Applied Sciences, The Netherlands
Arnold Heemink · Delft University of Technology, The Netherlands
Hai Xiang Lin · Delft University of Technology, The Netherlands
Miguel Mujica Mota · Amsterdam University of Applied Sciences, The Netherlands
Paolo Scala · KLM, The Netherlands

Program Committee

Margarita Bagamanova · Amsterdam University of Applied Sciences, The Netherlands
Abdel El Makhloufi · Amsterdam University of Applied Sciences, The Netherlands
Idalia Flores de la Mota · National Autonomous University of Mexico, Mexico
Arnold Heemink · Delft University of Technology, The Netherlands
Gieta Inderdjiet · Amsterdam University of Applied Sciences, The Netherlands
Hai Xiang Lin · Delft University of Technology, The Netherlands

Miguel Mujica Mota	Amsterdam University of Applied Sciences, The Netherlands
Alejandro Murrieta Mendoza	Amsterdam University of Applied Sciences, The Netherlands
Paolo Scala	KLM, The Netherlands

Additional Reviewers

Sameer Alam	Mercedes Perez
David Al-Dabass	Miquel Angel Piera
Asteris Apostolidis	Edgar Possani
Margarita Bagamanova	David Post
Sahil Belsare	Blaž Rodič
Agostino Bruzzone	Rodrigo Romero Silva
Roman Buil	Henrik Rothe
Erik Dahlquist	Christina Rott
Mohammad Dehghani	Cristina Ruiz
Bruno Desart	Yuri Senichenkov
Abdel El Makhloufi	Michael Schultz
Javier Faulin	Konstantinos Stamoulis
Luis Manuel Fernández-Ahumada	Gary Tan
Idalia Flores de la Mota	Susana C. Tellez
Jorge Luis Garcia-Alcaraz	Ricardo Torres
Egils Ginters	Fausto Vieira
Edmond Hajrizi	Edward Williams
Arnold Heemink	Richard Wu
Glenn Jenkins	Borut Zupančič
Emilio Jimenez	Raid Al-Aomar
Esko Juuso	Goran Andonovski
Gorazd Karer	Benedikt Badanik
Juan Ignacio Latorre	Markus Bans
Bernt Lie	Felix Breitenecker
Vito Logar	Branko Bubalo
Edgar Lopez-Rojas	Jordi Casas
José Antonio Marmolejo	Fabio De Felice
Marina Massei	Daniel Delahaye
Peter Meincke	Alejandro Di Bernardi
Miguel Mujica Mota	Rosario Enriquez
Gašper Mušič	Roberto Felix Patron
Allan Nõmmik	Jaume Figueras i Jové
Jorge Ochoa	Luiz Augusto Franzese
Stephan Onggo	Amir Ghasemi
Jose Maria Ortiz-Gomez	Antoni Guasch Petit
Antonio Padovano	Cathal Heavey

David Martin Herold
Edgar Jimenez
Angel Juan
Sofia Kalakou
Andreas Korner
Sanja Lazarova-Molnar
Hai Xiang Lin
Francesco Longo
David Lucido
Luis Martin-Domingo
Rina Mary Mazza
Yuri Merkurev
Alejandro Murrieta Mendoza
Letizia Nicoletti
Jenaro Nosedal
Elias Olivares Benitez
Angel Orozco
Taha Osman
Maria Papanikou

Antonella Petrillo
Jorge Pinho de Sousa
Nikki Popper
Juan Jose Ramos
Alicia Roman-Alonso
Kristjan Roosipold
Olivier Rose
Mario Ruz
Raquel Salamanca
Paolo Scala
Joh Shortle
Adriano Solis
Artis Teilans
Ali Torabi
Ashish Verma
Gabriel Wainer
Bogusz Wisnicki
Alejandro Zarate Perez

Sponsors

Contents – Part II

Logistics and Transportation Systems

Monitor, Control, and Theoretical Systems

Contents – Part I

Production Systems

Business and Industries

Error-Model Predictive Control
of Wheeled Mobile Robots
for Minimum-Time Trajectory Tracking

Martina Benko Loknar$^{(\boxtimes)}$ (ID), Andrej Zdešar (ID), Sašo Blažič (ID),
and Igor Škrjanc (ID)

Faculty of Electrical Engineering, University of Ljubljana,
Tržaška 25, 1000 Ljubljana, Slovenia
martina.benkoloknar@fs.uni-lj.si, {andrej.zdesar,saso.blazic,
iigor.skrjanc}@fe.uni-lj.si

Abstract. In this paper, we propose an error-based four state kinematic model of a wheeled mobile robot with tricycle drive for trajectory tracking control. The trajectory tracking algorithm was developed
for a discrete system and implemented using a model predictive control (MPC) approach. The objective function of the MPC is minimized
with particle swarm optimization (PSO). The minimum-time trajectory
used in the experiments satisfies velocity, acceleration, and jerk constraints, which we used to indirectly describe the dynamic properties
of the wheeled mobile robot (WMR). Simulation results showed robust
performance in the presence of various non-ideal conditions, such as measurement noise, delays, and constrained control velocities. Consequently,
we were also able to apply the approaches from the simulations to a
real robot platform to confirm the real-time applicability. Our proposed
control algorithm and trajectory generation approach are well suited for
automated guided vehicles (AGVs) used in logistics in industrial environments, where efficient operation depends on minimizing travel time.

Keywords: intelligent control · model predictive control · particle
swarm optimization · trajectory-tracking · automated guided vehicles ·
jerk constraints

1 Introduction

In recent decades, the field of autonomous robotic systems has attracted great
interest and made remarkable progress. The increasing demands of international
trade, e-commerce, and labor shortages in the logistics industry have sparked
commercial interest in automated guided vehicles (AGVs) and autonomous
mobile robots (AMRs). AGVs, in particular, have become a highly sought-after

Supported by Slovenian Research Agency under Grant P2-0219.

© The Author(s), under exclusive license to Springer Nature Switzerland AG 2024
M. Mujica Mota and P. Scala (Eds.): EUROSIM 2023, CCIS 2033, pp. 3–18, 2024.
https://doi.org/10.1007/978-3-031-68438-8_1

technology due to their ability to efficiently handle material supply and transportation in modern manufacturing, warehousing, and supply operations. Their features such as high maneuverability, agility, low cost, autonomy and flexibility have made them an attractive solution in the industry [19,20]. The movement of WMRs that have non-holonomic constraints, can be accomplished by start-to-goal pose control or by trajectory tracking control. According to Brockett's theorem, non-holonomic systems cannot be stabilized to a declared posture by differential or continuous time-invariant feedback, and for a driftless system to be stabilized by continuous time-invariant feedback, the number of control inputs must be greater or equal than the number of state variables [6]. To circumvent the constraints imposed by Brockett's conditions, the control action is very often split into two control parts: feedforward and feedback [13]. The goal of trajectory tracking control is generally to determine control inputs to guide the mobile robot along a predefined path, and can be divided into three tasks: *point-to-point motion* (also: *point stabilization*) [3,27], *path following* [9,12], and *trajectory tracking* [11,14,16]. The definition of trajectory tracking is that a wheeled mobile robot reaches a predefined path and travels along it while maintaining a desired velocity profile.

In the literature there are many examples of various control strategies that were proposed for trajectory tracking of AGVs. The authors in [2] divided these types of controllers in the following classes: classical [7], geometric and kinematic [24], dynamic, optimal [1], adaptive [8], and model based controllers. MPC is one of the most widely used control methods. It is an advanced control method otherwise known as moving/receding horizon control [17,22,23]. An attribute of MPC is that it can be designed for multi-input-multi-output (MIMO) systems – which include mobile systems. It also takes into account system constraints and includes upcoming reference information. The basic principle of MPC operation is that on a finite horizon the future behaviour of the system is predicted and, based on an overall mobile robot model, control inputs are calculated from minimization of a predefined cost function. This procedure often includes solving a real-time optimization problem to obtain the optimal value of the control input [25]. In [21], the authors compared three heuristics in solving the optimization problems that occured in MPC application on mobile robot systems, namely ant colony optimization (ACO), gravitational search algorithm (GSA), and PSO, which was concluded to be preferable.

This paper deals with a trajectory tracking controller for wheeled mobile robots with tricycle configuration. We derived an error-model for predictive control to gain the ability to deal with linear constraints and time-varying systems and achieve good performance on tracking problems. We coupled it with PSO to achieve fast search speed in global optimization. Other advantages of the PSO method include using very few algorithm parameters, insensitivity to scaling of design variables, and easy parallelization for concurrent processing [15]. To ensure efficient operation of the vehicle, we implemented minimum velocity planning based on the idea proposed in [4]. We validated the accuracy of our method on a real AGV. The main contributions of this paper can be summarized as follows: we derived a novel error-based four state kinematic model for WMR with tricycle drive and used it to implement predictive control for trajectory tracking. We also used the minimum-time trajectory approach (described in detail in [4]) to also take into account the dynamical properties of the vehicle, given in the form of constraints on velocity, acceleration and also jerk. Generally, constraints on velocity, acceleration and jerk prevent slipping of the wheels, tipping over of AGV or its load, and unwanted vibrations in mechanical structure caused by extreme values of jerk [4,26]. Additionally, we showed that the proposed methods, developed in simulation environment based on a discrete-time kinematic model, also achieve the desired performance on a real mobile robot.

The kinematics of the WMR is presented in Sect. 2. The MPC is based on the derived error-model of the tricycle, where the cost function is minimized with PSO (Sect. 3). The experiment setup and minimum-time trajectory generation are presented in Sect. 4. The proposed control algorithms were validated in simulation environment, considering some non-ideal conditions, and on a real AGV (Sect. 5). The proposed control law is computationally efficient and can be used in real-time applications. Conclusions are drawn in the final section.

2   WMR with Tricycle Kinematics

The mobile robot in Fig. 1 has four wheels: the front two wheels can be driven independently by the two motors and the back wheels are passive. Each motor is equipped with an encoder and has an internal velocity controller. The mobile robot can therefore be controlled by setting the desired velocity of the right and left wheel, v_R and v_L, respectively. The two front wheels are mounted on a drive cart that is connected to the main body of the mobile robot with a passive rotational joint. An additional absolute encoder measures the angle γ between the drive cart and the main body.

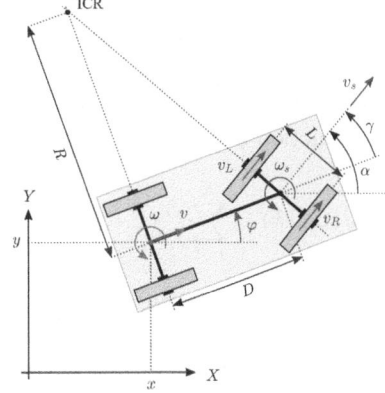

Fig. 1. The WMR used in the experiments (left) and drive mechanism of a WMR with tricycle drive (right)

Although the WMR has four wheels, it has kinematics of a tricycle. The drive mechanism of a WMR with a tricycle drive is shown in Fig. 1 (right). Every point on the rigid body rotates around a virtual axis through instantaneous center of rotation (ICR) with the same angular velocity ω. In a global coordinate frame let x and y be the coordinates of the center of the rear axis, φ the orientation of the robot body, and α the steering angle of the drive cart. WMR with tricycle drive has two non-holonomic constraints

$$
\begin{aligned}
\dot{x}\sin\varphi - \dot{y}\cos\varphi &= 0 \\
\dot{x}\sin\alpha - \dot{y}\cos\alpha - \dot{\varphi}D\cos\gamma &= 0
\end{aligned}
\tag{1}
$$

where D is the distance between the center of the drive cart and the rear axis. External kinematics of a WMR with tricycle drive is determined by

$$
\begin{bmatrix} \dot{x}(t) \\ \dot{y}(t) \\ \dot{\varphi}(t) \\ \dot{\gamma}(t) \end{bmatrix} = \begin{bmatrix} \cos\varphi(t)\cos\gamma(t) & 0 \\ \sin\varphi(t)\cos\gamma(t) & 0 \\ \frac{1}{D}\sin\gamma(t) & 0 \\ -\frac{1}{D}\sin\gamma(t) & 1 \end{bmatrix} \begin{bmatrix} v_{\mathrm{s}}(t) \\ \omega_{\mathrm{s}}(t) \end{bmatrix}.
\tag{2}
$$

The translational and the angular velocity $v_{\mathrm{s}}(t)$ and $\omega_{\mathrm{s}}(t)$ of the mobile robot are related to the speeds of the left $v_{\mathrm{L}}(t)$ and right $v_{\mathrm{R}}(t)$ wheels:

$$
v_{\mathrm{s}}(t) = \frac{v_{\mathrm{L}}(t) + v_{\mathrm{R}}(t)}{2}, \qquad \omega_{\mathrm{s}}(t) = \frac{v_{\mathrm{R}}(t) - v_{\mathrm{L}}(t)}{L}
\tag{3}
$$

In L is the distance between the front wheels. The angular velocity $\omega_{\mathrm{s}}(t)$ is defined according to the global coordinate system, i.e. $\omega_{\mathrm{s}}(t) = \dot{\alpha}(t) = \dot{\varphi}(t) + \dot{\gamma}(t)$.

3   MPC Based on Error-Model

In this section, we first summarize the main concepts of MPC and present the derivation of the error-model in Subsect. 3.1. The basic idea of model predictive control is well established [18,19]. The goal is to calculate the optimal control signals by minimization of the cost function over a finite horizon h. Let J be a cost function at sample k defined as

$$
J(\mathbf{u}_{\mathrm{fb}}, k) = \sum_{i=1}^{h} \epsilon^{T}(k+i)\mathbf{Q}\epsilon(k+i) + \mathbf{u}_{\mathrm{fb}}^{T}(k+i-1)\mathbf{R}\mathbf{u}_{\mathrm{fb}}(k+i-1),
\tag{4}
$$

where $\mathbf{u}_{\mathrm{fb}}$ are the future actions of the feedback control part, and i stands for the i-th prediction step. $\mathbf{Q}$ and $\mathbf{R}$ are symmetric matrices in which the elements represent weights, $\mathbf{e}_{r}(k+i)$ is the future reference tracking error, $\mathbf{e}(k+i|k)$ is the predicted tracking error, and $\epsilon(k+i) = \mathbf{e}_{r}(k+i) - \mathbf{e}(k+i|k)$. The predicted i-th step tracking error is determined by the error model

$$
\mathbf{e}(k+i|k) = \mathbf{A}(k+i-1)\mathbf{e}(k+(i-1)|k) + \mathbf{B}(k+i-1)\mathbf{u}_{\mathrm{fb}}(k+i-1),
\tag{5}
$$

where $\mathbf{A} \in \mathbb{R}^n \times \mathbb{R}^n$, n is the number of the state variables and $\mathbf{B} \in \mathbb{R}^n \times \mathbb{R}^m$, and m is the number of input variables. Substituting $\mathbf{e}(k + (i-1)|k)$ into (5) for $i \in \{1, \ldots, h-1\}$ yields the error-model prediction at the time instant h:

$$
\begin{aligned}
\mathbf{e}(k+h|k) = {}& \prod_{j=0}^{h-1} \mathbf{A}(k+j)\mathbf{e}(k) \\
& + \sum_{j=1}^{h-1} \left(\prod_{i=j}^{h-1} \mathbf{A}(k+i) \right) \mathbf{B}(k+j-1)\mathbf{u}_{\text{fb}}(k+j-1) \\
& + \mathbf{B}(k+h-1)\mathbf{u}_{\text{fb}}(k+h-1).
\end{aligned}
\tag{6}
$$

To shorten notation we let $\Lambda(k, i)$ to stand for the product of matrices $\mathbf{A}(k)$

$$
\begin{aligned}
\Lambda(k, i) = {}& \prod_{j=i}^{h-1} \mathbf{A}(k+j) \\
= {}& \mathbf{A}(k+h-1)\mathbf{A}(k+h-2)\ldots\mathbf{A}(k+i),
\end{aligned}
\tag{7}
$$

where $h-1 \geq i$. We will denote by $\mathbf{U}_{\text{fb}}(k)$ the vector that contains all the control actions and error predictions over the finite horizon h:

$$
\mathbf{U}_{\text{fb}}(k) = \begin{bmatrix} \mathbf{u}_{\text{fb}}^T(k) & \mathbf{u}_{\text{fb}}^T(k+1) & \ldots & \mathbf{u}_{\text{fb}}^T(k+h-1) \end{bmatrix}^T.
\tag{8}
$$

Let us denote by $\mathbf{E}^*(k)$ the predicted tracking error vector, and by $\mathbf{E}_r^*(k)$ the predicted reference tracking error vector:

$$
\mathbf{E}^*(k) = \begin{bmatrix} \mathbf{e}^T(k+1) & \mathbf{e}^T(k+2) & \ldots & \mathbf{e}^T(k+h) \end{bmatrix}^T,
\tag{9}
$$

$$
\mathbf{E}_r^*(k) = \begin{bmatrix} \mathbf{e}_r^T(k+1) & \mathbf{e}_r^T(k+2) & \ldots & \mathbf{e}_r^T(k+h) \end{bmatrix}^T.
\tag{10}
$$

Applying (8–9) we can rewrite Eq. (6) as

$$
\mathbf{E}^*(k) = \mathbf{F}(k)\mathbf{e}(k) + \mathbf{G}(k)\mathbf{U}_{\text{fb}},
\tag{11}
$$

where $\mathbf{F}(k)$ is

$$
\mathbf{F}(k) = \begin{bmatrix} \mathbf{A}^T(k) & (\mathbf{A}(k+1)\mathbf{A}(k))^T & \ldots & \Lambda^T(k,0) \end{bmatrix}^T,
\tag{12}
$$

and $\mathbf{G}(k)$ is

$$
\begin{bmatrix}
\mathbf{B}(k) & 0 & \ldots & 0 \\
\mathbf{A}(k+1)\mathbf{B}(k) & \mathbf{B}(k+1) & \ldots & \ldots \\
\vdots & \vdots & \ddots & \vdots \\
\Lambda(k,1)\mathbf{B}(k) & \Lambda(k,2)\mathbf{B}(k+1) & \ldots & \mathbf{B}(k+h-1)
\end{bmatrix}
\tag{13}
$$

and both $\mathbf{F}(k)$ and $\mathbf{G}(k)$ depend on vehicle kinematics. Similarly, applying (10) we make the following definition

$$
\mathbf{E}_r^*(k) = \mathbf{F}_r e(k),
\tag{14}
$$

where $\mathbf{F}_r$ is

$$\mathbf{F}_r = \begin{bmatrix} \mathbf{A}_r & \mathbf{A}_r^2 & \cdots & \mathbf{A}_r^h \end{bmatrix}^T, \tag{15}$$

and $\mathbf{A}_r$ is the reference model matrix that determines the time dependency of the reference error. Substituting (8–14) into (4) we can rewrite the cost function in a matrix form

$$J(\mathbf{U}_{\text{fb}}) = (\mathbf{E}_r^* - \mathbf{E}^*)^T \bar{\mathbf{Q}}(\mathbf{E}_r^* - \mathbf{E}^*) + \mathbf{U}_{\text{fb}}^T \bar{\mathbf{R}} \mathbf{U}_{\text{fb}}, \tag{16}$$

where $\bar{\mathbf{Q}}$ and $\bar{\mathbf{R}}$ are weighting matrices:

$$\bar{\mathbf{Q}} = \begin{bmatrix} \mathbf{Q} & 0 & \cdots & 0 \\ 0 & \mathbf{Q} & \cdots & 0 \\ \vdots & \vdots & \ddots & \vdots \\ 0 & 0 & \cdots & \mathbf{Q} \end{bmatrix} \qquad \bar{\mathbf{R}} = \begin{bmatrix} \mathbf{R} & 0 & \cdots & 0 \\ 0 & \mathbf{R} & \cdots & 0 \\ \vdots & \vdots & \ddots & \vdots \\ 0 & 0 & \cdots & \mathbf{R} \end{bmatrix}. \tag{17}$$

Finally, the optimal control vector as a solution to the optimal control problem is calculated by minimization $\left(\frac{dJ}{d\mathbf{U}_{\text{fb}}} = 0 \right)$ of the cost function

$$\mathbf{U}_{\text{fb}}(k) = \left(\mathbf{G}^T(k) \bar{\mathbf{Q}} \mathbf{G}(k) + \bar{\mathbf{R}}(k) \right)^{-1} \mathbf{G}^T(k) \bar{\mathbf{Q}} \left(\mathbf{F}_r - \mathbf{F}(k) \right) \mathbf{e}(k). \tag{18}$$

3.1   Derivation of the Error-Model for Tricycle Drive

Error model (5) in shorter notation reads:

$$\mathbf{e}(k+1) = \mathbf{A}(k)\mathbf{e}(k) + \mathbf{B}(k)\mathbf{u}_{\text{fb}}(k). \tag{19}$$

For a short sampling time T_s the discrete matrices A and B can be calculated as follows:

$$\mathbf{A}(k) = \mathbf{I} + \mathbf{A}_c T_s, \qquad \mathbf{B} = \mathbf{B}_c T_s, \tag{20}$$

where $\mathbf{A}_c(t)$ and $\mathbf{B}_c(t)$ are matrices of the continuous state-space model [19]. Let us start the derivation of a four state kinematic model for a tricycle by denoting the pose error $\mathbf{e}(t) = [e_x(t) \ e_y(t) \ e_\varphi(t) \ e_\gamma(t)]^T$, where e_x represents the error in the direction of driving, e_y the error in the perpendicular direction, e_φ the error in the orientation, and e_γ the error in the steering angle, all regarding the virtual reference robot (Fig. 2). The pose error $\mathbf{e}(t)$ is determined by the pose $\mathbf{q}_{\text{ref}}(t) = [x_{\text{ref}}(t) \ y_{\text{ref}}(t) \ \varphi_{\text{ref}}(t) \ \gamma_{\text{ref}}(t)]^T$ of the virtual reference robot and the actual pose $\mathbf{q}(t) = [x(t) \ y(t) \ \varphi(t) \ \gamma(t)]^T$ of the real robot:

$$\begin{bmatrix} e_x(t) \\ e_y(t) \\ e_\varphi(t) \\ e_\gamma(t) \end{bmatrix} = \begin{bmatrix} \cos\varphi(t) & \sin\varphi(t) & 0 & 0 \\ -\sin\varphi(t) & \cos\varphi(t) & 0 & 0 \\ 0 & 0 & 1 & 0 \\ 0 & 0 & 0 & 1 \end{bmatrix} (\mathbf{q}_{\text{ref}}(t) - \mathbf{q}(t)). \tag{21}$$

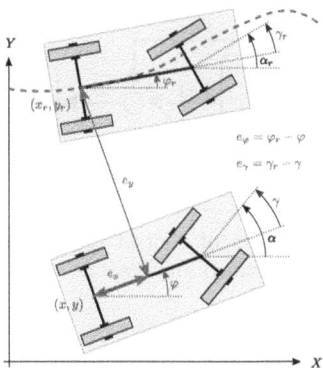

Fig. 2. The pose error $\mathbf{e}(t) = [e_x(t)\ \ e_y(t)\ \ e_\varphi(t)\ \ e_\gamma(t)]^T$ in local coordinates

For clarity we omit the time dependency notation in the following derivation. Assuming that the real robot and the reference virtual robot have the same kinematic model, we first calculate the time derivative of pose error by taking into account Eq. (21) to write the error model in the following form

$$
\begin{bmatrix} \dot{e}_x \\ \dot{e}_y \\ \dot{e}_\varphi \\ \dot{e}_\gamma \end{bmatrix} = \begin{bmatrix} \cos e_\varphi \cos \gamma_{\text{ref}} & 0 \\ \sin e_\varphi \cos \gamma_{\text{ref}} & 0 \\ \frac{\sin \gamma_{\text{ref}}}{D} & 0 \\ -\frac{\sin \gamma_{\text{ref}}}{D} & 1 \end{bmatrix} \begin{bmatrix} v_{\text{s,ref}} \\ \omega_{\text{s,ref}} \end{bmatrix} + \begin{bmatrix} -\cos \gamma_r + \frac{\sin \gamma_r}{D} e_y & 0 \\ -\frac{\sin \gamma_r}{D} e_x & 0 \\ -\frac{\sin \gamma_r}{D} & 0 \\ \frac{\sin \gamma_r}{D} & -1 \end{bmatrix} \begin{bmatrix} v_s \\ \omega_s \end{bmatrix} \tag{22}
$$

where $\gamma_r = \gamma_{\text{ref}} - e_\gamma$, and $v_{\text{s,ref}}$ and $\omega_{\text{s,ref}}$ are

$$
v_{\text{s,ref}} = \sqrt{D^2 \omega_{\text{ref}}^2 + v_{\text{ref}}^2}, \tag{23a}
$$

$$
\omega_{\text{s,ref}} = \omega_{\text{ref}} + D \frac{v_{\text{ref}} \dot{\omega}_{\text{ref}} - \omega_{\text{ref}} \dot{v}_{\text{ref}}}{v_{\text{ref}}^2} \cos^2(\gamma_{\text{ref}}). \tag{23b}
$$

The linear and the angular reference velocities v_{ref} and ω_{ref} are determined from the reference trajectory. Rearranging and linearizing Eq. (22) yields the following:

$$
\begin{bmatrix} \dot{e}_x \\ \dot{e}_y \\ \dot{e}_\varphi \\ \dot{e}_\gamma \end{bmatrix} = \underbrace{\begin{bmatrix} 0 & \frac{a}{D} & 0 & -c \\ -\frac{a}{D} & 0 & b & 0 \\ 0 & 0 & 0 & \frac{b}{D} \\ 0 & 0 & 0 & -\frac{b}{D} \end{bmatrix}}_{A_c} \begin{bmatrix} e_x \\ e_y \\ e_\varphi \\ e_\gamma \end{bmatrix} + \underbrace{\begin{bmatrix} -\cos \gamma_r & 0 \\ 0 & 0 \\ -\frac{\sin \gamma_r}{D} & 0 \\ \frac{\sin \gamma_r}{D} & -1 \end{bmatrix}}_{B_c} \begin{bmatrix} v_{\text{fb}} \\ \omega_{\text{fb}} \end{bmatrix} \tag{24}
$$

where $a = v_{\text{s,ref}} \sin \gamma_r$, $b = v_{\text{s,ref}} \cos(\gamma_{\text{ref}})$, and $c = v_{\text{s,ref}} \sin(\gamma_{\text{ref}})$. Our control system (Fig. 3) consists of a trajectory generator that provides reference pose $\mathbf{q}_{\text{ref}}$ and feedforward input vector $\mathbf{u}_{\text{ff}}$. The feedback input vector $\mathbf{u}_{\text{fb}}$ is the output of the MPC controller. The control action $\mathbf{u} = [v_s\ \omega_s]^T$ is fed to the WMR.

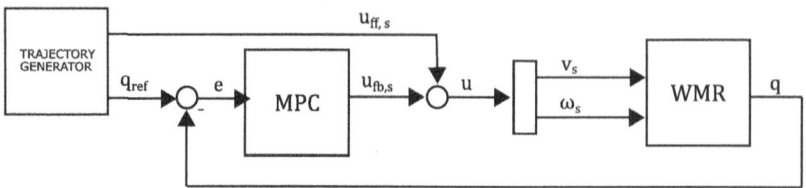

Fig. 3. Simplified schematic display of control signals calculation using MPC. The control action u consists of feedforward and feedback part and is fed to the system.

3.2   Optimal Control Based on PSO

An effective global search optimization technique that can be used when solving optimization problem of minimizing the cost function (Eq. 4) in MPC is PSO method. It uses a population of candidate solutions, the so-called particles. For each particle i in a swarm of n particles there is a parameter vector $x_i = [x_1 x_2 \dots x_n]^T$ that defines its position, and vector $v_i = [v_1 v_2 \dots v_n]^T$ that defines its velocity in the d dimensional parameter space. In our case x_i consisted of the control actions v_s and ω_s, and v_i was the change of the control velocities. Particles' best known parameters are stored in vector $\mathbf{pBest}_i = [p_{i1}, p_{i2}, \dots, p_{id}]^T$, and the swarm's best known parameters are stored in vector $\mathbf{gBest} = [g_1, g_2, \dots, g_d]^T$. The updates of velocity and position of the i-th particle in the r-th iteration are:

$$
\begin{aligned}
\mathbf{v}_i^{(r)} &\leftarrow \omega_{\text{in}} \mathbf{v}_i^{(r-1)} + c_1 r_1 (\mathbf{pBest}_i^{(r-1)} - \mathbf{x}_i^{(r-1)}) \\
&+ c_2 r_2 (\mathbf{gBest}_i^{(r-1)} - \mathbf{x}_i^{(r-1)}),
\end{aligned}
\tag{25}
$$

$$
\mathbf{x}_i^{(r)} \leftarrow \mathbf{x}_i^{(r-1)} + \mathbf{v}_i^{(r)}.
\tag{26}
$$

The parameter ω_{in} is the inertia weight. The cognitive constant c_1 and parameter r_1 weigh the stochastic attraction towards the best parameter of the particle $\mathbf{pBest}_i$ (local search weight). r_1 has a random value on the interval $[0, 1]$ and increases the probability of finding the global optimum or avoiding a convergence being too fast. The constant of social behavior in the swarm c_2 and the parameter r_2 weigh the stochastic attraction to the best parameter of the swarm $\mathbf{gBest}$ (global search weight). r_2 has a random value in the interval $[0, 1]$ and has the same role as r_1. We have assumed that control actions are constant over the forecast horizon and during the optimization process we explicitly ensured that velocity constraints are fulfilled.

4   Experimental Setup

We calculated a minimum-time velocity profile on the reference path

$$
x_{\text{ref}}(u) = x_0 + A \sin(u), \qquad y_{\text{ref}}(u) = y_0 + B \sin(2u)
\tag{27}
$$

for $u \in [0, 4\pi]$, $x_0 = 1.3 \, \mathrm{m}$, $y_0 = 0.92 \, \mathrm{m}$, $A = 0.7 \, \mathrm{m}$ and $B = 0.7 \, \mathrm{m}$. The chosen curve in Eq. (27) belongs to a family of curves that are often used for comparison between different trajectory-tracking algorithms, for example in [11, 14, 20].

For a mobile system driving along a three times continuously differentiable plane curve, the velocity vector $\mathbf{v}(t)$, the acceleration vector $\mathbf{a}(t)$, and the jerk vector $\mathbf{j}(t)$ in the tangential-normal form are:

$$
\begin{aligned}
\mathbf{v}(t) &= v(t) \cdot \hat{\mathbf{T}}, \\
\mathbf{a}(t) &= a_{\mathrm{T}}(t) \cdot \hat{\mathbf{T}} + a_{\mathrm{R}}(t) \cdot \hat{\mathbf{N}} = \dot{v} \cdot \hat{\mathbf{T}} + v^2 \kappa \cdot \hat{\mathbf{N}}, \\
\mathbf{j}(t) &= j_{\mathrm{T}}(t) \cdot \hat{\mathbf{T}} + j_{\mathrm{R}}(t) \cdot \hat{\mathbf{N}} = \left(\ddot{v} - v^3 \kappa^2 \right) \cdot \hat{\mathbf{T}} + \tfrac{1}{v} \left(\tfrac{\mathrm{d}}{\mathrm{d}t}(v^3 \kappa) \right) \cdot \hat{\mathbf{N}},
\end{aligned}
\tag{28}
$$

where $\hat{\mathbf{T}}$ and $\hat{\mathbf{N}}$ are the unit tangential and the unit normal vector, respectively, and κ is the curvature of the path at time t. A detailed explanation of minimum-time velocity profile calculation is presented in [4]. The velocity, acceleration and jerk constraints are imposed in the following form:

$$
\begin{aligned}
0 &\leq |v(t)| \leq v_{\mathrm{MAX}}, \\
\frac{a_{\mathrm{R}}^2(t)}{a_{\mathrm{R_{MAX}}}^2} &+ \frac{a_{\mathrm{T}}^2(t)}{a_{\mathrm{T_{MAX}}}^2} \leq 1, \\
\frac{j_{\mathrm{R}}^2(t)}{j_{\mathrm{R_{MAX}}}^2} &+ \frac{j_{\mathrm{T}}^2(t)}{j_{\mathrm{T_{MAX}}}^2} \leq 1.
\end{aligned}
\tag{29}
$$

The calculated minimum-time reference trajectory is shown in Fig. 4.

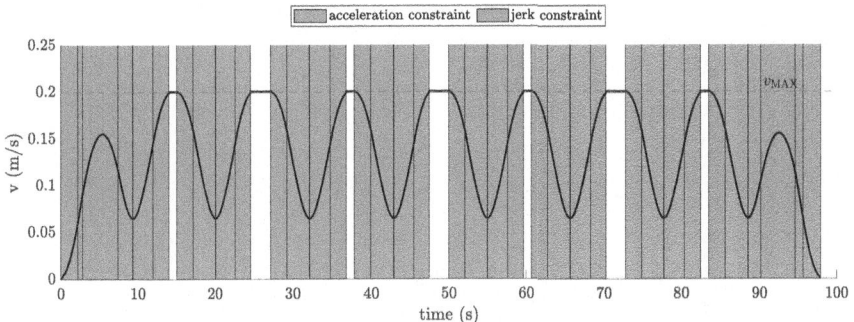

Fig. 4. Minimum-time velocity profile for the selected path. The colored bands indicate the intervals in which acceleration and jerk reach the threshold values. (Color figure online)

The simulation experiments were performed in MATLAB, using Runge-Kutta integration method ode45 to simulate a continuous state-space kinematic model of a tricycle drive and the controller was implemented in a discrete form. The noise was estimated to be normally distributed with zero mean value and standard deviation $0.001 \, \mathrm{m}$ in the position data, $0.01 \, \mathrm{rad}$ for measurements of φ and $0.001 \, \mathrm{rad}$ for measurements of γ. These values were determined based on the properties of the measure system of the real WMR.

The initial robot pose $[x(0)\ y(0)\ \varphi(0)\ \gamma(0)]^T = [1.29\ 0.87\ 0\ 0]^T$ both in the simulation and experiment was slightly different than the starting reference pose. The diagonal matrices with weights were $\mathbf{Q} = \mathrm{diag}([8\ 40\ 0.1\ 0])$, $\mathbf{R} = \mathrm{diag}([0.1\ 5 \cdot 10^{-4}])$ (in simulation experiments), and $\mathbf{R} = \mathrm{diag}([0.4\ 5 \cdot 10^{-4}])$ (in experiments on a real robot). Other parameters used in the simulation and the experiment together with some basic geometric parameters of the WMR are listed in Table 1.

Table 1. Parameters in the experiments.

Parameter	Symbol	Value	Units
distance between the rear axis and center of the drive cart	D	0.1207	m
distance between the front wheels	L	0.0430	m
horizon	h	$\{5, 10, 15\}$	/
sample time	T_s	$\frac{1}{15}$	s
time delay	t_D	$2T_s$	s
maximum value of velocity	v_{MAX}	0.2	$\mathrm{m\,s^{-1}}$
maximum value of radial acceleration	$a_{R_{MAX}}$	0.05	$\mathrm{m\,s^{-2}}$
maximum value of tangential acceleration	$a_{T_{MAX}}$	0.05	$\mathrm{m\,s^{-2}}$
maximum value of radial jerk	$j_{R_{MAX}}$	0.04	$\mathrm{m\,s^{-3}}$
maximum value of tangential jerk	$j_{T_{MAX}}$	0.02	$\mathrm{m\,s^{-3}}$
maximum value of angular velocity	ω_{MAX}	2.5	$\mathrm{rad\,s^{-1}}$

We also applied additional velocity and angular velocity constraints of the AGVs. If the value of angular velocity is larger than the allowed value, it is set to ω_{MAX}. The speeds of left and right wheels are calculated according to Eq. (2). If either v_L or v_R exceeds the allowed value v_{MAX}, a scale factor k_s is calculated:

$$k_s = \max\left(\left|\frac{v_L}{v_{MAX}}\right|, \left|\frac{v_r}{v_{MAX}}\right|, 1\right). \tag{30}$$

The composite control signal $\mathbf{u} = [v_s, \omega_s]$ is then scaled down with the same factor in order to preserve the curvature $\kappa = \omega_s/v_s$ of the path: $\mathbf{u}_{constr} = 1/k_s\,[v_s, \omega_s]$.

5   Results

In all experiments, WMR drove two laps. Figures 5, 6 and 7 show the results of the simulation experiments for the first lap, where we took into account some non-ideal conditions such as noise and delay that could be expected for a real robot.

Figures 8, 9 and 10 show the results of the experiments on the real robot for the first lap. The experimental results were obtained on a real mobile robot platform (Fig. 1, left). An overhead vision-based object-tracking system measured pose (position and orientation) of the WMR. The WMR has an artificial marker from the Ar-Uco library [10], a distinctive pattern with an encoded identification information, which enables tracking and distinction of multiple markers through a sequence of images that are obtained with Raspberry Pi Camera Module V2. We used Robot Operating System (ROS), a set of software libraries and

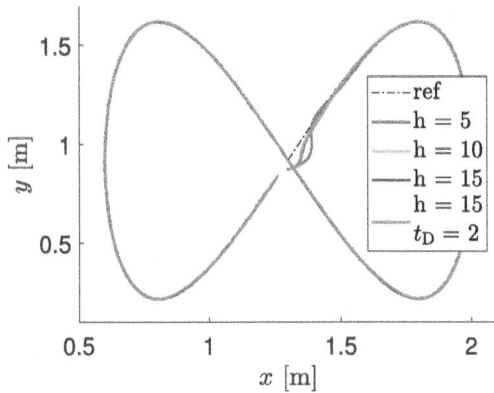

Fig. 5. Reference path (dashed) and control tracking results for different values of h and t_D in simulation

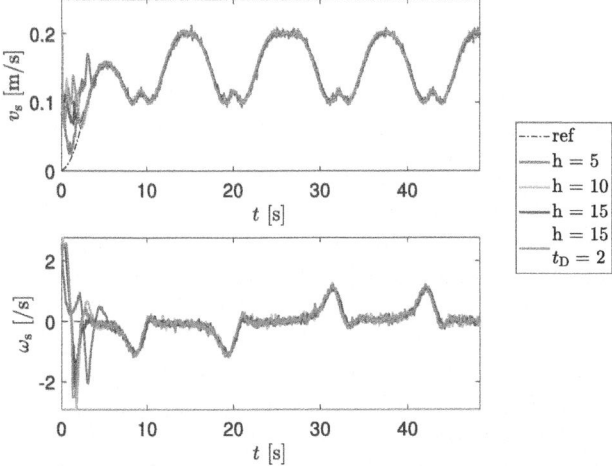

Fig. 6. Reference control signals $v_\mathrm{s,ref}(t)$ and $\omega_\mathrm{s,ref}(t)$ (dashed) and control actions $v_\mathrm{s}(t)$ and $\omega_\mathrm{s}(t)$ for different values of h in simulation

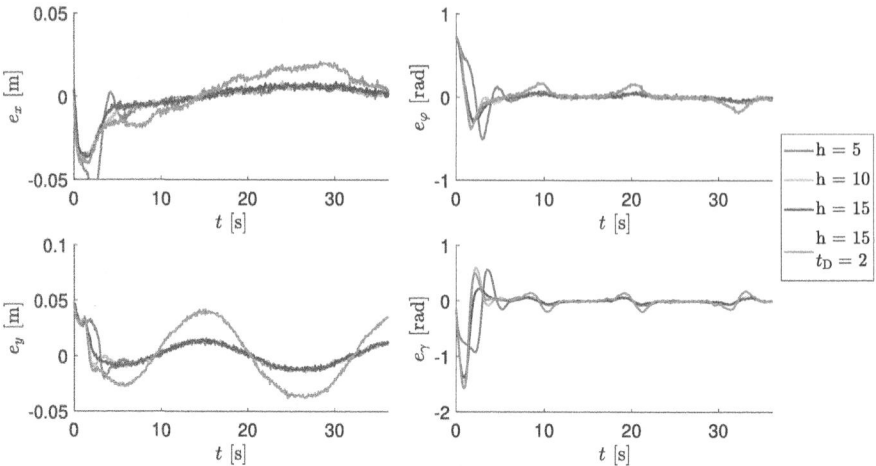

Fig. 7. Pose error vectors $e_x(t)$, $e_y(t)$, $e_\varphi(t)$, and $e_\gamma(t)$. The simulation experiment was conducted for different values of h.

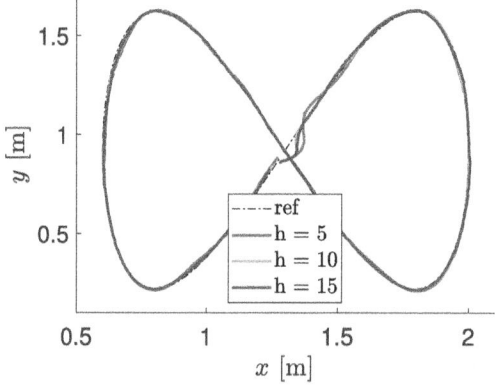

Fig. 8. Reference path (dashed) and control tracking results for different values of h and t_D in experiment on a real robot

tools, to enable data exchange and to establish a communication between the WMR, PC and camera. Additionally, all experiments also show how the prediction horizon influences the trajectory-tracking performance. Overall, we have confirmed (Fig. 5) that the presented controller not only works satisfactorily when a mobile robot is near the reference trajectory, but also reduces the initial error and the tracking errors (disturbances). The experiments have shown that the proposed implementation of the predictive controller achieves good tracking performance (Fig. 7). With proper tuning of the control parameters (which is intuitive), it is possible to achieve robust performance, even in the presence of measurement noise and constrained control velocities. The results in Fig. 6 show that the velocity constraints have been satisfied. If we compare the simulation

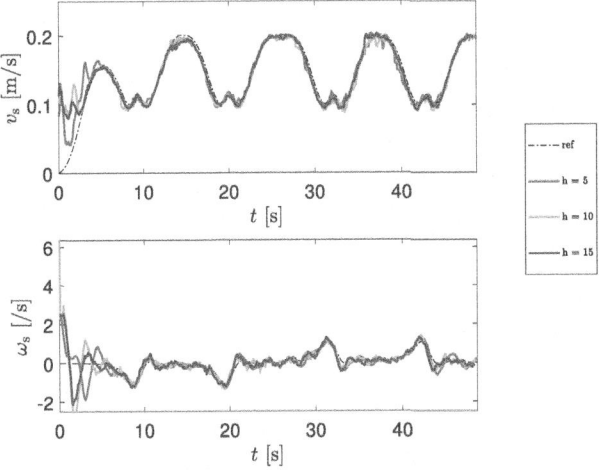

Fig. 9. Results from the experiment on a real robot: reference control signals $v_{s,ref}(t)$ and $\omega_{s,ref}(t)$ (dashed) and control actions $v_s(t)$ and $\omega_s(t)$ for different values of h

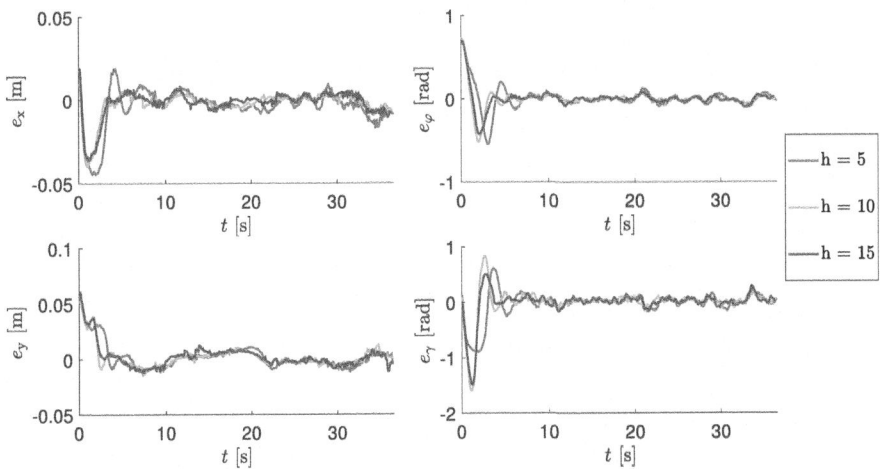

Fig. 10. Pose error vectors $e_x(t)$, $e_y(t)$, $e_\varphi(t)$, and $e_\gamma(t)$. The experiment on a real robot was conducted for different values of h.

experiments with the experiments on the real mobile robot, we can see that the transfer of the controller developed on the basis of the kinematic error model in the simulation environment was successful. Our findings can be highly valuable for optimizing the time-efficient trajectories of autonomous mobile robots in automated warehouses, taking into account dynamic constraints that may vary based on the type of load being transported [5].

6 Conclusion

In this paper, we proposed an error-based four state kinematic model for WMR with tricycle drive and used it to implement predictive control for trajectory tracking. Additionally, we performed simulation experiments and experiments on a real robot traveling along a minimum-time trajectory with constraints on velocity, acceleration and jerk. In this way, the study indirectly considered the dynamical properties of the mobile system to avoid wheel slipping or tipping over.

The proposed approach, which is designed for non-holonomic mobile systems, can also be applied on holonomic mobile systems. Another feature is that the controllers implicitly include constraints on velocity and angular velocity. Additionally, small measurement delays (as expected with physical WMRs) can also be handled and therefore a seamless transition of the controller to real WMRs can be achieved. Our approach has wide practical applications. The proposed trajectory-tracking control can be especially useful for automated guided vehicles (AGVs) that operate in industrial environments, for example in production halls or in automated warehouses, where trajectories are known in advance and it is important to ensure a safe, effective and comfortable ride.

Acknowledgements. This research was funded by Slovenian Research Agency under Grant P2-0219.

References

1. Abbasi, A., Moshayedi, A.J.: Trajectory tracking of two-wheeled mobile robots, using LQR optimal control method, based on computational model of KHEPERA IV. J. Simul. Anal. Novel Technol. Mech. Eng. **10**(3), 41–50 (2018)
2. Amer, N.H., Zamzuri, H., Hudha, K., Kadir, Z.A.: Modelling and control strategies in path tracking control for autonomous ground vehicles: a review of state of the art and challenges. J. Intell. Robo. Syst.: Theory Appl. **86**(2), 225–254 (2017). https://doi.org/10.1007/s10846-016-0442-0
3. Azizi, M.R., Keighobadi, J.: Point stabilization of nonholonomic spherical mobile robot using nonlinear model predictive control. Robot. Auton. Syst. **98**, 347–359 (2017). https://doi.org/10.1016/j.robot.2017.09.015
4. Benko Loknar, M., Blažič, S., Klančar, G.: Minimum-time velocity profile planning for planar motion considering velocity, acceleration and jerk constraints. Int. J. Control **96**(1), 251–265 (2023). https://doi.org/10.1080/00207179.2021.1987526
5. Benko Loknar, M., Klančar, G., Blažič, S.: Minimum-time trajectory generation for wheeled mobile systems using bézier curves with constraints on velocity, acceleration and jerk. Sensors **23**(4) (2023). https://doi.org/10.3390/s23041982, https://www.mdpi.com/1424-8220/23/4/1982
6. Brockett, R.: Asymptotic stability and feedback stabilization. Differ. Geom. Control Theory **27**, 181–191 (1983)
7. Cui, M., Liu, W., Liu, H., Jiang, H., Wang, Z.: Extended state observer-based adaptive sliding mode control of differential-driving mobile robot with uncertainties. Nonlinear Dyn. **83**(1–2), 667–683 (2016). https://doi.org/10.1007/s11071-015-2355-z

8. Ding, L., Li, S., Liu, Y.J., Gao, H., Chen, C., Deng, Z.: Adaptive neural network-based tracking control for full-state constrained wheeled mobile robotic system. IEEE Trans. Syst. Man Cybern.: Syst. **47**(8), 2410–2419 (2017). https://doi.org/10.1109/TSMC.2017.2677472

9. Fernandez, B., Herrera, P.J., Cerrada, J.A.: A simplified optimal path following controller for an agricultural skid-steering robot. IEEE Access **7**, 95932–95940 (2019). https://doi.org/10.1109/ACCESS.2019.2929022

10. Garrido-Jurado, S., Muñoz-Salinas, R., Madrid-Cuevas, F., Marín-Jiménez, M.: Automatic generation and detection of highly reliable fiducial markers under occlusion. Pattern Recogn. **47**(6), 2280–2292 (2014). https://doi.org/10.1016/j.patcog.2014.01.005

11. Guechi, E.H., Belharet, K., Blažič, S.: Tracking control for wheeled mobile robot based on delayed sensor measurements. Sensors **19**(23), 5177 (2019). https://doi.org/10.3390/s19235177

12. Kapitanyuk, Y.A., Proskurnikov, A.V., Cao, M.: A guiding vector-field algorithm for path-following control of nonholonomic mobile robots. IEEE Trans. Control Syst. Technol. **26**(4), 1372–1385 (2018). https://doi.org/10.1109/TCST.2017.2705059

13. Kara-Mohamed, M.: Advanced trajectory tracking for UAVs using combined feed-forward/feedback control design. Robot. Auton. Syst. **96**, 143–156 (2017). https://doi.org/10.1016/j.robot.2017.07.009

14. Kayacan, E., Chowdhary, G.: Tracking error learning control for precise mobile robot path tracking in outdoor environment. J. Intell. Robot. Syst.: Theory Appl. **95**(3–4), 975–986 (2019). https://doi.org/10.1007/s10846-018-0916-3

15. Kennedy, J., Eberhart, R.: Particle swarm optimization. In: Proceedings of ICNN 1995 - International Conference on Neural Networks, vol. 4, pp. 1942–1948 (1995). https://doi.org/10.1109/ICNN.1995.488968

16. Keymasi Khalaji, A., Jalalnezhad, M.: Robust forward\backward control of wheeled mobile robots. ISA Trans. **115**, 32–45 (2021). https://doi.org/10.1016/j.isatra.2021.01.016

17. Khan, S., Guivant, J., Li, X.: Design and experimental validation of a robust model predictive control for the optimal trajectory tracking of a small-scale autonomous bulldozer. Robot. Auton. Syst. **147**, 103903 (2021). https://doi.org/10.1016/j.robot.2021.103903

18. Klančar, G., Škrjanc, I.: Tracking-error model-based predictive control for mobile robots in real time. Robot. Auton. Syst. **55**(6), 460–469 (2007). https://doi.org/10.1016/j.robot.2007.01.002

19. Klančar, G., Zdešar, A., Blažič, S., Škrjanc, I.: Wheeled Mobile Robotics: From Fundamentals Towards Autonomous Systems. Butterworth-Heinemann (2017)

20. Kokot, M., Miklic, D., Petrovic, T.: A unified MPC design approach for AGV path following. In: 2022 IEEE/RSJ International Conference on Intelligent Robots and Systems (IROS), vol. 2022-Octob, pp. 4789–4796. IEEE (2022). https://doi.org/10.1109/IROS47612.2022.9981575

21. Merabti, H., Belarbi, K., Bouchemal, B.: Nonlinear predictive control of a mobile robot: a solution using metaheuristcs. J. Chin. Inst. Eng. Trans. Chin. Inst. Eng. Ser. A **39**(3), 282–290 (2016). https://doi.org/10.1080/02533839.2015.1091276

22. Sekiguchi, S., Yorozu, A., Kuno, K., Okada, M., Watanabe, Y., Takahashi, M.: Human-friendly control system design for two-wheeled service robot with optimal control approach. Robot. Auton. Syst. **131**, 103562 (2020). https://doi.org/10.1016/j.robot.2020.103562

23. Škrjanc, I., Klančar, G.: A comparison of continuous and discrete tracking-error model-based predictive control for mobile robots. Robot. Auton. Syst. **87**, 177–187 (2017). https://doi.org/10.1016/j.robot.2016.09.016

24. Wang, R., Li, Y., Fan, J., Wang, T., Chen, X.: A novel pure pursuit algorithm for autonomous vehicles based on salp swarm algorithm and velocity controller. IEEE Access **8**, 166525–166540 (2020). https://doi.org/10.1109/ACCESS.2020.3023071

25. Xiao, H., et al.: Robust stabilization of a wheeled mobile robot using model predictive control based on neurodynamics optimization. IEEE Trans. Industr. Electron. **64**(1), 505–516 (2017). https://doi.org/10.1109/TIE.2016.2606358

26. Zdešar, A., Škrjanc, I.: Optimum velocity profile of multiple Bernstein-Bézier curves subject to constraints for mobile robots. ACM Trans. Intell. Syst. Technol. **9**(5), 1–23 (2018). https://doi.org/10.1145/3183891

27. Zhang, Y., Zhao, X., Tao, B., Ding, H.: Point stabilization of nonholonomic mobile robot by Bézier smooth subline constraint nonlinear model predictive control. IEEE/ASME Trans. Mechatron. **26**(2), 990–1001 (2021). https://doi.org/10.1109/TMECH.2020.3014967

Augmented Reality Applied to the "Balneario Paso de la Danta" in Manaure Cesar, an Interactive Experience of ICT and Cultural Heritage

Paola Patricia Ariza-Colpas[1,3]([✉]), Marlon-Alberto Piñeres-Melo[2,3],
Roberto Cesar Morales-Ortega[1,4], Andres -Felipe Rodriguez-Bonilla[3],
Shariq Butt-Aziz[5], Ronald Alexander Vacca Ascanio[6], Yuneidis Morales Ortega[1],
and Sumera Naz[7]

[1] Department of Computer Science and Electronics, Universidad de la Costa CUC,
080002 Barranquilla, Colombia
{pariza1,rmorales1}@cuc.edu.co, rmorales@certika.co
[2] Department of Systems Engineering, Universidad del Norte, 081001 Barranquilla, Colombia
pineresm@uninorte.edu.co
[3] Blazing Soft Company, 081001 Barranquilla, Colombia
andres.rodriguez@blazingsoft.com
[4] Certika Company, 081001 Barranquilla, Colombia
[5] Department of Computer Science, University of South Asia, Lahore, Pakistan
[6] Faculty of Engineering and Technology, Universidad Popular del Cesar, 200004 Valledupar,
Cesar, Colombia
ronaldalexandervacca@unicesar.edu.co
[7] Department of Mathematics, Division of Science and Technology, University of Education,
Lahore, Pakistan
sumera.naz@ue.edu.pk

Abstract. This scientific article focuses on showcasing the technological novelty and the post-pandemic economic impact of implementing augmented reality at Paseo de la Danta, a tourist resort located in the Cesar department. The research was conducted through a case study, analyzing the use of augmented reality in tourism promotion and its effect on generating income at the resort. The results show that the implementation of this technology has allowed for an improvement in the tourist experience, increasing the influx of visitors and, consequently, the economic income generated. Furthermore, in the context of the pandemic, augmented reality becomes a safe alternative for tourism promotion and entertainment, allowing for the economic reactivation of the sector during crisis. In conclusion, this study highlights the importance of technology and innovation in the economic and tourism development of a region, and the need to continue exploring new tools to improve the competitiveness of tourism destinations.

Keywords: Augmented Reality · Digital Media · Balneario Paso de la Danta · cultural heritage

© The Author(s), under exclusive license to Springer Nature Switzerland AG 2024
M. Mujica Mota and P. Scala (Eds.): EUROSIM 2023, CCIS 2033, pp. 19–30, 2024.
https://doi.org/10.1007/978-3-031-68438-8_2

1 Introduction

The implementation of new technologies in the tourism sector has become a relevant topic in recent years, as their use can significantly improve the tourist experience and increase the competitiveness of tourist destinations. However, their adoption is not without challenges and limitations, as it can generate digital and access gaps for certain population groups, as well as issues related to technology investment and staff training. In this context, the case of Paseo de la Danta spa, located in the department of Cesar, Colombia, is presented, where augmented reality technology has been implemented with the aim of improving the tourist experience and generating a positive economic and social impact on the local community.

The implementation of augmented reality at the spa has generated a significant economic impact, as it has increased the influx of visitors and, consequently, generated economic income. Additionally, its use has also generated a positive social and cultural impact on the community, as it has promoted knowledge and ownership of technology among residents and improved the tourist offering of the destination. The problematizing question posed in this study is how can the implementation of augmented reality at Paseo de la Danta spa improve the tourist experience and generate a positive economic and social impact on the local community?

This project is based on two components: technological innovation and social innovation. Regarding the first, the implementation of augmented reality at the spa represents an important technological innovation, as it improves the tourist experience and makes the destination more attractive and competitive in an increasingly demanding market. Regarding the second component, social innovation, it is expected that the implementation of augmented reality will have a positive impact on the local community, as it can generate new employment opportunities and improve the quality of life of residents.

The implementation of augmented reality at Paseo de la Danta spa represents a unique opportunity to improve the tourist experience, generate a positive economic impact, and promote technological and social innovation in the region. This research seeks to analyze the results of the implementation of this technology and its impact on the local community, with the aim of providing valuable information for future similar projects in other tourist destinations.

2 Conceptual Information

2.1 Augmented Reality

The term was coined in 1992 by the scientist and researcher Thomas P. Caudell while he was developing one of the most famous airplanes in the world: the Boeing 747. Caudell observed that the operators in charge of assembling the new aircraft wasted too much time interpreting the instructions and thought: What if they had access to a screen that would guide them through the installation? The invention failed, but at that precise moment, the concept of Augmented Reality was born.

It arose then and not in the summer of 2016, as many of us believe when we were infected by Pokémon GO fever, a video game that consisted of searching for and capturing different characters from the Japanese saga and which, at its peak, reached the astronomical figure of 45 million daily active users. Pokémon GO popularized AR, and brought it closer to the general public—to all audiences—, but by then there were already many companies from very different sectors (health, education, architecture, services, retail, etc.) that were beginning to use it with the aim of creating valuable experiences for its customers.

There are different applications of augmented reality in our daily lives, among which we can highlight. In medicine, doctors can view the different organs of the patient through tablets or glasses and consult their history during surgeries. Regarding the automotive sector, the manufacturers offer their future clients the possibility of being able to configure their new car according to their preferences. In education, the way of learning has been transformed through notebooks with markers that allow 3D images to be viewed [1, 2]. Other users who are fond of sports can program their routes on a 3D surface to learn about the challenges offered by the adventure they are going to embark on.

Other applications that also stand out are simultaneous translations, which energizes the tourism sector since it allows a translation into the desired language through a photograph. Participation in social networks has evolved through the implementation of facial scanners that allow users to be identified and related to their profile on social networks. In the same way, many advertising agencies use applications based on augmented reality to be able to display information about product brands in public places. Regarding tourism and architecture, the evolution of an important cultural place can be observed through the development of time capsules and allows to improve the experience of travelers in a certain place.

2.2 Municipality of Manaure

In South America, the Republic of Colombia, to the north of the Department of Cesar, with an area of 127 square kilometers, is one of the most picturesque towns rich in a variety of landscapes and fauna embedded in the center of the Eastern Cordillera. As our old minstrels describe it, Manaure is a beautiful savannah surrounded by green mountains, the Serranía de Perijá that surrounds it from south to east, presenting itself imposing, inviting tourists to scrutinize its sites that, in addition to being reserve areas, are very striking. Manaure Balcón del Cesar has been par excellence the Tourist Balcony of Cesar, it is characterized by the great beauty of its vegetation, the freshness of its environment, the beauty of its river and above all by the friendliness of its people. The municipality of Manaure has a great tourist infrastructure to provide and please the visitor, it has spas with waterfalls of crystalline waters, stadiums with swimming pools and kiosks at the exit of the municipality and very close to the Manaure River.

3 Literature Review

There are several countries around the world that are actively developing and implementing augmented reality (AR) technology, see Fig. 1. Among them are the United States, Japan, China, South Korea, and the United Kingdom. In the United States, companies such as Google, Apple, and Microsoft are leading the way in AR development. Google's ARCore platform and Apple's ARKit are two of the most popular AR development tools, while Microsoft's HoloLens is a well-known AR headset used in various industries.

Japan has a strong AR industry, with companies such as Sony and Panasonic developing AR applications for entertainment and business purposes. One notable example is Sony's "PlayStation VR" headset, which utilizes AR technology to create an immersive gaming experience. China has also made significant advancements in AR technology, with companies like Huawei and Xiaomi releasing AR-enabled smartphones and tablets. Additionally, Chinese tech giant Alibaba has developed an AR shopping platform, allowing customers to virtually try on clothes and accessories before making a purchase.

South Korea is another country that has embraced AR technology, with Samsung releasing AR-enabled smartphones and tablets, and LG developing AR applications for use in education and healthcare. The United Kingdom is home to several AR development studios, including Zappar and Blippar, which have created AR applications for advertising and marketing campaigns, as well as educational purposes.

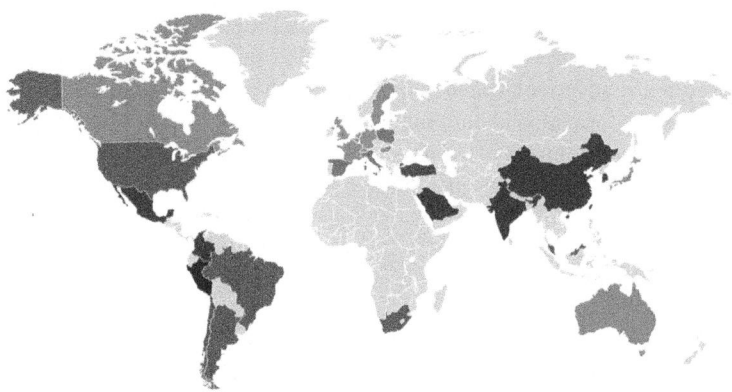

Fig. 1. Countries that develop augmented reality.

Touristic augmented reality has a wide range of applications in different sectors, some of which include:

3.1 Cultural Tourism

Augmented reality can be used to enhance the experience of tourists visiting cultural sites and historical landmarks. For example, interactive applications can be created that allow visitors to explore ancient ruins or museums and obtain additional information about the objects and the history behind them.

Many studies have focused on exploring the potential of augmented reality for enhancing the tourist experience, providing interactive and immersive experiences, and promoting cultural heritage. One example is a study by Mellis and Del Chiappa [3], which explored the use of augmented reality in the context of cultural tourism in Italy. The study examined the impact of an augmented reality application on tourists' behavior and perception of cultural heritage sites. Another study by Guttentag [4] investigated the use of augmented reality for enhancing museum experiences. The study found that augmented reality improved visitors' engagement with the museum exhibits and provided a more immersive and interactive experience.

Similarly, a study by Yuu [5], explored the use of augmented reality for promoting Chinese cultural heritage. The study developed an augmented reality application that allowed users to interact with Chinese cultural artifacts and provided additional information about the history and significance of the artifacts. These studies demonstrate the potential of augmented reality in enhancing the tourist experience and promoting cultural heritage. As technology continues to advance, it is likely that we will see even more innovative and immersive applications of augmented reality in cultural tourism.

3.2 Gastronomic Tourism

Augmented reality can also be used to enhance the gastronomic experience of tourism. For instance, applications can be created that enable users to scan restaurant menus and obtain detailed information about dishes and ingredients.

The application of augmented reality in gastronomic tourism has been the subject of study in several recent scientific articles. For example, Wang [6] developed a mobile augmented reality application to enhance diners' experience in a restaurant. The application allowed users to scan the menu and view 3D images of dishes and nutritional information. The authors found that the application improved user experience and increased customer satisfaction.

Another study by Wei [7] focused on the application of augmented reality in wine tasting. The researchers developed a mobile application that allowed users to scan wine labels and obtain information about the wine, its taste, pairing, and origin. The results indicated that the application significantly improved the user experience in wine tasting. Penco [8] also explored the use of augmented reality in gastronomic tourism, particularly in food selection at markets. They developed a mobile application that allowed users to scan products in the market and obtain detailed information about their origin, taste, and nutritional characteristics. The authors found that the application increased consumers' confidence in selecting food and improved their shopping experience at the market. These studies demonstrate that the application of augmented reality in gastronomic tourism can significantly enhance user experience and increase customer satisfaction. Additionally, it is suggested that augmented reality can be a useful tool in consumer education about food and the promotion of local gastronomic culture.

3.3 Adventure Tourism

Augmented reality can be a useful tool in adventure tourism to provide information and guidance in unfamiliar environments. For example, applications can be created for hikers that provide information about the route and the hazards along the way.

The application of augmented reality in adventure tourism has been a topic of interest in several recent scientific articles. For example, Huang [9] developed an augmented reality-based navigation system for hikers. The system allowed users to scan the surrounding area and receive real-time guidance on the trail and potential obstacles. The

authors found that the system improved navigation accuracy and increased user satisfaction. Another study by Fanini [10] focused on the application of augmented reality
in rock climbing. They developed an augmented reality-based training tool that allowed
climbers to practice and improve their skills in a virtual environment. The authors found
that the tool improved climbers' technical abilities and their confidence in real-world
climbing situations.

In a different study, Marino [11] explored the use of augmented reality in the promotion of cultural heritage sites in adventure tourism. They developed an augmented
reality-based mobile application that provided users with information about the history and cultural significance of the site they were visiting. The authors found that the
application enhanced visitors' engagement and learning experience, leading to a deeper
appreciation of the cultural heritage site. Augmented reality can be a valuable tool in
adventure tourism, improving navigation, skill-building, and cultural education. The
potential for augmented reality to enhance the adventure tourism experience remains
largely untapped, and further research is needed to explore its full potential.

Nature Tourism: Augmented reality can be used to enrich the experience of tourists
exploring nature. For instance, applications can be created that allow users to identify
the flora and fauna of a region and obtain detailed information about them.

The use of augmented reality (AR) in natural tourism has been the subject of study
in several recent scientific articles. For instance, Bashir [12] developed an AR-based
mobile application to enhance the visitor experience at a wildlife sanctuary. The app
provided information on flora and fauna and allowed users to interact with 3D models
of animals. The authors found that the app increased visitor engagement and learning.
Another study by Sánchez-Ancajima [13] explored the application of AR in hiking
trails. The researchers developed an app that provided interactive information on the
trail, such as the difficulty level, distance, and elevation, as well as AR-based virtual
signs and markers. The study found that the app improved the overall hiking experience
and increased visitor satisfaction.

Bohn [14] also investigated the use of AR in natural tourism, focusing on the interpretation of archaeological sites. The authors developed a mobile app that displayed AR
overlays of the site and provided historical and archaeological information. The study
found that the app increased visitor engagement and understanding of the site. Additionally, Masneri [15] developed an AR-based app for birdwatching, which provided
interactive information about bird species and their habitats. The authors found that the
app increased visitor engagement and learning about birdwatching.

4 Methodology

The project development methodology is detailed below with the purpose of achieving
the specific objectives of the research. Each phase of the project development is outlined
below.

Phase 1 Analysis of the Current Situation: A detailed evaluation of the current state of
the technological infrastructure in the different secretariats of economic development,
environment, and tourism of the municipalities of the department of Cesar will be carried

out, as well as the technological resources available in the tourism sector. Existing municipal development plans and previous efforts to promote social and cultural tourism will also be analyzed.

Phase 2 Definition of Strategies: The objective of this phase is to gather detailed information about the necessary requirements for the construction of an augmented reality software that meets the project's objectives. To achieve this. The users' needs and specific requirements for the augmented reality software was be identified. An investigation into the different types of augmented reality technology and their applications in the tourism sector will be conducted.

The specific objectives of the augmented reality software were be established, and the functionalities it should have to meet the project's objectives will be defined. The appropriate technology for the construction of the augmented reality software was selected, considering the previously defined needs and objectives. The interface of the augmented reality software was designed, considering usability and the user experience.

Phase 3 Implementation: The strategies defined in the previous phase were put into action, by assigning the necessary human, technical, and financial resources for its development. The software development team was formed, and the selected technology was utilized to start the software development process. The interface design was implemented, considering the usability and user experience. Testing was performed at different stages of development to ensure that the software met the defined objectives and requirements. Once the software was fully developed and tested, it was launched and made available to the public. Ongoing maintenance and support were also provided to ensure that the software continued to function effectively and efficiently.

5   Augmented Reality Software in the "Balneario Paso de la Danta" in Manaure, Cesar

5.1   Balneario Paso de la Danta

The Balneario Paseo de la Danta is a tourist destination located in the municipality of Manaure, in the department of Cesar, Colombia. It is a place that offers visitors a relaxing experience and contact with nature thanks to its tranquil atmosphere and its location amidst vegetation. The resort features a natural pool fed by a nearby river and has an artificial waterfall that adds to the ambiance of the place. Visitors can enjoy the tranquility of the resort while swimming in the pool or resting in the green areas of the place, see Fig. 2.

In addition to its natural beauty, the Balneario Paseo de la Danta also offers food services with options for typical regional cuisine, as well as a camping area for those who wish to spend the night at the site, is an ideal tourist destination for those seeking a tranquil and natural environment in the department of Cesar. It offers a unique experience of contact with nature and an opportunity to relax and disconnect from daily routine.

5.2   Innovative Character and Components of the Software

See (Fig. 3).

Fig. 2. The "Balneario Paseo de la Danta"

Component of Augmented Reality Software

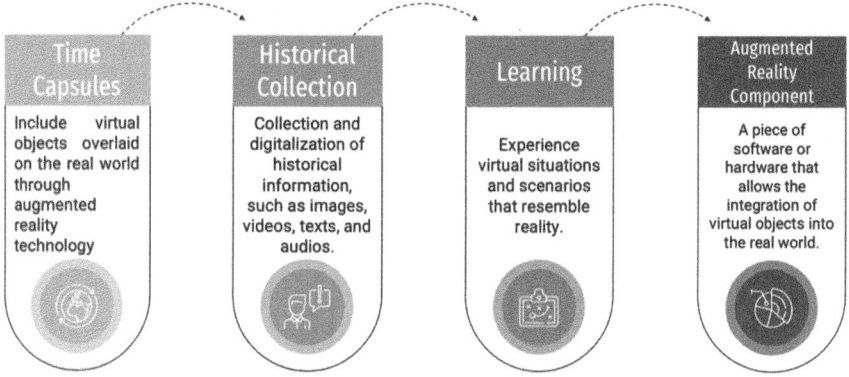

Fig. 3. Component of Augmented Reality Software.

5.2.1 Time Capsules

In the application, the creation of time capsules was used, which allows for the creation of an interactive experience for users where they can identify virtual objects overlaid on the real world. This allows them to see a futuristic proposal of the place they are visiting, enabling them to interact with the different components that are part of the capsule, enriching their knowledge about the place they are visiting, see Fig. 4.

5.2.2 Historical Information

The historical collection component used for the development of the augmented reality software constitutes an essential part of providing a complete immersive experience to

Fig. 4. Visualization of the "Balneario Paseo de la Danta"

users. This component allowed for the collection and digitization of historical information about Balneario Paseo de la Danta, including images, videos, texts, and audios, which were then integrated into the augmented reality application.

The historical collection was carried out through research and information gathering from different sources such as books, historical archives, museums, and personal testimonies of people who lived during that period. Once collected, the information was digitized and integrated into the application. In the augmented reality application, the historical collection component was used to superimpose virtual objects that represent the past of a place and its historical significance, including 3D models, historical photos superimposed on the current location, videos, and audios that tell the story of this important natural site.

The historical collection component allows users to have an enriching experience by learning about the history of the place or event they are visiting. This not only provides them with a deeper understanding of local history and culture but also allows them to connect with the place in a more personal and emotional way. Therefore, it is an important component in augmented reality software as it allows users to experience a place or event from a historical and enriching perspective, contributing to a complete and authentic immersive experience, see Fig. 5.

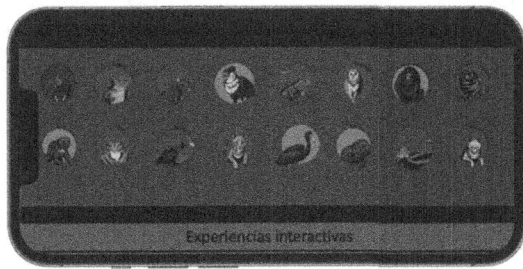

Fig. 5. Learning based on gamification of fauna and flora in Manaure.

5.2.3 Learning

The learning in the application was carried out through different approaches and strategies, focused on different users and audiences, see Fig. 6 and 7, which are detailed below:

Exploration-based learning: This is a learning strategy that allows users to explore the real world and digital information in an interactive and autonomous way. In the software, users can explore virtual objects and scenarios superimposed on the real world and learn about them by interacting with them.

Game-based learning: Games can be an attractive and motivating way to learn. In the application, this includes virtual elements superimposed on the real world, challenges and goals that must be achieved to progress in the game and learn new concepts.

Simulation-based learning: Simulators allow users to experience virtual situations and scenarios that resemble reality. Users can experience virtual situations superimposed on the real world and learn about different topics in a practical and safe way.

Instruction-based learning: In the application, users can receive information and guidance through virtual elements superimposed on the real world, such as step-by-step instructions, tips, and recommendations.

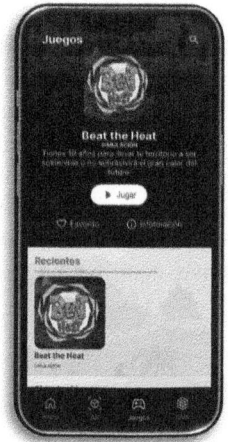

Fig. 6. Game-based learning.

Fig. 7. OVA interface.

5.2.4 Augmented Reality Component

The augmented reality component is an essential element in the development of applications that combine the real world with virtual elements. Among the most important features that can be seen in the application are its ability to integrate virtual objects into the real world, allowing users to experience them as if they were part of their environment. Additionally, these components can detect and tracking real-world elements such as surfaces, objects, and people, to overlay virtual objects in a precise and coherent manner. They are also able to adapt to different lighting and perspective conditions to ensure a fully immersive and uninterrupted experience for the user and can be easily integrated and used on different platforms and devices such as smartphones and tablets, ensuring a satisfying and high-quality augmented reality experience for users.

6   Results and Implementation of Augmented Reality Software in the "Balneario Paso de la Danta" in Manaure, Cesar

The implementation of augmented reality software in the "Balneario Paso de la Danta" in Manaure, Cesar, has shown promising results in terms of enhancing the user experience and improving the attraction of visitors to the site. The project consisted of developing an application that combines the real-world environment of the natural park with virtual elements, such as interactive maps, educational content, and digital games. The software was designed to be easy to use, and compatible with a range of mobile devices, including smartphones and tablets. Visitors can download the application for free, and use it to explore the park, learn about its flora and fauna, and engage in a variety of interactive experiences. One of the key features of the software is its ability to overlay virtual objects onto the real world, using the camera and sensors of the mobile device to detect and track the user's location and orientation.

The implementation of the augmented reality software has led to several positive outcomes for the Balneario Paso de la Danta. Firstly, it has increased the attractiveness of the site to visitors, particularly younger audiences who are more likely to be engaged by interactive digital experiences. The software has also been successful in promoting educational content, such as information about the park's biodiversity and conservation efforts, and in encouraging visitors to explore areas of the park that they might otherwise overlook.

The implementation of the software has also had a positive impact on visitor engagement and satisfaction. Feedback from users has been overwhelmingly positive, with many expressing enthusiasms for the interactive experiences and the innovative use of technology. Additionally, the software has helped to create a sense of community among visitors, who can share their experiences and compete in digital games and challenges.

The implementation of augmented reality software in the "Balneario Paso de la Danta" in Manaure, Cesar, has been a success, both in terms of enhancing the visitor experience and promoting educational content. The software has proven to be an effective tool for engaging visitors and increasing the attraction of the site. With further development and refinement, it has the potential to become an asset for the promotion of tourism and conservation efforts in the region.

7   Conclusion

This application allows the generation of nature tourism application scenarios, generates new opportunities for the economic revitalization of the municipality and positions the department of Cesar as a benchmark in the processes of implementing new technologies. It locates the users in the natural diversity of the municipality, the gastronomic offer, socialize the customs of the municipality and allows the exchange of knowledge and social appropriation of the technologies in all the users.

References

1. Ariza-Colpas, P., et al.: Cyclon language first grade app: technological platform to support the construction of citizen and democratic culture of science, technology and innovation in children and youth groups. In: Tiwary, U.S., Chaudhury, S. (eds.) IHCI 2019. LNCS, vol. 11886, pp. 270–280. Springer, Cham (2019). https://doi.org/10.1007/978-3-030-44689-5_24

2. Ariza-Colpas, P., et al.: Glyph reader app: multisensory stimulation through ICT to intervene literacy disorders in the classroom. In: Tiwary, U.S., Chaudhury, S. (eds.) IHCI 2019. LNCS, vol. 11886, pp. 259–269. Springer, Cham (2019). https://doi.org/10.1007/978-3-030-44689-5_23

3. Melis, G., McCabe, S., Atzeni, M., Del Chiappa, G.: Collaboration and learning processes in value co-creation: a destination perspective. J. Travel Res. **62**(3), 699–716 (2023)

4. Guttentag, D., Smith, S., Potwarka, L., Havitz, M.: Why tourists choose Airbnb: A motivation-based segmentation study. J. Travel Res. **57**(3), 342–359 (2018)

5. Yung, R., Khoo-Lattimore, C.: New realities: a systematic literature review on virtual reality and augmented reality in tourism research. Curr. Issue Tour. **22**(17), 2056–2081 (2019)

6. Wang, S.W., Hsu, M.K.: Airline co-branded credit cards—An application of the theory of planned behavior. J. Air Transp. Manag. **55**, 245–254 (2016)

7. Wei, W.: Research progress on virtual reality (VR) and augmented reality (AR) in tourism and hospitality: A critical review of publications from 2000 to 2018. J. Hospitality Tourism Technol. **10**(4), 539–570 (2019). https://doi.org/10.1108/JHTT-04-2018-0030

8. Penco, L., Serravalle, F., Profumo, G., Viassone, M.: Mobile augmented reality as an internationalization tool in the "Made In Italy" food and beverage industry. J. Manage. Governance **25**, 1179–1209 (2021)

9. Huang, W., Sun, M., Li, S.: A 3D GIS-based interactive registration mechanism for outdoor augmented reality system. Expert Syst. Appl. **55**, 48–58 (2016)

10. Fanini, B., Pagano, A., Pietroni, E., Ferdani, D., Demetrescu, E., Palombini, A.: Augmented Reality for Cultural Heritage. In: Springer Handbook of Augmented Reality, pp. 391–411. Springer International Publishing, Cham (2023)

11. Marino-Romero, J.A., Palos-Sanchez, P.R., Velicia-Martin, F.: Improving KIBS performance using digital transformation: study based on the theory of resources and capabilities. J. Serv. Theory Pract. **33**(2), 169–197 (2023)

12. Bashir, A.K., et al.: A Survey on Federated Learning for the Healthcare Metaverse: Concepts, Applications, Challenges, and Future Directions. arXiv preprint arXiv:2304.00524 (2023)

13. Sánchez-Ancajima, R.A., et al.: Applications of intelligent systems in tourism: relevant methods. J. Internet Serv. Inform. Secur. **13**(1), 54–63 (2023)

14. Bohn, D., Carson, D.A., Demiroglu, O.C., Lundmark, L.: Public funding and destination evolution in sparsely populated Arctic regions. Tourism Geographies **25**(8), 1833–1855 (2023)

15. Masneri, S., Domínguez, A., Sanz, M., Zorrilla, M., Larrañaga, M., Arruarte, A.: CleAR: an interoperable architecture for multi-user AR-based school curricula. Virtual Reality **27**, 1813–1825 (2023)

Simulation Analysis of Flow Patterns Inside a Cyclone Bag Separator at Increasing Clogging Levels

Federico Solari, Natalya Lysova[✉], Federico Iasoni, Giovanni Paolo Tancredi, Roberto Montanari, and Andrea Volpi

Deparment of Engineering for Industrial Systems and Technologies, University of Parma, Parco Area delle Scienze 181/a, 43124 Parma, Italy
natalya.lysova@unipr.it

Abstract. Milling plants often employ pneumatic conveying systems that use air to transport grains and flour within the plant. Before the particulate-laden air can be released into the atmosphere, however, it must be purified to meet strict environmental regulations. Cyclones and bag filters are commonly adopted for this purpose.

In this study, CFD simulation was used to reproduce the functioning of a cyclone separator equipped with a fabric bag filter. The digital model was used to simulate different clogging conditions of the fabric bags and evaluate their impact on the system functioning in terms of flow rate distribution and pressure drop. Based on the results of the simulations, additional sensor locations could be defined where the calculated velocities exhibited a clear trend as the clogging levels increased. In future research activities, additional sensors will be installed on the plant at the identified locations to validate the results obtained.

The results of the simulations performed allowed for gathering important insights for the optimization of the maintenance of the device. Indeed, in future research activities, the digital model developed aims to be implemented for predictive maintenance purposes and the generation of a Digital Twin of the pilot plant.

Keywords: cyclone separator · bag filter · CFD simulation · predictive maintenance

1 Introduction

Bag filters are widespread components in all industrial processes that generate dust, particulate matter, and other air pollutants. Indeed, they are commonly used in milling farms to control dust emissions from grain (wheat) handling, milling, and other operations. They are designed to capture and remove these contaminants from the air before it is released into the environment. The crucial issue associated with bag filters is clogging. As the filter media captures dust and other particles, it becomes increasingly clogged, reducing filtering efficiency and increasing pressure drop across the filter. This led to

© The Author(s), under exclusive license to Springer Nature Switzerland AG 2024
M. Mujica Mota and P. Scala (Eds.): EUROSIM 2023, CCIS 2033, pp. 31–43, 2024.
https://doi.org/10.1007/978-3-031-68438-8_3

decreased airflow, increased energy consumption, and reduced equipment life. There-fore, understanding the factors that influence filter clogging and developing effective strategies for maintaining filter performance is critical for ensuring the efficient opera-tion of milling farms. In this context, Computational Fluid Dynamics (CFD) modelling can represent a valuable and powerful tool for understanding the behaviour of fluids and particulate matter in complex systems.

This study focuses on the simulation of a cyclone bag separator commonly used in milling plants to abate the particulate content of the airflow (mainly flour and bran) before releasing it into the atmosphere. In the plant considered, separation is accomplished by exploiting two physical phenomena: (*i*) cyclonic effect and (*ii*) filtration effect of the fabric bags [1, 2].

To maximize the first effect and reduce the number of solid particles that have to be separated by filtration, cyclone separators are designed with a cylindrical section and tangential airflow input, to create a swirling motion that forces the heavier particles to adhere to the walls of the cylinder, dissipating energy and accumulating in the conical bottom. The air and the fraction of solid particles that are not separated by cyclonic effect, flow through the filtering bags, which, depending on the material porosity, retain most of the suspended particles so that purified air can be released into the atmosphere. As a result, solid particles tend to accumulate on the bags, progressively reducing their permeability. This phenomenon is slowed down by a bag cleaning system that, at regular intervals, flushes an amount of compressed air in countercurrent to remove part of the solid particles deposited on the surface of the bags. Such a system is able to slow down the clogging phenomenon, and thus delay bag replacement, but it is not able to eliminate it. Indeed, the flow of compressed air can remove the solid particles that weakly and superficially adhere to the bags, but it is unable to remove those solid particles which have penetrated deep into the fabric.

To maximize the filter performance the following objectives have to be achieved: (*i*) minimize the pressure drops; (*ii*) maximize the useful life of the bags.

Regarding the first objective, the filtering surface area has to be maximized and has to be properly chosen to have the highest possible permeability given the cut-off dimension.

Regarding the second objective, bag layout is another crucial aspect addressed during the design phase: it is essential to distribute the flow inside the device as evenly as possible to ensure a uniform flow rate of the particle-laden air, thus preventing excessive clogging and early failures of specific bags. Indeed, it is important that the bag clogging is as uniform as possible, and that no bag is impacted by a high flow of dirty air. Excessive impact velocity against the bag surface would, indeed, cause deeper penetration of solid particles, into the porous matrix, which would then be much more difficult to remove during the cleaning phase with compressed air.

Consequently, as time goes on, the bags become more and more clogged, offering more and more resistance to the passage of air, and the cleaning system is no longer able to restore an adequate permeability value. When most bags are clogged, it is necessary to proceed with their replacement. When an intervention of this type is scheduled, all bags are replaced.

Currently, bag replacement is carried out when a pressure gauge, installed on the machine, measures a pressure drop across the bags that is greater than a target value.

Several studies have been conducted over time to optimize the functioning of either cyclone separators [3–6] or fabric filters [7–10], by acting on geometric features, materials, and operating parameters. Some recent studies have evaluated the performances of particle separators combined with filters [11, 12].

In this study, CFD simulation was used to reproduce the functioning of a cyclone separator equipped with a bag fabric filter in standard operating conditions and at increasing levels of clogging due to the accumulation of particle matter in the porous material constituting the filtering bags. Using a digital model of the plant validated in a previous study [13], the normal functioning of the separator with clean bags was simulated to evaluate the entities of the flow rate across the device.

The results allowed for the definition of four sectors of bags, based on the flow rate passing through them. In this way, it was possible to determine which groups of bags were most exposed to the contaminated flow, thus establishing a clogging sequence. Then, the simulations of the failure conditions were performed, reproducing increasing levels of clogging of the bag filter by modifying the set-up accordingly.

The results obtained allowed for the characterization of the airflow inside the device at different clogging conditions. By leveraging this information, it is possible to predict the degree and uniformity of bag clogging and then optimize the maintenance and cleaning operations, resulting in a more efficient filtration power, longer useful life of filtration bags, as well as fewer stops and interventions. Furthermore, the validated digital model of the cyclone can be used to develop a Digital Twin of the system, which could be exploited for predictive maintenance purposes [14–17].

2   Materials and Methods

2.1 Pilot Plant

The plant analyzed in this study is a laboratory-scale reproduction of a mill separation system, consisting of a bag filter with 31 fabric bags, installed inside a cyclone separator. The pilot plant is installed at the University of Parma, in a laboratory of the Department of Engineering and Architecture. An overview of the dimensions of the filter is reported in Table 1. The plant is equipped with a suction fan, which draws the air from the environment to the inlet of the device. The air stream then flows inside the filter and

Table 1. Geometrical features of the cyclone bag separator

Dimension	Value	Units
Total separator height	5.308	m
Separator diameter	1.300	m
Conical bottom height	1.550	m
Bags length	3.000	m
Bags diameter	0.123	m

through the bags. After passing through the separation system, the purified air is released back into the atmosphere.

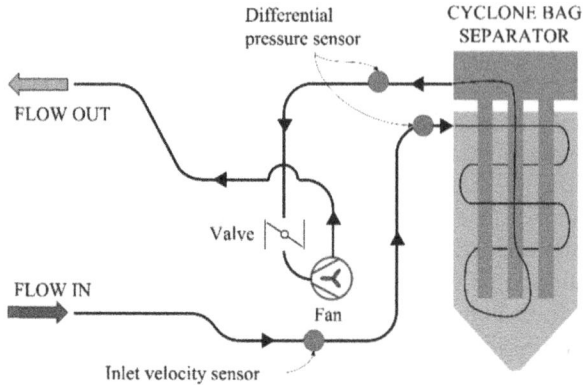

Fig. 1. Schematic representation of the pilot plant

The pilot plant is currently equipped with several sensors for the evaluation of the velocity magnitude and pressure drop inside the system.

The velocities are evaluated with hot-wire anemometers (E + E EE650-T3A6L300K2) and a Pitot tube flowmeter (KIMO Instrument CP213), while the pressure values are measured with differential pressure sensors (Endress Hauser Deltabar S PMD75). A schematic representation of the pilot plant is presented in (Fig. 1).

With regards to the fabric bags, the permeability α of the material was determined using the air permeability curve provided by the manufacturer, resulting in $\alpha = 6.72 \times 10^{-11}$ m^2. The thickness of the bag fabric was equal to 1.4 mm.

As said in the Introduction, in industrial settings, these filters are treated as "black boxes", with only the global pressure drop across the bags being monitored to evaluate the conditions of the device. In this way, the insights about what is happening inside the filter are very limited, and all the filtering bags are usually substituted at the same time when the pressure drop reaches a predefined critical value.

To gather more information about the effects of increasing clogging on the fluid dynamics inside the filter, several simulations were performed under different operating conditions.

2.2 Simulation Settings

To carry out the simulations, the 3D geometry of the filter was generated with Ansys® SpaceClaim, reproducing the real design and dimensions of the pilot plant, starting from the model used and validated in a previous study [13]. The length of both inlet and outlet ducts was increased to allow for the flow to fully develop. The geometry of the filter is presented in Fig. 2.

The 31 filtering bags were modelled inside the cyclonic separator. Based on the information provided by both the manufacturer and the end users of the system, who have observed uneven clogging of the bags along their length, each bag geometry was divided into 3 sectors: top, middle and bottom (Fig. 3). This division logic, that was based on the hypothesis of non-uniform clogging along the bag length, allowed for the simulation of different clogging conditions at specific locations. The thickness of the bags was not included in the 3D geometry but was accounted for in the simulation settings.

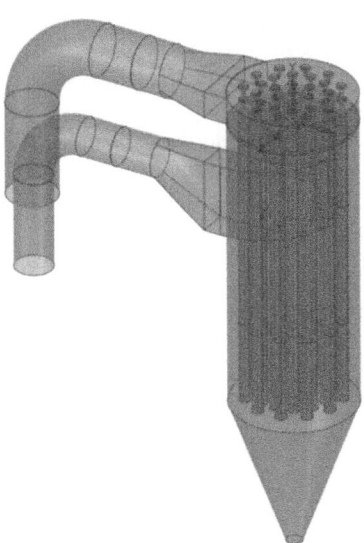

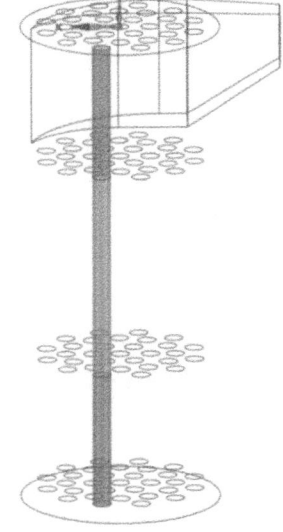

Fig. 2. 3D geometry of the filter   **Fig. 3.** Example of a bag divided into 3 sectors along its length

The domain was then discretized, based on the finite volume method, with Ansys® Meshing. The volume was divided into tetrahedral elements, which were then grouped into polyhedra. In this way it was possible to increase the overall quality of the mesh and decrease the number of elements, thus the computation time. Appropriate mesh refinement in the narrowest regions was performed to guarantee adequate precision of the calculated velocity field.

The numerical simulations were performed with Ansys® Fluent 2022 R2 software, to solve the flow and turbulence equations. "Velocity inlet" boundary condition was defined at the separator inlet, assuming an air velocity of 20 m/s. "Pressure outlet" with null gauge pressure, was defined at the outlet. The filtrating bags were modelled using the "porous jump" setting, thus modelling the filtering bags as thin, porous membranes with known permeability and thickness. Porous jump is a 1D simplification for modelling porous media that allows for the calculation of the pressure drop introduced by a medium without modelling the momentum field inside the porous volume.

A first simulation at standard operating conditions, i.e., with completely clean bags, was carried out. The air velocity streamlines, as well as the flow rate processed by each of the 31 bags, were analysed to confirm the assumption of non-uniform clogging along the bag length, determine the preferred direction of the flow, and estimate a possible "clogging order" based on the flow rate elaborated by each bag. In this way, it was possible to divide the bags into 4 groups based on their tendency to obstruct, from the most to the least prone to clogging. To reproduce the clogging, the permeability of the filtering medium was decreased proportionally to the obstruction level ($c_\%$), setting uniform boundary conditions for the bags of the same group. The decreased permeability values (α_c) were determined with (1).

$$\alpha_c = \alpha - \alpha \cdot c_\%$$ (1)

A simulation campaign was carried out by gradually increasing the clogging levels of the fabric bags (Table 2). In particular, the clogging of the 2nd case was defined based on the flow rates processed in the 1st case with clean bags; the clogging of the 3rd case was defined based on the 2nd one, and so on.

The results of the simulations were finally analysed to determine a way of detecting the clogging in a non-invasive way, without opening the separator chamber and extracting the bags, but through the values measured by sensors installed at properly defined locations of the device.

Table 2. Simulated conditions, in terms of the percentage of clogging of the three sectors of the 31 filtering bags divided into four groups

GROUP	SECTOR	CASE							
		1	2	3	4	5	6	7	8
1	top	0	20	40	55	70	80	95	95
	middle	0	0	0	5	10	20	30	50
	bottom	0	0	0	5	10	10	15	20
2	top	0	20	40	55	70	85	95	95
	middle	0	5	10	15	25	35	50	70
	bottom	0	0	0	0	5	5	10	15
3	top	0	15	30	45	60	75	90	95
	middle	0	5	10	15	25	30	40	55
	bottom	0	0	0	0	5	10	15	20
4	top	0	15	30	45	60	75	90	95
	middle	0	5	10	15	25	30	40	55
	bottom	0	0	0	0	5	10	10	15

3 Results and Discussion

The results of the simulations of the separator functioning in normal operating conditions allowed for the characterization of the air flow inside the device, highlighting a preferred path (Fig. 4).

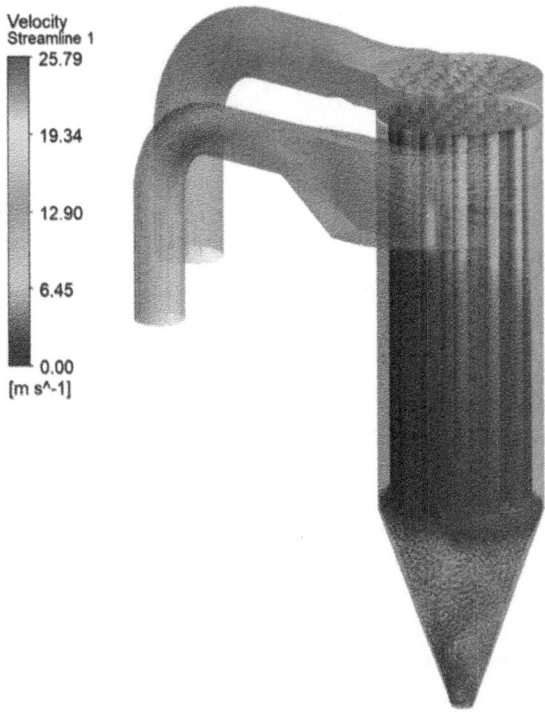

Fig. 4. Direction of the fluid flow inside the cyclonic separator

To identify the bags most subjected to contamination with solid particles, the flow rate through each bag was calculated. This allowed to sort the bags and divide them into four groups: the first group was characterized by the maximum flow rate, while the last one by the minimum. As expected, the highest flow rate occurred in correspondence with the top sectors of the bags, with lower flow rates passing through the middle and bottom sectors (Fig. 5). It was assumed that the group with the highest flow rate was the one that was most prone to clogging. This assumption was confirmed by the fact that the first group was close to the inlet of the contaminated air (Fig. 6).

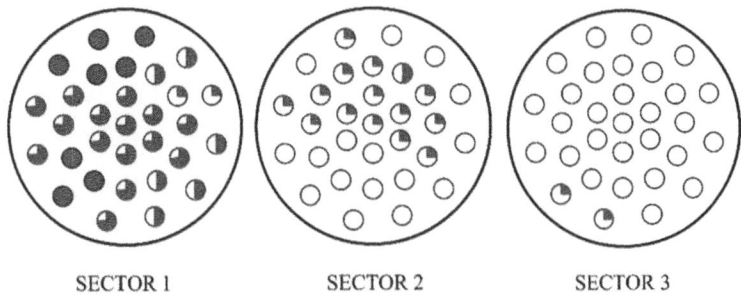

Fig. 5. Conditional formatting of the flow rate magnitudes passing through each bag

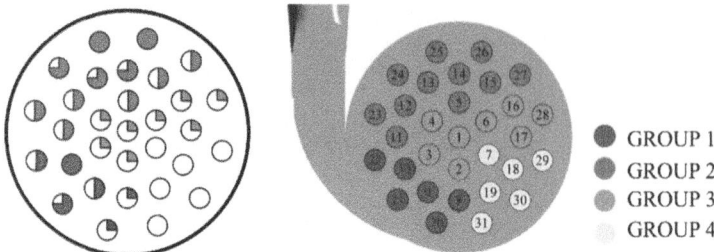

Fig. 6. Conditional formatting of the flow rates passing through the top sectors of the bags (left), and the division into 4 groups defined based on it.

Several clogging conditions were simulated according to the logics described. The results of the simulations exhibited, as expected, an increase in the system pressure drop as the clogging level increased (Fig. 7).

The results were then analyzed to identify the trends that the flow assumes in the different zones of the filter at different clogging conditions. In particular, the analysis focused on the velocity components that characterize the most swirling flows: (*i*) the tangential component and (*ii*) the ascending-descending components. These velocity values were monitored at properly defined locations, reported in Table 3. Locations where the velocity values were monitored are represented in Fig. 8.

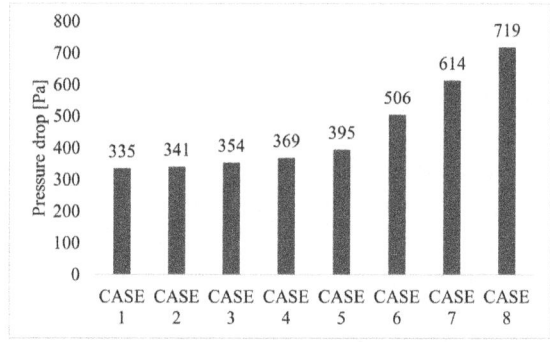

Fig. 7. Pressure drop across the separator for increasing clogging levels

Table 3. Locations where the velocity values were monitored

Location	Distance from the bag top [m]	Angular distance from the inlet section [°]
Plane 1	0.6	-
Plane 2	1.2	-
Plane 3	1.8	-
Plane 4	2.4	-
Line 1	-	0
Line 2	-	90
Line 3	-	180
Line 4	-	270

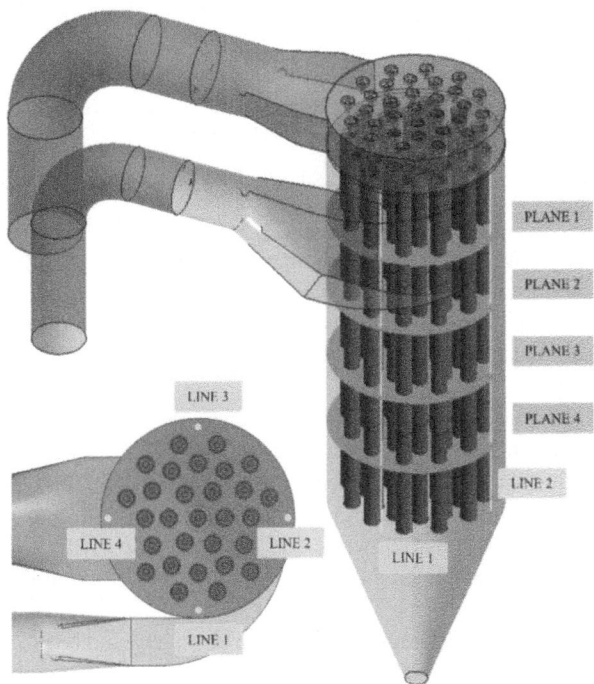

Fig. 8. Location of the line where tangential velocity was monitored

The ascending and descending components of fluid velocity were evaluated on the four transversal planes. The results are represented in Fig. 9. It can be seen from the results that on planes 2, 3 and 4 the ascending component of velocity increases, as clogging increases, while on plane 1 an opposite behaviour can be observed.

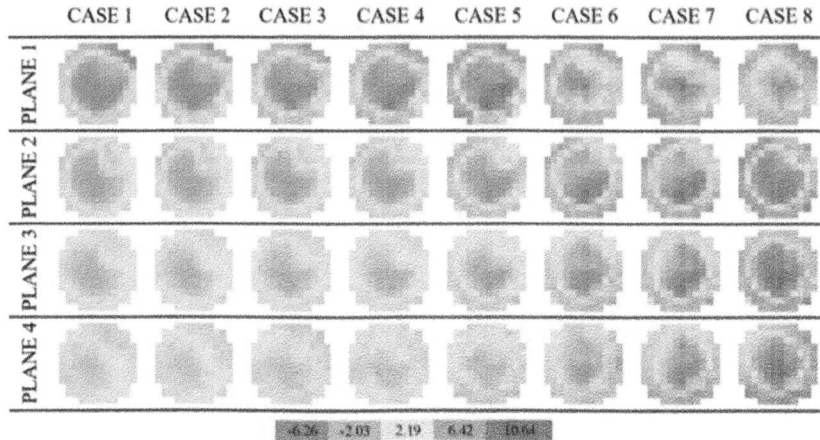

Fig. 9. Ascending velocity values on the transversal planes, formatted according to their values

As regards tangential components they were evaluated on four vertical lines placed at 90° to each other, and at a 5 cm distance from the wall. The length of the lines was chosen to detect the tangential velocity component, as a function of position, over the entire length of the bags. The values of tangential velocity calculated on the 4 lines are presented in Fig. 10.

On lines 1 and 4, no regular trends can be seen as the bag clogging changes. On lines 2 and 3, on the other hand, particularly at some points, well-defined trends can be observed as clogging increases. This means that, by installing additional sensors and performing measurements of the time gradient of the velocity at these points, it would be possible to indirectly estimate how the bag clogging is evolving. In particular, the two identified positions are located at 1.60 m from the sleeve top on line 2, and 2.40 m on line 3.

As a result, by knowing the clogging rate of the bags, it would be possible to predict, in real-time, when the next bag replacement should be planned.

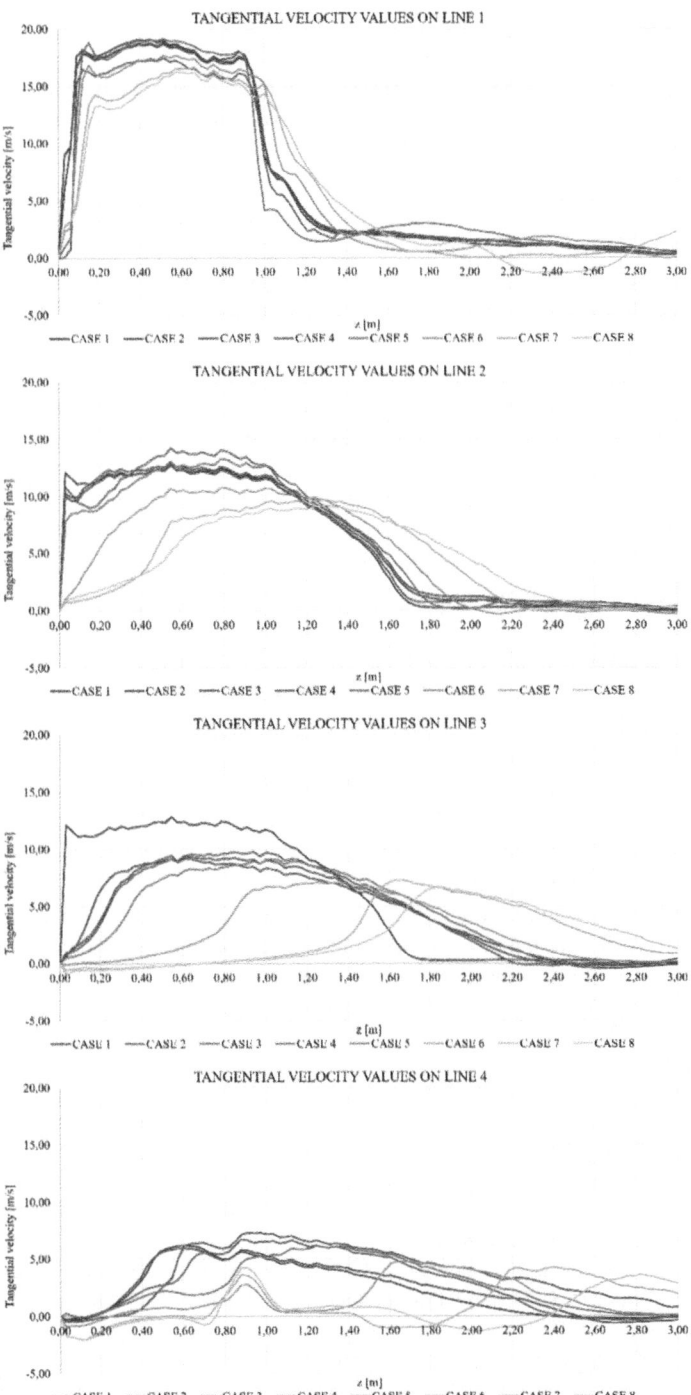

Fig. 10. Values of tangential velocity on the four lines for the 8 operating conditions simulated

4 Conclusions

Predictive maintenance is becoming more and more important, as evidenced by the growing number of studies in recent years that aim to fill the gaps in the literature, showing how, among the application fields of Industry 4.0, it was the last one in chronological order to be undertaken and deepened.

It is well known that downtimes due to maintenance and failures heavily affect the productivity and costs of industrial systems. An optimized approach to plant mainte-nance, therefore, is essential to predict the evolution in the health status of components or even entire plants, allowing to estimate the remaining useful life and prevent failure events.

This study proposes an innovative methodology, based on the results of fluid dynamic simulations, which aims to improve the current preventive maintenance planning, which is solely based on the pressure drop across the separator.

The results of the simulations showed that there is a correlation between the evolution of bag clogging and velocity components at certain points within the filter.

The results obtained allow for the characterization of the system functioning at differ-ent clogging conditions. The information acquired is of great value for the optimization of the performance and the maintenance of the filter, since clogging is one of the main issues with these devices.

Experimental tests should be conducted to validate the results obtained.

Using the methodology presented in this study, further research could be performed to identify innovative ways to identify the outbreak of filter failures, such as uneven clogging of the bags. Such an occurrence would cause a significant reduction in the service life of the bags. Being able to identify such failures in a timely manner would allow early intervention, thus quickly restoring optimal working conditions and significantly increasing the useful life of the bags, reducing maintenance interventions, and increasing the efficiency of the industrial plant.

Furthermore, the digital model developed could be used to generate a Digital Twin of the system, to be implemented on the pilot plant for real-time control and system optimization, as well as for predictive maintenance purposes.

References

1. Ogawa, A.: Mechanical separation process and flow patterns of cyclone dust collectors. Appl. Mech. Rev. **50**(3), 97–130 (1997). https://doi.org/10.1115/1.3101697
2. Löffler, F.: Fundamental principles of particle separation in fabric filters. In: Löffler, F., Diet-rich, H., Flatt, W. (eds.) Dust Collection with Bag Filters and Envelope Filters, pp. 1–54. Vieweg+Teubner Verlag, Wiesbaden (1988). https://doi.org/10.1007/978-3-663-07900-2_1
3. Misiulia, D., Antonyuk, S., Andersson, A.G., Lundström, T.S.: High-efficiency industrial cyclone separator: A CFD study. Powder Technol. **364**, 943–953 (2020). https://doi.org/10.1016/j.powtec.2019.10.064
4. Singh Brar, L.: Application of response surface methodology to optimize the performance of cyclone separator using mathematical models and CFD simulations. Mater. Today: Proc. **5**(9), 20426–20436 (2018). https://doi.org/10.1016/j.matpr.2018.06.418

5. Safikhani, H., Hajiloo, A., Ranjbar, M.A.: Modeling and multi-objective optimization of cyclone separators using CFD and genetic algorithms. Comput. Chem. Eng. **35**(6), 1064–1071 (2011). https://doi.org/10.1016/j.compchemeng.2010.07.017

6. Elsayed, K., Lacor, C.: Optimization of the cyclone separator geometry for minimum pressure drop using mathematical models and CFD simulations. Chem. Eng. Sci. **65**(22), 6048–6058 (2010). https://doi.org/10.1016/j.ces.2010.08.042

7. Krammer, G., Kavouras, A., Anzel, A.: Optimization of pulse cleaning frequency during bag filter operation. Chem. Eng. Technol. **26**(9), 951–955 (2003). https://doi.org/10.1002/ceat.200303032

8. Andersen, B.O., Nielsen, N.F., Walther, J.H.: Numerical and experimental study of pulse-jet cleaning in fabric filters. Powder Technol. **291**, 284–298 (2016). https://doi.org/10.1016/j.powtec.2015.12.028

9. Tanabe, E.H., Barros, P.M., Rodrigues, K.B., Aguiar, M.L.: Experimental investigation of deposition and removal of particles during gas filtration with various fabric filters. Sep. Purif. Technol. **80**(2), 187–195 (2011). https://doi.org/10.1016/j.seppur.2011.04.031

10. Pereira, T.W.C., Marques, F.B., Pereira, F.A.R., Ribeiro, D.C., Rocha, S.M.S.: The influence of the fabric filter layout of in a flow mass filtrate. J. Cleaner Product. **111**, 117–124 (2016). https://doi.org/10.1016/j.jclepro.2015.09.070

11. Zhang, Z., et al.: Experimental and numerical study of a gas cyclone with a central filter. Particuology **63**, 47–59 (2022). https://doi.org/10.1016/j.partic.2021.04.014

12. Xie, B., Li, S., Jin, H., Hu, S., Wang, F., Zhou, F.: Analysis of the performance of a novel dust collector combining cyclone separator and cartridge filter. Powder Technol. **339**, 695–701 (2018). https://doi.org/10.1016/j.powtec.2018.07.103

13. Solari, F., Tagliavini, G., Montanari, R., Bottani, E., Malagoli, N., Armenzoni, M.: CFD model validation of a bag filter for air filtration in a milling plant. Paper presented at the International Food Operations and Processing Simulation Workshop, FoodOPS 2017, Held at the International Multidisciplinary Modeling and Simulation Multiconference, I3M 2017, pp. 73–81 (2017)

14. Liu, H., Xia, M., Williams, D., Sun, J., Yan, H.: Digital twin-driven machine condition monitoring: a literature review. J. Sensors **2022**, 1–13 (2022). https://doi.org/10.1155/2022/6129995

15. van Dinter, R., Tekinerdogan, B., Catal, C.: Predictive maintenance using digital twins: a systematic literature review. Inf. Softw. Technol. **151**, 107008 (2022). https://doi.org/10.1016/j.infsof.2022.107008

16. You, Y., Chen, C., Hu, F., Liu, Y., Ji, Z.: Advances of digital twins for predictive maintenance. Procedia Comput. Sci. **200**, 1471–1480 (2022). https://doi.org/10.1016/j.procs.2022.01.348

17. Melesse, T.Y., Pasquale, V.D., Riemma, S.: Digital twin models in industrial operations: a systematic literature review. Procedia Manuf. **42**, 267–272 (2020). https://doi.org/10.1016/j.promfg.2020.02.084

Inventory Control in MSMEs Using Monte Carlo Simulation

Juan Genaro Chip Domínguez[✉], Guillermo Pérez Camacho[✉],
and Sonia Karina Pérez Juárez[✉]

Faculty of Engineering, UNAM, Ciudad Universitaria, Ciudad de México, Coyoacán 66455,
México
juan.genaro.chip@gmail.com, memo060412@gmail.com,
ing.karinaperezj@gmail.com

Abstract. In this article, the optimal inventory levels for a beer retailer are determined by using the Economic Order Quantity (EOQ) model with the help of @Risk tools to generate a Monte Carlo simulation. In this sense, the product catalog was classified according to the ABC system for its analysis, the products with the highest rotation were selected, limiting them to the seven most sold beers; the estimated demand was determined through probabilistic models and the total inventory costs in order to establish the optimal level of orders, the safety inventory and the reorder points. For each product selected, a test was carried out with three inventory policies in order to choose the most efficient one according to its total costs and service level.

This model has the particularity of being able to be applied in companies that identify difficulties in inventory management and thus identify areas for improvement.

Keywords: Economic Quantity Order Model (EOQ) · ABC System · Reorder Point · Demand · Costs · Monte Carlo Simulation Model

1 Introduccion

Poor warehouse management can lead to losses of more than 20% of the company's profits, approximately. It is an estimate, in addition, there are issues such as pilferage, not having the product on time or over-inventory, said Raul Avila, Commercial Director in Latin America of Datalogic. Pilferage is the fourth biggest crime suffered by Mexican companies, according to data from the National Business Victimization Survey 2016 of the National Institute of Geography (INEGI). In Mexico City alone, 5% of micro, small and medium-sized enterprises (MSMEs) claim to have suffered from it, reports the Chamber of Commerce, Services and Small Tourism (Canacope).

Commercial companies are dedicated to the purchase and sale of finished products, so part of their financial success lies in inventory control and management. The analysis of sales requires the study of turnover, identifying those items that generate more movement and profits, in order to implement strategic plans to improve investment in the procurement of merchandise.

© The Author(s), under exclusive license to Springer Nature Switzerland AG 2024
M. Mujica Mota and P. Scala (Eds.): EUROSIM 2023, CCIS 2033, pp. 44–58, 2024.
https://doi.org/10.1007/978-3-031-68438-8_4

Currently, there are still problems in inventory management in small and medium-sized enterprises. For medium-sized companies that already have knowledge in the determination of an inventory policy, the constant search for an inventory policy that minimizes total costs and satisfies demand, within optimal conditions, is a constant challenge. For small establishments, where there is little knowledge of inventory management and the economic repercussions it can have in the long term, it can be more difficult to know or decide what quantity of products to keep in inventory and when to order them so as not to experience unsatisfied demand or have expired products in their inventories. Given these current inventory problems, specifically in small establishments where it is more difficult to decide the number of products to order and to keep in inventory to minimize total costs, we intend to determine an optimal inventory policy in their products of higher rotation so that they can greatly reduce their operating costs and have the possibility of increasing the satisfaction of their customers by increasing the level of service provided. With this we could reduce the competitive disadvantage they may have with respect to medium and larger enterprises.

We focus on a company in charge of selling and providing a service to customers of various products (in total there are 508 different items), where we prioritize the first 77 best-selling products within the company, in order to determine through an analysis that 7 products contain the 10% of potential sales within the company. A series of models with a determined sequence is proposed as a proposal to have a process to follow specifically aimed at solving the problem of costs and customer service that triggers poor inventory management in small and medium enterprises.

The inventory policy was determined to better fit their resources and needs, taking as decision factors the level of service desired for their customers and the shortages allowed per cycle. For this purpose, the Monte Carlo simulation tool was used to represent the total cost of inventory management given the demand in a specific period of time for each of the selected products. The products were selected by the ABC model. For the analysis, three service levels and the Economic Order Quantity (EOQ) model were used to obtain the reorder point and the optimal replenishment quantity, respectively.

Service levels using the Monte Carlo simulation model for the products that representing 10% of total sales to minimize total inventory management cost. If an analysis of the control in the inventory system is made from different inventory policies proposed by means of a Monte Carlo simulation, we will be able to determine an convenient service level policy for its management, where it minimizes the total cost and the number of breaks.

Reach an efficient inventory management in the company could be linked to these main goals that are develop a system in order to improve inventory management, find a useful inventory policy for the selected products and establish an suitable stock point for the products.

2 State of the Art and Theoretical Framework

The objective in inventory problems is to minimize the costs of the system, subject to the restriction that a known or random demand must be satisfied, if you have a very large inventory, you will be using a lot of storage space and you will have a lot of money without producing, on the contrary, if you have a very poor inventory, this will cause continuous losses of time in the production, repair or restock of items due to lack of spare parts.

Inventories have different forms, depending on the company in question, they can be brass bars, steel sheets, iron castings, etc. However, all of them, no matter what form they take, are assets and the fact of keeping them in stock without rotation is money that costs the company interest instead of producing them.

Therefore, the problems that excess inventory can cause affect different variables within the company. They can be summarized as follows:

- Financial costs.
- Logistic costs.
- Obsolescence.
- Service level.

Likewise, the Economic Order Quantity (EOQ) model can be considered as the simplest and most fundamental of all inventory models, since it describes the important trade-off between fixed costs and inventory holding costs. This is the basis for the implementation of much more complex systems, and is the fundamental model of inventory models (Harris, 1915).

Zapata (2003) commented that there is no single methodology for implementing Monte Carlo simulation, since it is a tool that can be used in many fields and is adaptable to the type of situation being applied.

Monte Carlo simulation uses distribution functions with the purpose of performing an experiment whose results come, after a convenient number of trials (iterations), to simulate what would happen in a real system. This tool combines statistical concepts with the capacity of computer programs capable of generating random numbers and automating calculations, in order to reduce the total cost of the company's inventory (Coss, 1999).

Some other criteria that are applied as a measure of value are profit, unit cost or some measure of risk. The ABC product classification method assumes the statistical property known as the Pareto principle, which is a way of classifying products preliminarily according to certain criteria such as important impact on total value, whether inventory, sales, or costs (Dickie, 1951).

The procedure to perform the ABC classification model, based on some value criterion, is summarized in the following steps (Herron, 1976):

1. Select the value criterion (e.g., sales).
2. Sort the items in order of the importance of their value.
3. Calculate, for each item, its cumulative percentage of value and its cumulative percentage of the number of items.
4. Construct a graph of the cumulative percentage of the number of items as a function of the cumulative percentage of value.
5. Classify the items into categories A, B or C. The A items are those few "relevant" items and the C items are the many "non-relevant" items. The B's fall between the A's and C's.

3   Materials and Methods

The proposed phases for develop the simulation study were six which begin with a complete analysis of the company and its current situation, as well as identifying its main areas of opportunity within its operations that had to do with the control of its inventory system, and ends with an analysis of objective results to assess the advantages of determining the best inventory policy for each product (Fig. 1).

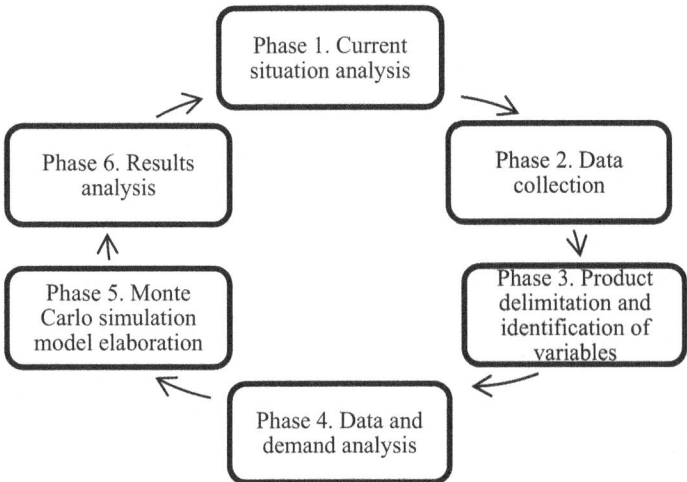

Fig. 1. The phases of research work.

3.1   Phase 1. Current Situation Analysis

A diagnostic evaluation was carried out to determine the current situation of the company and to determine whether it was feasible to implement the control of its inventory system by means of Monte Carlo simulation. The main problems were identified and followed up with the knowledge of the current state of the system. With this, access to the warehouse and it database was request to carry out the pertinent analysis.

In this phase it was found that the company does had problems within its inventory control which were the following:

- Lack of knowledge of the period for reordering without having a shortage of a product.
- Limitation in the total capacity of the warehouse in question.
- Unsatisfied demand for the products they claimed to sell the most.
- Lack of knowledge of the sales of all products from the most demanded to the one that does not generate much investment.

3.2   Phase 2. Data Collection

- The company was asked to provide a database on its sales for a period of no less than six months, including daily sales of each product.

- Initial inventory of each of the selected products was determined. To do this, the establishment was visited and a count was made of the products in stock.
- The average consumption of the refrigerators where the products were also stored was requested.
- The delivery times of each product from the time a new order was placed were consulted.
- Measurements were taken of the dimensions of the warehouse and of the products.

3.3   Phase 3. Product Delimitation and Identification of Variables

To take the significant products for this exercise, the ABC product classification method were used based on the data collected. It was decided to work with the first seven products of category A, which represented approximately 10% of the company's total sales in the study period (six months). The following Table 1 shows the selected products for analysis.

Table 1. List of the first seven products that belong to category A.

#	Product	Sales (MXN)	Accumulated sales (MXN)	% Accumulated sales
1	Bohemia Cristal 355 ml	7762.05	7762.05	2%
2	Indio Botella 1.2 L	6235.4	13997.5	4%
3	Carta Blanca 473 ml	5355	19352.5	5%
4	XX Lager 1.2 L	4339.8	23692.3	6%
5	Hidromiel Druida 355 ml	4111	27803.3	7%
6	Straffe Hendrik 330 ml	4100	31903.3	9%
7	X-Ray Imperial Stout 355 ml	3996.93	35900.2	10%

It was decided to analyze the inventory policies for each of them through simulation, and thus generate strategies to improve the costs of each of them and to be able to scale this control to the other products on sale in the same company.

The measured and collected data with the company carried to the input variables necessaries for calculating the economic quantity to be ordered are shown below (Table 2).

All values have an annualized period. The cost per unit and the restocking time were consulted with the manager on duty. The results of the costs for placing an order are the same between products because it costs the same to order one or the other. Fixed costs such as rent, electricity and the volume occupied by each unit were considered to determine the cost of keeping a product in inventory. The results of the annual demand were obtained by analyzing the establishment's sales history by product.

Table 2. Input variables.

#	Product	Cost per unit (MXN)	Cost of placing an order (MXN)	Annual inventory holding cost (MXN)	Restock time (days)	Annual demand (units)
1	Bohemia Cristal 355 ml	$ 15.13	$ 1.28	$ 0.90	3	361.74
2	Indio Botella 1.2 L	$ 35.75	$ 1.28	$ 3.85	3	232.87
3	Carta Blanca 473 ml	$ 8.50	$ 1.28	$ 1.01	3	291.65
4	XX Lager 1.2 L	$ 39.42	$ 1.28	$ 3.85	3	153.74
5	Hidromiel Druida 355 ml	$ 75.00	$ 1.28	$ 1.36	3	61.04
6	Straffe Hendrik 330 ml	$ 90.00	$ 1.28	$ 1.36	1	74.61
7	X-Ray Imperial Stout. 355 ml	$ 51.00	$ 1.28	$ 2.10	2	61.04

3.4   Phase 4. Data and Demand Analysis

For a statistical analysis of input data, we used the version 3 of Statistically Fit Software (Stat::Fit) tool to ensure that the net sales data for each product were independent of each other and could be treated within the simulation afterwards. The software is very intuitive, easy to use and adapts to the data available for the simulation, which is why it was used for this phase. Data from product 3 of category A are presented as an example.

The scatter plot was used as a test of independence in which it was expected that in all cases the points were randomly scattered within the coordinate system and to detect that the data had an independent behavior.

The autocorrelation plot was also used as a test of independence in which it was expected that the data were not autocorrelated, observing within the autocorrelation plot an oscillation between the positive and negative regions approaching the value of zero (Fig. 2).

The test of spurts was used as a test of independence to determine if there is any correlation between them. Within this test, the median test and the rotating point test are performed in which the initial hypothesis is expected to be accepted, being the null hypothesis H0 = the data set is random. As a result, in this test it was expected that the tool would yield for all cases the result of acceptance of the null hypothesis for both the median test and the rotating point test (Fig. 3).

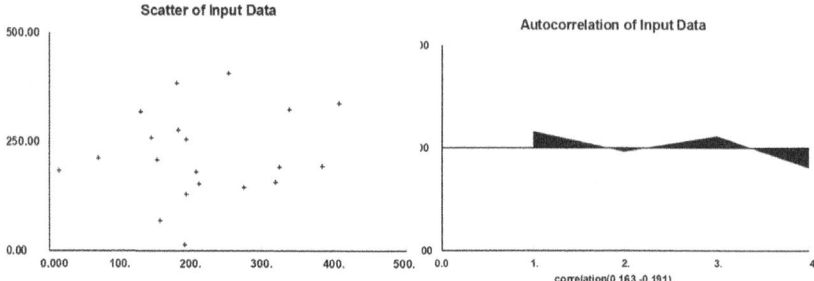

Fig. 2. Scatter plot and autocorrelation plot for product 3 of category A.

runs test on input

runs test (above/below median)

data points	21
points above median	10
points below median	10
total runs	14
mean runs	11
standard deviation runs	2.17643
runs statistic	1.3784
level of significance	0.05
runs statistic(0.025)	1.95996
p-value	0.168078
result	DO NOT REJECT

runs test (turning points)

data points	21
turning points	13
mean turnings	13.6667
standard deviation turnings	1.84692
turnings statistic	0.360961
level of significance	0.05
turnings statistic(0.025)	1.95996
p-value	0.718128
result	DO NOT REJECT

Fig. 3. Results of the tests of spurts for product 3 of category A.

The results showed that all the data analyzed for the type A products accepted the null hypothesis and it was concluded that they were independent of each other. Thus, it is certain that these values can be used as input data within the simulation and the methodology was continued.

A descriptive analysis of demand based on the historic sales database for each of the selected products was performed to determine the average demand and the standard deviation of demand in the periods necessary to calculate the economic quantity to order (S) and the reorder point (s). The results obtained are shown in the Table 3.

Table 3. Demand for type A products at delivery period.

#	Product	$\overline{D_\tau}$	σ
1	Bohemia Cristal 355 ml	29.91	20.58
2	Indio Botella 1.2 Lt	19.26	7.95
3	Carta Blanca 473 ml	24.12	12.95
4	XX Lager 1.2 Lt	12.71	8.02
5	Hidromiel Druida 355 ml	5.05	1.67
6	Straffe Hendrik 330 ml	6.17	3.72
7	X-Ray Imperial Stout. 355 ml	5.05	1.67

Using the Economic Order Quantity (EOQ) inventory control model, the size of the order to be placed each time the reorder point is reached was determined. For the application of the model, the following conditions were assumed for its application: it does not allow shortages, the ordered quantity arrives at the same time and a periodic review system is considered.

Having the data for each of the variables the quantity to order (S) was determined for each product using the following equation:

$$Q = \sqrt{(2 \cdot D \cdot CPO)/IHC} \tag{1}$$

Q Quantity to order (S)
D Demand
IHC Inventory holding cost
CPO Cost of placing an order

The results obtained of this exercise are shown at Table 4.

Three different service levels were determined to control the number of expected shortages per period from a beta level (β) to obtain three different reorder points for each product. Finally, to have three inventory policies that will be analyzed in the simulation.

A service level policy was used for reorder point where it is desired to control the amount of expected shortages per period from a beta level (β), where β is assortment rate or fraction of demand served. This level refers to the number of customers to be satisfied during delivery times. In order to obtain three inventory policies as analysis, we choose three possible service levels for each product.

A reorder point (s) was determined for each chosen policy. The service level selected for each policy was: $\beta 1 = 90\%$, $\beta 2 = 95\%$ and $\beta 3 = 98\%$. Once obtained the average demand for each product and their standard deviation of demand, the following equations were applied.

$$L(z) = ((1 - \beta) \cdot Q)/\sigma) \tag{2}$$

$$s = \overline{D}_\tau + z\sigma_\tau \tag{3}$$

$\overline{D}_\tau$ Expected value of demand at delivery period.
σ Standard deviation of demand in the delivery period.

To determine it, $L(z)$ is calculated using S (calculated above as Q) and the level β selected. Subsequently with use of the table of the Integral of the unit normal linear loss, z is obtained. Finally, the reorder point (s) is evaluated using z as the equations follow (3). The results obtained of reorder point for each product are grouped in Table 4.

The three inventory policies for each selected product were obtained through the results of the match of the optimal order quantity (S) and optimal reorder points (s) for each service level. The following table shows the results of the different inventory policies with the different service levels selected.

In order to simulate a demand closest to reality, we use the inverse transform method for discrete random variables in order to represent an aleatory demand. For each product selected, all the values of the probability distribution p(x) were calculated from the

Table 4. Inventory policies for each product with the different service levels selected.

#	Product	Quantity to order (S)	Reorder point (s) with $\beta = 90\%$	Reorder point (s) with $\beta = 95\%$	Reorder point (s) with $\beta = 98\%$
1	Bohemia Cristal 355 ml	33	2	6	10
2	Indio Botella 1.2 L	13	2	3	5
3	Carta Blanca 473 ml	28	2	4	7
4	XX Lager 1.2 L	11	2	4	5
5	Hidromiel Druida 355 ml	11	0	1	2
6	Straffe Hendrik 330 ml	12	0	1	2
7	X-Ray Imperial Stout. 355 ml	9	1	1	2

sample; subsequently, the values of the cumulative distribution P(x) were calculated. In this way, when generating pseudo-random values r ~ U (0,1), a correspondence is made to determine the value of x corresponding to P(x) and thus have the demand for an event or iteration.

3.5 Phase 5. Monte Carlo Simulation Model Elaboration

Torres & Téllez, 2022 suggest a Monte Carlo simulation modeling methodology, and this was used in this simulation study. The methodology includes the following steps:

1. Identification of the decision variables and the response variables of the system.

Following the methodology to develop the simulation model, the input variables and the performance variables were defined (Table 5).

Table 5. Variables for simulation.

	VARIABLES
TC	TOTAL COST
TCPO	TOTAL COST OF PLACING AN ORDER
VOC	VARIABLE ORDER COST
TIHC	TOTAL INVENTORY HOLDING COST
TCEO	TOTAL COST OF EMERGENCY ORDER

(continued)

Table 5. (*continued*)

	VARIABLES
QO	QUANTITY ORDERED
UC	UNIT COST
IHC	INVENTORY HOLDING COST
CPO	COST OF PLACING AN ORDER
IA	INITIAL AVAILABLE
FA	FINAL AVAILABLE
EO	EMERGENCY ORDERS
EC	EMERGENCY COST

2. Elaboration of an influence diagram with the variables.

The Influence diagram was constructed, wich allowed us to know the variables with which we were going to work for the elaboration of the simulation, as follows (Fig. 4):

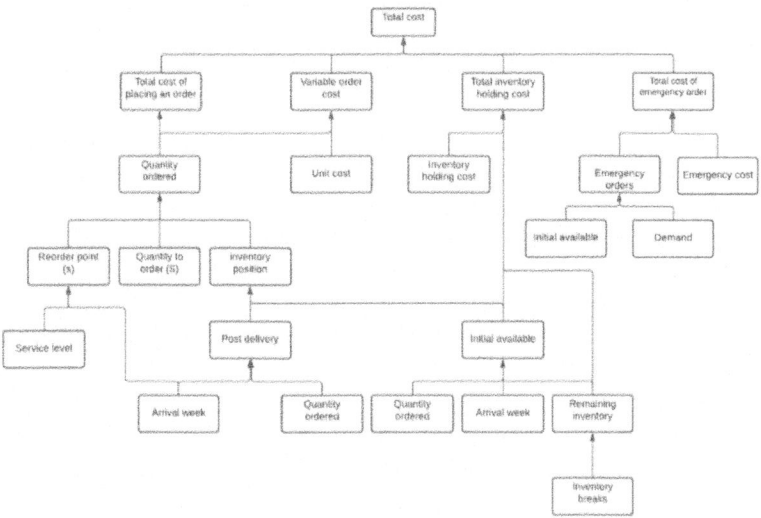

Fig. 4. Influence diagram.

3. Establishment of a mathematical representation showing the relationship between decision/input variables with the performance variables (KPI's).

Thus, from the Influence diagram we have the following mathematical representations of the relationship among variables.

$$VOC = QO(UC) \tag{4}$$

$$TIHC = IHC(IA + FA)/2 \tag{5}$$

$$TCPO = CPO; \quad with \ QO > 0 \tag{6}$$

$$TC = TCPO + TIHC + VOC \tag{7}$$

With these mathematical representations, the Monte Carlo simulation model can be developed, where the response variable is the total cost (7). For the number of order breaks per order, it means there was no product available in inventory for sale and it was denied.

4. Elaboration of a simulation model in a spreadsheet from the influence diagram and the mathematical representation. Here a deterministic model is elaborated to make a verification and validation of the model.
5. Change the decision or input variables by the distribution functions that represent them.
6. Establishment of the number of iterations necessary to guarantee the representativeness of the model.

After the success of the verification and validation of the model at step 4, the simulation was carried out with the @Risk tool in which the all input data were introduced for the analysis of each product with 10,000 iterations in order to obtain significant results. As response variables the total cost and the number of breakages per period.

The following image shows the representation of the Monte Carlo simulation in which, starting from the input values of each of the products we feed it with, it will perform the exercise for a total of 6 months and calculate the total cost and the amount of breakage in that period (Figs. 5 and 6).

CARTA BLANCA 473 ml							
Costs		**Demand distribution**			**Different inventory policies**		
Cost of placing an order	$1,28	Demand (units)	Probability	Accumulated	Policy	s	S
Unit cost	$8.50	0	0.5912	0.5912	1	2	28
Inventory holding cost	$1,01	1	0.1887	0.7799	2	4	28
Demand		2	0.1195	0.8994	3	7	28
Mean of weekly demand	6	3	0.0566	0.9560			
Standard deviation weekly demand	3	4	0.0189	0.9748		**Results**	
Initial inventory	52	5	0.0126	0.9874	Breaks		
		6	0.0126	1.0000	Total cost		

Fig. 5. Input data for the simulation of product 3 of category A.

Simulation	Variables								Costs			Random	
Day	Initial available	Post delivery	Inventory position	Quantity ordered	Arrival week	Demand	Remaining inventory	Breaks	Cost of placing an order	Variable order cost	Inventory holding cost	Random	Demand
1	52	0	52	0	0	0	52	0	0	0	$52,52	0,045	0
2	52	0	52	0	0	0	52	0	0	0	$52,52	0,499	0

Fig. 6. Example of simulation using the @Risk tool for the product 3 of the category A.

7. Elaboration of an analysis of the results of both the input variables but focusing on the response variables (KPI's).

A simulation supported by the @Risk program was run for each selected product. As a result, the application produced a series of graphs as the one shown below (Fig. 7).

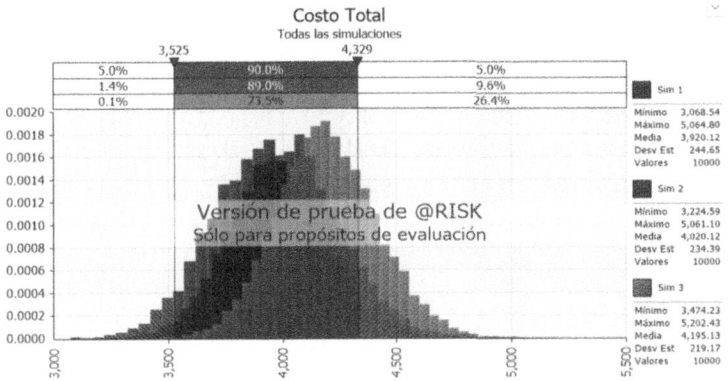

Fig. 7. Example of a graph obtained from the total costs of each policy by product.

3.6 Phase 6. Results Analysis

• The results obtained for each of the selected products were compiled and the best replenishment point was determined together with its reorder point.
• The expected future state of the company was established considering the best inventory policy for each product and comparing it with the current state.
• Recommendations were given to the company on the actions to be taken to control its products.

4 Results and Discussion

During the process, it became known issues that the company did not have, such as the knowledge of its best-selling products.

Figure 8 shows the results of a product and the impact that the decision to adopt one or another policy can have on inventory control.

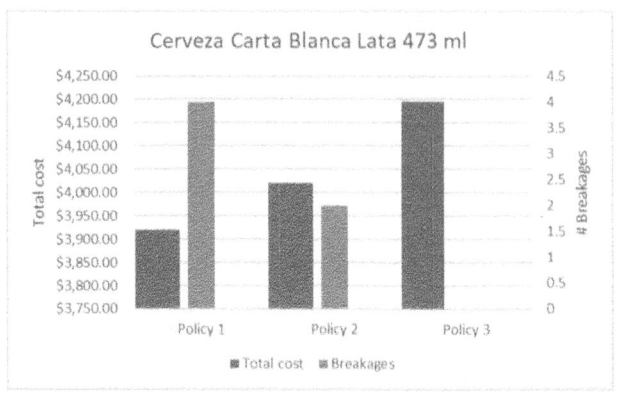

Fig. 8. Results obtained for product 3 of category A.

Table 6. Results through simulation by product.

#	Product	Policy 1 Total cost	Policy 1 Breakage	Policy 2 Total cost	Policy 2 Breakage	Policy 3 Total cost	Policy 3 Breakage
1	Bohemia Cristal 355 ml	$4.919,40	9.45	$ 5.219,09	5.28	$ 5.518,29	3.5
2	Indio Botella 1.2 Lt	$ 7.560,96	8	$ 7.560,96	5	$ 8.478,24	2
3	Carta Blanca 473 ml	$ 3.920,12	4	$ 4.020,12	2	$ 4.195,13	0
4	XX Lager 1.2 Lt	$ 6.239,57	5.32	$ 6.833,73	1.51	$ 7.139,86	0.7
5	Hidromiel Druida 355 ml	$ 2.039,05	5.36	$ 3.053,54	2.05	$ 3.204,31	0.64
6	Straffe Hendrik 330 ml	$ 4.651,34	3	$ 4.895,07	0.88	$ 5.116,89	0.25
7	X-Ray Imperial Stout. 355 ml	$ 9.327,35	0.037	$ 9.307,82	0.01	$ 9.328,31	0

For a better presentation of the results, the Table 6 shows the total costs and number of breakages for each inventory policy by product.

Through the graphs obtained with the help of @RISK software, as shown in the previous example, we were able to obtain the total costs of the policies presented, thus showing an appropriate option for each product and the company opted to take an improvement action.

As can be seen in the Table 6, the higher the service level, the more products should be in stock to protect against possible unsatisfied demand during delivery time that means the higher the total costs will be, but the average number of breakdowns in the period will be lower.

There is a relationship between the total cost and the level of service provided in which there must be a healthy balance within the company where the decision includes profitability and development. It can be seen that for all products there are less than ten units of breakage in any inventory policy. The company´s priorities must be taken in consideration for the choice of the policy to be adopted, which can be service or cost oriented.

5 Conclusions and Recommendations

It was possible to obtain through a Monte Carlo simulation the identification of the best inventory policy of a selected product that fits with the company´s necessities. With this, we have an inventory control that satisfies the current real market demand.

The results obtained from the application of a Monte Carlo simulation, managed to determine an inventory policy for each product where it has the opportunity to provide a better service while minimizing total costs and increasing the level of service provided to customers. Thus, achieving a satisfied customer demand by almost always having the product of their choice, without neglecting the other products in the same way. By applying the proposed inventory control and expecting a satisfied demand as a consequence, it would also be expected an increase in sales within the company so it is recommended to perform the proposed method periodically in order to ensure the healthy development of the company.

Monte Carlo simulation is an effective and efficient tool for correct decision making based on hard data such as output variables. In our application, it resulted in a very relevant improvement in the internal management of inventories because it had a direct impact on the total operating costs of the company and on the quality of customer service, where, as far as possible, the goal is to have inventory available for sale and at the same time not to have an over-inventory. Having an inventory policy based on hard data reduces the disadvantages of small companies with respect to medium and large ones because better service is provided and prices can be lowered without affecting profits.

Knowledge of optimal inventory levels allows the warehouse to be managed efficiently. In the case of the company, it has limited space for product storage and delicate management during replenishment is crucial. Knowing when to order and how much to order helps considerably to make the best use of the available space.

Some answers to the problems faced by the company were solved during the application of the method. One of them was the lack of knowledge of the best-selling products, which in turn had an unsatisfied demand. During the elaboration of the project we were given the task of investigating the inventory control since the company had a series of problems to work with, thus achieving that the work team has a broad knowledge of their products in terms of supply and demand, making clear which are the products that must be in constant observation according to their reorder with suppliers, minimizing the costs of ordering and maintaining.

If we apply in the future the ISO 26000, Social Responsibility Guide within the company for its inventory planning and control practices, we will be able to contribute directly and positively to sustainable development, as is the Sustainable Development Goal (SDG) 12: Responsible production and consumption; Specifically, if an optimal inventory policy is implemented for each product within the company with the proposed model, we would expect a sustainable use of resources such as economic, energy, physical space, transportation and fair competition where demand can be met at the required level, which could favorably impact the environment and consumer issues.

References

Sipper, D., Bulfin Jr., R.L.: Planeación y control de la producción (Primera Edición). McGraw-Hill Interamericana Editores (1998)

Ruíz, H.: Exceso de inventarios y la problemática de gestión (2018). https://www.teamnet.com.mx/blog/2018/04/exceso-de-inventarios

Fuentes Hernández, E.: Ampliación de la simulación Montecarlo para el control de inventarios [Tesis de Titulación]. Universidad Nacional Autónoma de México (1986)

Dongo Chira, D.F.: Desarrollo y optimización de la gestión y control del planeamiento de la producción para mejorar los indicadores de gestión en una empresa procesadora de alimentos cárnicos en la ciudad del cusco. [Tesis de Titulación]. Universidad católica de santa maría (2019)

Rodríguez, E.C.: Modelo de inventarios para control económico de pedidos en empresa comercializadora de alimentos. Revista Ingenierías Universidad de Medellín **14**(27), 163–178 (2015). https://doi.org/10.22395/rium.v14n27a10

Muñoz Sanabria, C.G., Villegas Aldana, G.A.: Diseño de un modelo de control de inventarios de producto terminado para un ingenio azucarero ubicado en el norte del valle del cauca [tesis]. Universidad del valle (2017)

Viera & Calderón: Políticas de Inventarios, Mantenimiento y Pronósticos. En Facultad de Ingeniería Universidad de la República Oriental del Uruguay. Dpto. De Investigación Operativa, In. Co., Facultad de Ingeniería. Recuperado 10 de diciembre de 2022 (2002). https://www.colibri.udelar.edu.uy/jspui/bitstream/20.500.12008/3048/1/tg-ferrer.pdf

ISO 26000 Y LOS ODS. (s. F.). En ISO.ORG. Recuperado 11 de diciembre de 2022. https://www.iso.org/files/live/sites/isoorg/files/store/sp/PUB100401_sp.pdf

Hernández Sampieri. (s. F.). Metodología de la Investigación, En ic.mujeres. Recuperado 18 de diciembre de 2022. https://www.icmujeres.gob.mx/wp-content/uploads/2020/05/Sampieri.Met.Inv.pdf

Hurtado de Barrera. (s. F.). Capítulo 3. La investigación aspectos preliminares. En Guía para la comprensión holística de la ciencia y Tipos de investigación holística. Recuperado 18 de diciembre de 2022. https://dariososafoula.files.wordpress.com/2017/01/hurtado-de-barrera-metodologicc81a-de-la-investigaciocc81n-guicc81a-para-la-comprensiocc81n-holicc81stica-de-la-ciencia.pdf

Torres & Téllez: Modelado en simulación con hojas de cálculo I [presentación de diapositivas]. Curso Simulación, FI UNAM, México (2022)

A Monte Carlo Simulation Methodology for Uncertainty Analysis in Product Recall Management

Jean P. Morán-Zabala[iD] and Juan M. Cogollo-Flórez[(✉)][iD]

Instituto Tecnológico Metropolitano – ITM, 050034 Medellín, Colombia
jeanmoran281289@correo.itm.edu.co, juancogollo@itm.edu.co

Abstract. Product recall campaigns are performed when defective or unsafe products are in the market or another supply chain stage. Product recall management is an uncertain issue for estimating the operations or processes that cause product failure. It is necessary to use quality engineering techniques and tools for analyzing uncertainty in product recall management. Therefore, we propose a Monte Carlo Simulation for Uncertainty Analysis in Product Recall (MCS-UAPR) methodology to improve decision-making in product recall management. It was applied in a company in the automotive sector and made it possible to identify the operations with the highest impact on the total unit recall cost.

Keywords: Automotive industry · Product recall management · Monte Carlo Simulation applications · Quality costs · Quality engineering

1 Introduction

Product Recall is the collection or replacement of defective products with a potential or actual risk to consumer health. Product recalls are frequent and can have adverse consequences for companies and consumers [1]. Product recall management focuses on collecting all the defective or unsafe products in the production process, at points of sale, or at consumer facilities [2]. Product recall management involves identifying and managing supply chain risks by jointly considering the recall total cost and the product quality guarantee [3].

Product recall management is a current and complex issue to ensure safety, quality, and consumer protection [4, 5]. Product recall aims to prevent problems caused by the failure of some part or the whole product. All manufacturers are responsible for the product quality delivered to customers [6, 7]. In the last decade, there has been a trend for outsourcing New Product Development (NPD) activities, so it is necessary to strengthen product quality beyond providing warranties [8].

Product recall is a complex phenomenon with adverse consequences for companies and supply chains [9, 10]. However, decisions can be made to reduce the likelihood of product recall [11, 12]. Regardless of the product recalls severity, there will be opportunities to improve processes and gain consumer confidence [13]. Product recalls are not

© The Author(s), under exclusive license to Springer Nature Switzerland AG 2024
M. Mujica Mota and P. Scala (Eds.): EUROSIM 2023, CCIS 2033, pp. 59–70, 2024.
https://doi.org/10.1007/978-3-031-68438-8_5

theoretical possibilities, but they are a frequent phenomenon in the business activities of manufacturers and supply chains [2]. Ben-Shahar [14] states that damages to the product quality may appear over time, and the manufacturer will have to decide whether to continue selling the product or to recall it from the market.

Although the main goal of product recall is to improve consumer safety, other drivers also need to be considered [15]. Research in product recall has focused on the requirement of running the process without examining all the statutory provisions [16]. There is a lack of jurisprudence on product liability [17–20], and attempts have been made to include these provisions in product safety [21].

A product recall can affect the brand value and reputation of the company [22]. Although the financial impact is known as a research issue [23], one of the biggest problems is compromising the health or safety of users with the product when the failures are identified in the consumption stage [24–26]. For example, the automotive industry stands out for the following needs: diversification, competitive prices, quality, short launching times, and market expansion. Several studies have identified that the final assembly process in the automotive manufacturing process requires more attention from the engineering area since it is important for the final quality product [24].

The business sectors with the highest occurrence of product recall levels are the toy industry, the automotive, electrical supplies, and the food industry [27]. The current challenges are centered on the follow-up and recalled products evaluation on the number of registered databases through the implementation of simulation tools to know the behavior of the process uncertainty, thus improving the decision-making.

Therefore, this work develops a methodology to analyze the uncertainty of recall processes through MCS to identify the variables with the highest impact and estimate their variability. The main contribution of this work is developing a flexible simulation methodology to solve complex quality engineering problems. Thus, it is possible to predict the behavior of the product recall variables and eliminate or reduce the deviations concerning the fulfillment of the target and improve the decision-making process.

2 Monte Carlo Simulation (MCS)

Simulation is based on building a logical-mathematical model of system or decision processes, conducting experiments to understand the behavior system, or assisting in decision-making. A significant element in simulation processes is the identification of the appropriate probability distributions for the data. The main objective of the simulation is to conduct experiments with the model and analyze the results.

MCS (or Multiple Probability Simulation) is a mathematical technique for estimating the possible outcomes of an event [28, 29]. MCS is based on the differential principle, where a and b are the limits and m is a finite parameter that increases as a function of the desired accuracy allowing to predict the results further in time with greater precision:

$$\int_a^b f(x)dx = \frac{(b-a)}{m} \sum_{n=1}^{m} f(x_n) \tag{1}$$

MCS has been applied for assessing the impact of risk in many real-life scenarios, such as artificial intelligence (AI), sales forecasting, project management, pricing, and stock prices [30, 31]. It also provides advantages for predictive models with fixed inputs, such as the ability to perform sensitivity analysis or calculate the input correlations.

Sensitivity analysis allows making decisions from the impact of individual inputs on an outcome [32, 33]. MCS seeks to predict outcomes set based on an estimated range of values against a set of fixed input values [34]. Regardless of the tool used, the MCS consists of 3 basic steps:

1. Set up the predictive model: the dependent variable (Y) is to be predicted, and the independent variables (X_n) are identified.
2. Specify the probability distributions of the independent variables (X_n). In addition, historical data should be used to define a range of probable values.
3. Run simulations repeatedly to generate random values of the independent variables (X_n).

Therefore, the application of MCS requires having enough historical information to establish the variables behavior and how they affect or are affected by other ones. Process errors may occur due to both bias and imprecision. Therefore, MCS models allow for predicting the risk proportion in response to the measurement results of determined bias and imprecision values [35]. MCS modeling is a statistical technique for assessing risks and uncertainty in complex systems. This technique is based on multiple random scenarios generation to simulate the behavior of a system under different conditions and parameters [36, 37].

The main contribution of Monte Carlo simulation modeling is to understand the uncertain nature of a system. Patterns and trends can be identified by generating multiple scenarios, and the probability of different outcomes occurring can be measured [38]. In addition, MCS modeling assesses the system sensitivity to some variables and parameters, and it allows the identification of areas of the system most susceptible to uncertainty and focuses on them to mitigate risks [39, 40].

3 Data Analysis

The analysis was carried out with a database of an automotive industry company. For this purpose, five input variables and one output variable are identified and codified, as shown in Table 1. The data are within two years and twenty-nine periods are classified for each of the variables to determine the behavior of these through the classifications (See Fig. 1). X_1 is the dark blue line, X_2 is the red line, X_3 is the green line, X_4 is the purple line, X_5 is the light blue line, and Y is the orange line.

Figure 1 shows that the variable with the lowest trend in all periods is X_2, and the variables with the highest trend throughout the twenty-nine periods relative to Y are X_1 and X_4. In addition, the highest total unit recall cost was in period twenty-six, reaching a value of $6,790, and the lowest total unit recall cost is in period four, reaching a value of $648.

Table 1. Model Variables. (Own Elaboration).

Variable	Description	Unit
X_1	*Unit Development Cost*	USD
X_2	*Unit Cost per Shipment*	USD
X_3	*Unit Logistics Cost*	USD
X_4	*Unit Production Cost*	USD
X_5	*Unit External Failures Cost*	USD
Y	*Total Unit Recall Cost*	USD

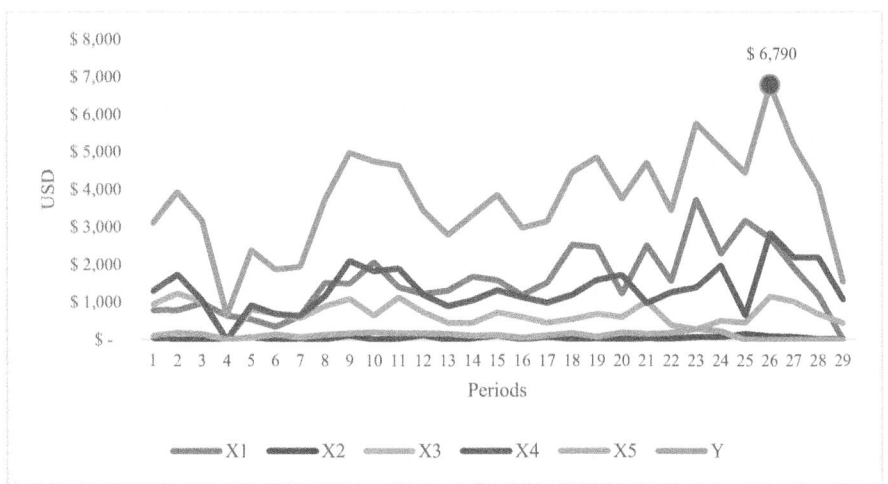

Fig. 1. Data behavior of the model variables. (Own elaboration).

4   MCS-UAPR Methodology

The Monte Carlo Simulation for Uncertainty Analysis in Product Recall (MCS-UAPR) methodology consists of four steps (Fig. 2). First, probability distributions were established to model the uncertainty or variability in the input data or model parameters and obtain multiple possible scenarios to calculate the probability of certain events or outcomes occurring. The mathematical model is then established to represent the system and is used to generate random values and calculate simulation results. It allows the probability of different outcomes to be estimated and informed decisions to be made based on these results.

Once the mathematical model and the probability distributions of the input data and model parameters have been defined and established, the simulations are run. During the simulation runs, random values of the probability distributions allow us to obtain multiple scenarios of the model.

Analysis of the simulation results provides valuable information about the system or process being simulated, such as the probability of certain events or outcomes being, or the model sensitivity to different input parameters. It aims to make informed decisions, such as optimizing the design of a product or process, evaluating the risks of strategies, or reducing the total costs of a company, product, or service. The results of applying the MCS-UAPR methodology are detailed in the following section.

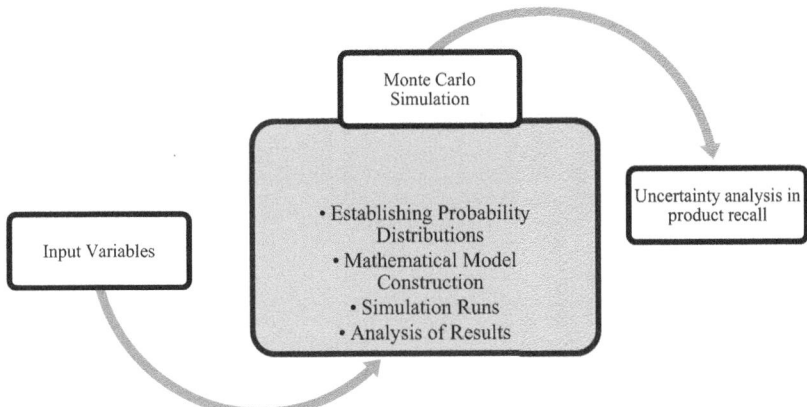

Fig. 2. Methodology steps. (Own elaboration).

5   Results

5.1   Establishing Probability Distributions

Seventy-seven probability distributions were tested using the Anderson-Darling goodness-of-fit test (AD test) to determine the extent to which the data fit a specific distribution (see Table 2). Anderson-Darling is a statistical goodness-of-fit test used to assess whether a data sample follows a specific distribution when high sensitivity in detecting deviations in the distribution tails is required [41]. This test compares the theoretical cumulative distribution function (CDF) and the empirical CDF of the data. In addition, the Anderson-Darling test is also helpful when working with moderate or small sample sizes, as it is less susceptible to type II errors compared to other goodness-of-fit tests [42, 43]. Graphic results of establishing the probability distributions are shown in Figs. 3, 4, 5, 6 and 7.

Table 2. Variables best-fit distribution adjustment. (Own Elaboration).

Variable	Best-fit distribution	Parameters
X_1	*Beta Binomial*	[431769;0.7701;4.2079]
X_2	*Beta*	[0.752;4.3577; −0.0001;466190.1575]
X_3	*Beta Binomial*	[0.2623;3.0366; −0.0001;246291.3044]
X_4	*Beta Binomial*	[409782;1.5033;5.1128]
X_5	Beta	[431769;0.7721;4.226]

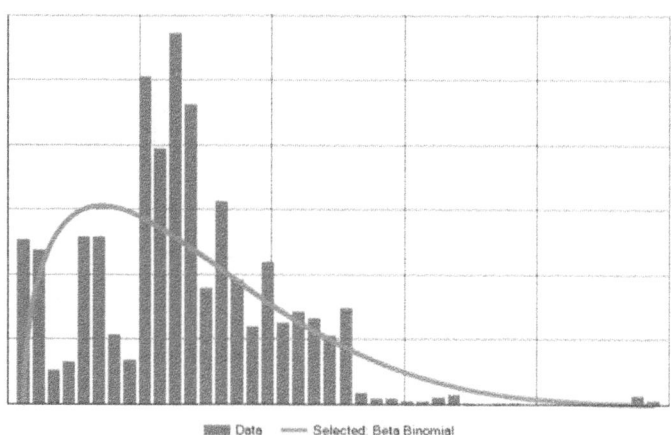

Fig. 3. X_1 variable best-fit distribution adjustment. (Own elaboration).

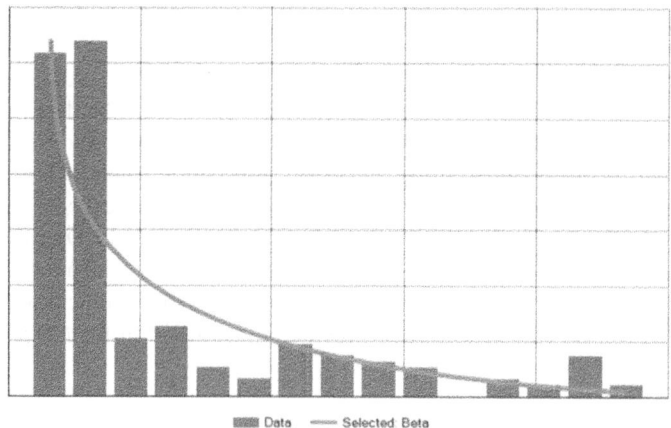

Fig. 4. X_2 variable best-fit distribution adjustment. (Own elaboration).

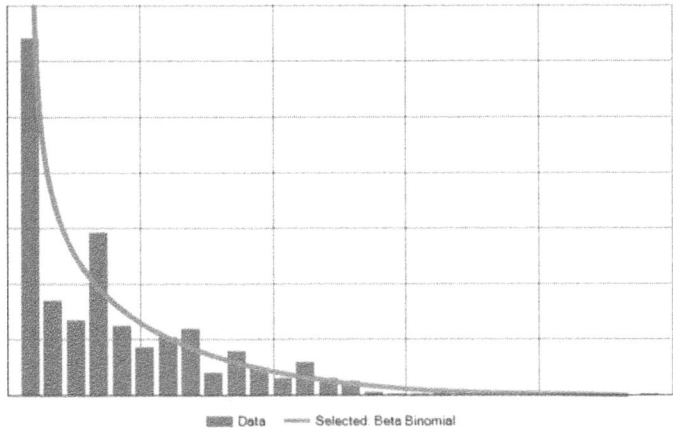

Fig. 5. X_3 variable best-fit distribution adjustment. (Own elaboration).

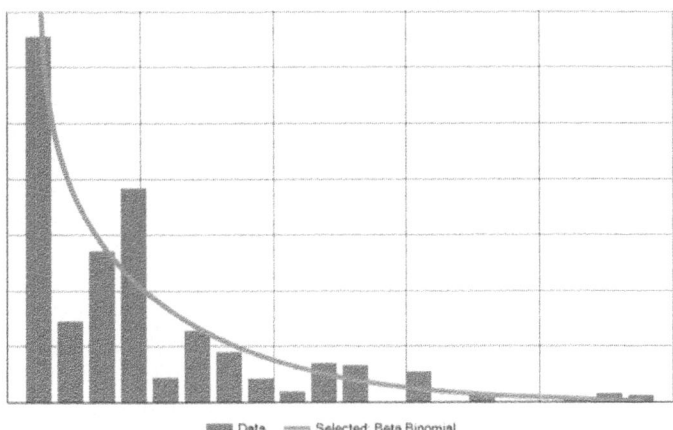

Fig. 6. X_4 variable best-fit distribution adjustment. (Own elaboration).

5.2 Mathematical Model Construction

The MCS mathematical model is detailed in (2) and is based on a multiple regression analysis. The probability distributions are noted as follows: E is Exponential, B is Beta and BeB is Beta Binomial. Thus, considering the relationship coefficients between the variables detailed in Table 3 and the probabilistic distributions (See Table 2), 10,000 simulation runs were performed.

$$
\begin{aligned}
Y = \{X_1 \rightarrow E[0; 68240.7357]\} + \{X_2 \rightarrow B[0.752; 4.3577; -0.0001; \\
466190.1575]\} + \{X_3 \rightarrow BeB[431769; 0.7721; 4.226]\} + \{X_4 \rightarrow BeB \\
[431769; 0.7701; 4.2079]\} + \{X_5 \rightarrow B[0.2623; 3.0366; -0.0001; \\
246291.3044]\}
\end{aligned}
\tag{2}
$$

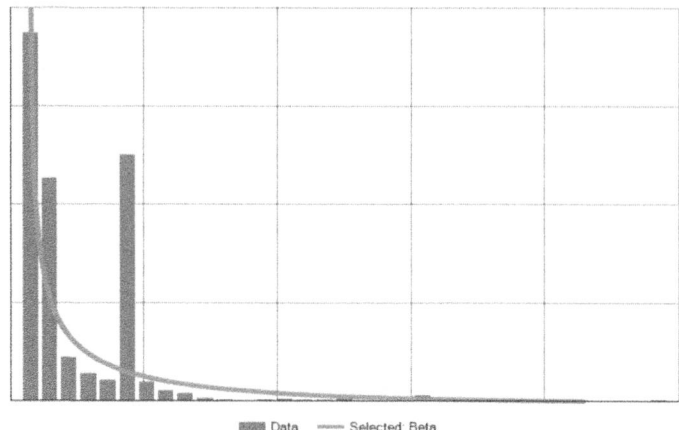

▨▨▨ Data   ——— Selected: Beta

Fig. 7. X_5 variable best-fit distribution adjustment. (Own elaboration).

Table 3. Correlation Matrix. (Own elaboration).

Origin Variable	Destination Variable	Correlation Coefficient
X_1	X_2	0.4392
X_1	X_3	−0.0576
X_1	X_4	0.3169
X_1	X_5	0.2652
X_1	Y	0.8081
X_2	X_3	0.0891
X_2	X_4	0.1868
X_2	X_5	−0.0539
X_2	Y	0.4161
X_3	X_4	0.5680
X_3	X_5	0.0226
X_3	Y	0.4379
X_4	X_5	0.0924
X_4	Y	0.7861
X_5	Y	0.2751

5.3 Simulation Runs

According to the simulation results, with 95% confidence, the values of X_1, X_2, X_3, X_4, X_5 and Y will be close to \$51.2, \$40.8, \$18.5, \$37.2, \$21, and \$109 respectively. Figure 8 shows the histogram of simulation results for the total recall cost (Y), where the green bars represent the frequency of the data, the blue line represents the cumulative frequency, and the red bar represents the mean value of the simulated data.

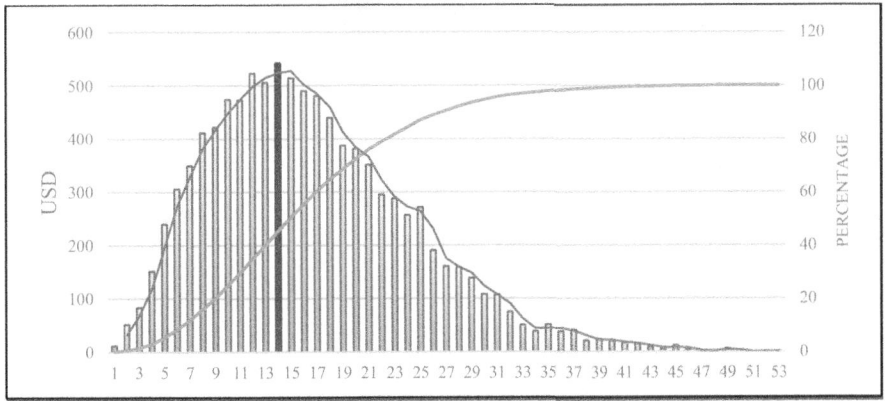

Fig. 8. Histogram of simulation result for total unit recall cost (Y). (Own elaboration).

5.4 Analysis of Results

The reliability percentage selected for the confidence intervals is 95%. Thereby, there is a 95% probability that the total unit recall cost is between \$100 and \$120. A sensitivity analysis was performed to predict the optimal results of the response variable (Y),

Fig. 9. Pearson sensitivity analysis for total unit recall cost. (Own elaboration).

considering the uncertainty conditions, and using Pearson's coefficient as a statistic to determine the strength of the relationship between the variables (See Fig. 9).

In Fig. 9, it is possible to state all the model variables are relevant to the MCS-UAPR methodology application proposed in this work. However, some variables have a higher impact on the output variable. Thus, the variables that most influence the total unit recall cost are unit cost per shipment (X_2) and unit production cost (X_4). Therefore, it is necessary to focus recall cost reduction decisions on these two variables.

6 Conclusions

Product recall is an important managerial decision to provide consumer safety when potentially hazardous products are identified. Therefore, it is essential to improve quality control processes to identify hazards that may occur in the market and establish control processes to ensure product reliability, and in turn, gain consumers trust. At this goal, MCS facilitates uncertainty analysis to know the risks by permuting ranges of values.

This MCS-UAPR methodology is a novel approach to the integration and application of simulation tools in product recall management. It allows a complete mapping of the performance of operations that could determine the presence of defective products in the market. The MCS-UAPR methodology has a generalized approach that allows its application in other business sectors interested in making decisions to reduce product recall.

The main contribution of this work is the implementation of a novel methodology for analyzing recall uncertainty in the automotive industry. Future work will focus on the development of lexicographic models combining optimization and simulation for solving complex problems in other supply chains. This topic is a relevant research gap in quality engineering.

References

1. Kumar, S., Schmitz, S.: Managing recalls in a consumer product supply chain – root cause analysis and measures to mitigate risks. Int. J. Prod. Res. **49**(1), 235–253 (2011). https://doi.org/10.1080/00207543.2010.508952
2. Copeland, T., Jackson, G., Morgan, F.: An update on product recalls. J. Mark. Channels **11**(2–3), 103–121 (2004). https://doi.org/10.1300/J049v11n02_06
3. Schniederjans, D., Khalajhedayati, M.: Product recall strategy in the supply chain: utility and culture. Int. J. Qual. Reliab. Manag. **38**(1), 195–212 (2021). https://doi.org/10.1108/IJQRM-03-2019-0077
4. Restrepo-Hincapié, M., Cogollo-Flórez, J.M.: The product recall problem: academic and legal approaches. Revista Lasallista de Investigación **18**(1), 114–133 (2021). https://doi.org/10.22507/rli.v18n1a8
5. Xie, X., Dai, B., Du, Y., Wang, C.: Contract design in a supply chain with product recall and demand uncertainty. IEEE Trans. Eng. Manage. **70**(1), 232–248 (2023). https://doi.org/10.1109/TEM.2021.3062279
6. Vizcarra, R.: Cars, recalls and INDECOPI: comments through the light of recent decisions and precedents of INDECOPI. IUS ET VERITAS **46**, 400–418 (2013)

7. Rodriguez, G., Miró, M.: Peltzman effect peruvian style: the law and economics of recalls for defective products. IUS ET VERITAS **60**, 134–144 (2020)
8. Kalaignanam, K., Kushwaha, T., Nair, A.: The product quality impact of aligning buyer-supplier network structure and product architecture: an empirical investigation in the auto-mobile industry. Cust. Need. Solut. **4**(1), 1–17 (2017). https://doi.org/10.1007/s40547-017-0074-y
9. Zhao, X., Li, Y., Flynn, B.: The financial impact of product recall announcements in China. Int. J. Product. Econ. **142**(1), 115–123 (2013). https://doi.org/10.1016/j.ijpe.2012.10.018
10. Mayo, K., Ball, G., Mills, A.: CEO tenure and recall risk management in the consumer products industry. Product. Operat. Manage. **31**(2), 743–763 (2021). https://doi.org/10.1111/poms.13576
11. Gibson, D.C.: Public relations considerations of consumer product recall. Public Relat. Rev. **21**(3), 225–240 (1995). https://doi.org/10.1016/0363-8111(95)90023-3
12. Quanhong, L., Xin, Z.: Corporate product recall and its influence on corporate performance. In: 3d International Conference on Advanced Information and Communication Technology for Education (ICAICTE-2015), pp. 241–246. Atlantis Press (2015).
13. Chandran, R., Lancioni, R.A.: Product recall: a challenge for the 1980s. Int. J. Phys. Distrib. Mater. Manage. **11**(8), 46–55 (1981). https://doi.org/10.1108/eb014520
14. Ben-Shahar, O.: How liability distorts incentives of manufacturers to recall products. In: Law & Economics Working Papers (2005). https://doi.org/10.2139/ssrn.655804. Last accessed 30 Dec 2022
15. Kalaignanam, K., Kushwaha, T., Eilert, M.: The impact of product recalls on future product reliability and future accidents: evidence from the automobile industry. J. Market. **77**(2), 41–57 (2013). https://doi.org/10.1509/jm.11.0356
16. Choudhary, A.: A proposal for introducing mandatory product recall in the consumer protection act. Int. J. Consum. Law Pract. **3**(7), 145–161 (2015)
17. Forgays, J.A.: The food and drug administration's power to recall a harmful product and other remedial actions: the powerless consumer. Vermont Law Rev. **10**, 129 (1985)
18. Warmer, R.C.: Judges as regulators: using injunctive relief to recall products. Defense Counsel J. **68**, 299 (2001)
19. Lens, J.W.: Product recalls: why is tort law deferring to agency inaction. John's Law Rev. **90**, 329 (2016)
20. Schwartz, T.M., Adler, R.S.: Product recalls: a remedy in need of repair. Case Western Reserve Law Rev. **34**, 401 (1983)
21. Venugopal, R., Tollefson, L., Hyman, F.N., Timbo, B., Joyce, R.E., Klontz, K.C.: Recalls of foods and cosmetics by the US food and drug administration. J. Food Protect. **59**(8), 876–880 (1996). https://doi.org/10.4315/0362-028X-59.8.876
22. Wei, J., Wang, Q., Yu, Y., Zhao, D.: Public engagement in product recall announcements: an empirical study on the Chinese automobile industry. J. Market. Commun. **25**(4), 343–364 (2019). https://doi.org/10.1080/13527266.2016.1251487
23. Ni, J.Z., Flynn, B.B., Jacobs, F.R.: Impact of product recall announcements on retailers finan-cial value. Int. J. Product. Econ. **153**, 309–322 (2014). https://doi.org/10.1016/j.ijpe.2014.03.014
24. Baraldi, E.C., Kaminski, P.C.: A study on the causes of recall in automotive vehicles marketed in brazil. SAE Tech. Paper (2016). https://doi.org/10.4271/2016-36-0169
25. Pyke, D., Tang, C.S.: How to mitigate product safety risks proactively? process, challenges and opportunities. Int. J. Logistics: Res. Appl. **13**(4), 243–256 (2010). https://doi.org/10.1080/13675561003720214
26. Kumar, S.: A knowledge based reliability engineering approach to manage product safety and recalls. Expert Syst. Appl. **41**(11), 5323–5339 (2014). https://doi.org/10.1016/j.eswa.2014.03.007

27. Cogollo-Flórez, J.M., Restrepo-Hincapié, M.: A taxonomical classification proposal for product recalls. Revista UIS Ingenierías **20**(3), 111–120 (2021). https://doi.org/10.18273/revuin.v20n3-2021007

28. Salazar, E., Alzate, W.: Application of the Monte Carlo simulation in the projection of statement of profit or loss: a case study. Espacios **39**(51), 11–18 (2018)

29. Jacobs, F.R., Chase, R.B.: Operations and Supply Chain Management, 16th edn. McGraw-Hill, New York, NY (2021)

30. Ji, W., AbouRizk, S.M.: Simulation-based analytics for quality control decision support: pipe welding case study. J. Comput. Civ. Eng. **32**(3), 1–9 (2018). https://doi.org/10.1061/(asce)cp.1943-5487.0000755

31. Shangguan, J., Guo, H., Yue, M.: Robust energy management of plug-in hybrid electric bus considering the uncertainties of driving cycles and vehicle mass. Energy **203**, 117836 (2020). https://doi.org/10.1016/j.energy.2020.117836

32. Saviano, A., Lourenço, F.: Measurement uncertainty estimation based on multiple regression analysis (MRA) and Monte Carlo (MC) simulations – Application to agar diffusion method. Measurement **115**, 269–278 (2017). https://doi.org/10.1016/j.measurement.2017.10.057

33. Wen, X.L., Zhao, Y.B., Wang, D.X., Pan, J.: Adaptive Monte Carlo and GUM methods for the evaluation of measurement uncertainty of cylindricity error. Precision Eng. **37**(4), 856–864 (2013). https://doi.org/10.1016/j.precisioneng.2013.05.002

34. Ioannides, A.M., Tingle, J.S.: Monte carlo simulation for flexible pavement reliability. In: International Airfield and Highway Pavements Conference 2021, pp. 13–25. American Society of Civil Engineers (2021). https://doi.org/10.1061/9780784483503.002

35. Inman, M., Parker, K., Strueby, L., Lyon, A.W., Lyon, M.E.: A simulation study to assess the effect of analytic error on neonatal glucose measurements using the canadian pediatric society position statement action thresholds. J. Diab. Sci. Technol. **14**(3), 519–525 (2019). https://doi.org/10.1177/1932296819884923

36. Dehghani, M., Saghafian, B., Saleh, F.N., Farokhnia, A., Noori, R.: Uncertainty analysis of streamflow drought forecast using artificial neural networks and Monte-Carlo simulation. Intl. J. Climatol. **34**(4), 1169–1180 (2013). https://doi.org/10.1002/joc.3754

37. Biwer, A., Griffith, S., Cooney, C.: Uncertainty analysis of penicillin V production using Monte Carlo simulation. Biotechnol. Bioeng. **90**(2), 167–179 (2005). https://doi.org/10.1002/bit.20359

38. Hofer, E.: (2008): How to account for uncertainty due to measurement errors in an uncertainty analysis using Monte Carlo simulation. Health Phys. **95**(3), 277–290 (2008). https://doi.org/10.1097/01.HP.0000314761.98655.dd

39. Thompson, K.M., Burmaster, D.E., Crouch, E.A.C.: Monte Carlo techniques for quantitative uncertainty analysis in public health risk assessments. Risk Anal. **12**(1), 53–63 (1992). https://doi.org/10.1111/j.1539-6924.1992.tb01307.x

40. Arunraj, N.S., Mandal, S., Maiti, J.: Modeling uncertainty in risk assessment: an integrated approach with fuzzy set theory and Monte Carlo simulation. Accid. Anal. Prev. **55**, 242–255 (2013). https://doi.org/10.1016/j.aap.2013.03.007

41. Wijekularathna, D.K., Manage, A.B., Scariano, S.M.: Power analysis of several normality tests: a Monte Carlo simulation study. Commun. Stat.-Simul. Comput. **51**(3), 757–773 (2020). https://doi.org/10.1080/03610918.2019.1658780

42. Razali, N.M., Wah, Y.B.: Power comparisons of shapiro-wilk, kolmogorov-smirnov, lilliefors and anderson-darling tests. J. Stat. Model. Analytics **2**(1), 21–33 (2011)

43. Li, Y., Singh, R.S., Sun, Y.: Goodness-of-fit tests of a parametric density functions: Monte Carlo simulation studies. J. Stat. Res. **39**(2), 111–133 (2005)

The Use of Simulation Throughout the Lifecycle of Advanced Software Algorithms – A Case Study

Yvo A. Saanen[✉], Zack Lu, and Gijsbert Bast

Portwise, Lange Kleiweg 12, 2288 GK Rijswijk, The Netherlands
{yvo.saanen,zack.lu,gijsbert.bast}@portwiseconsultancy.com

Abstract. Software development is a complex process. It is often faced with various challenges such as complexity, integration, scalability, reliability, and security. To address these issues, a stepwise approach is developed making use of dynamic simulation models in the development lifecycle from initial prototyping, implementation, testing, to operation. Such stepwise approach from rapid prototyping to testing control algorithms in a simulation environment to testing the entire system against a virtual environment ("digital twin") enables control over the development and insight in the expected behaviors as early in the process as possible. This paper, using a real case study in a container terminal, demonstrates the benefits of enabling such control and insight by using simulation being the essential key to a successful development and implementation of software. However, it should be acknowledged that such approach making use of simulation is still not a solution to everything. Drawbacks and limitations of its use in both prototyping and testing phases have been also observed from our experience. Examples of such drawbacks and limitation include the representation of simulation conditions vs. reality, happy-day operation vs. exception handling, unpredictable use cases, etc. This paper also presents these findings from our real case study experience and the lessons learnt that could be possibly utilized to mitigate and overcome (part of) these issues.

Keywords: Software · Simulation · Emulation · Digital twin

1 Introduction

In 2018, TBA initiated the development of a new Terminal Operating System (TOS), for large-scale, automated container terminals. A TOS (see Grifo, 2008) can be considered the ERP of a terminal, managing all activities from order management, planning, scheduling, dispatching, and allocation of resources. It's an essential asset of a container terminal. Activities managed by a TOS range from weeks in advance (e.g., berth scheduling), to hours in advance (e.g., vessel planning), to real-time scheduling and dispatching of equipment (see Grifo, 2008). Moreover, a TOS is highly interconnected with surrounding systems, such as Port Community Systems (e.g. customs), Billing systems, and operational systems such as Gate Operating Systems (GOS), and equipment

© The Author(s), under exclusive license to Springer Nature Switzerland AG 2024
M. Mujica Mota and P. Scala (Eds.): EUROSIM 2023, CCIS 2033, pp. 71–83, 2024.
https://doi.org/10.1007/978-3-031-68438-8_6

control software. In automated terminals, also automation control layers are connected to the TOS. Finally, to make the complexity puzzle complete, the TOS must deal with various external information flows from multiple sources. Information is, however, often incomplete, erroneous and internally conflicting.

Meanwhile, this new TOS has been operational for more than one year in a largely automated terminal. The path towards the go-live was not easy, as there were quite some challenges to overcome. To name a few:

- Complexity: As systems become larger and more interconnected, it becomes more difficult to ensure that all components are working together seamlessly and in real-time. Moreover, as systems become more automated, the exception cases (where automation does not provide the solution) become more difficult.
- Integration: Large-scale systems often require multiple components, each with its own operating system, to be integrated into a single cohesive system.
- Scalability: As systems grow in size, they must be able to handle increased loads, including more data and more users.
- Reliability: Real-time control systems must be highly reliable. Yet they operate in a dynamic environment, depend on input from users and external systems, and run on infrastructure and communicate via wired and wireless networks. All these factors can affect reliability.
- Security: As these systems are often connected to the Internet and other networks, they must be designed with security in mind to prevent unauthorized access and cyberattacks. Security systems themselves affect connectivity and operating speed.

A Development Approach Addressing Complexity

To address these issues, we followed a development approach that addressed the complexity. The approach relies heavily on the use of dynamic simulation models, enabling near-to-live testing (see Boer and Saanen, 2017) early in the development process. We will elaborate on the approach in the next section. An overview of the methodology is provided as follows.

First, we developed the architecture of the entire system – a multi-layered and highly modularized architecture based on microservices. Microservices is a software architecture style that structures an application as a collection of small independent services communicating with each other, rather than as a monolithic whole. Each microservice runs a unique process and communicates through APIs (Application Programming Interface). This approach allows for faster development and deployment, less internal interdependency, and easier maintenance and testing. Benefits include increased reliability, greater scalability, and the ability to upgrade individual services without affecting the overall system. This architecture allowed us to develop many separate modules (services) in parallel, within a strict architecture to foster a large degree of independence.

Within the architecture, the modules themselves were designed and specified, based on the operating procedures and the use cases thereof created. Here the first part of the development approach came in handy, with rapidly prototyping in a simulation environment (see Fig. 1). The prototypes represent the entire set of processes in a container terminal, including the TOS. The prototypes built and tested in the simulation environment, were the basis of the specification for the actual software development.

In the subsequent software implementation, critical modules, such as the "Yard module" (*responsible for real-time placement of containers and finding the best spot inside the automated yard, not only based on here and now, but also considering longer-term strategies*) and the "Scheduler" (*responsible for the timing and assignment of container moves to the various types of equipment*) (see also Xi, 2009), were tested in the same simulation environment, replacing the simulation prototypes but linking the real modules to the simulation environment. As such, these modules could be tested under "near-to-live" circumstances.

The final step of the development approach was to test the entire software package (including some of the surrounding software, such as the reefer management system and the gate operating system). All the components were tested in a dynamic emulation environment, where all equipment was modelled as digital twins of their physical counterparts.

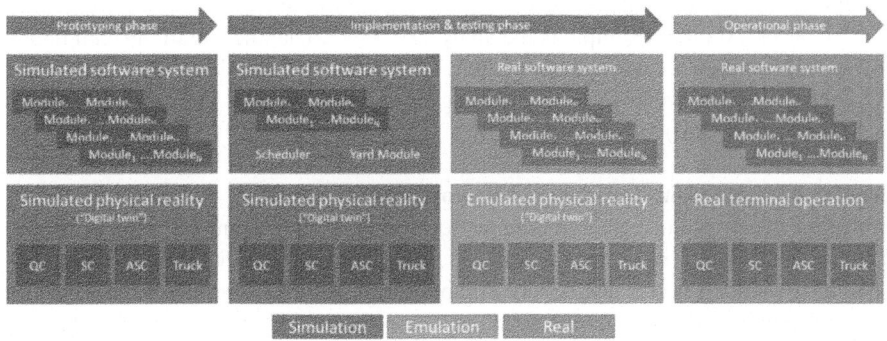

Fig. 1. Four phases of software development cycle from prototyping, to testing, to live operations (see also Boer and Saanen, 2008, 2017)

2   The Stepwise Approach: Elaboration

In the following sections, we will explore the stepwise approach and the different phases in more detail.

2.1   Rapid Prototyping Using Simulation Models

Commonly, dynamic simulation is used in the planning phase of container terminals (also in this case) (see for instance Agerschou et al. (2004), Rusca et al. (2018), and Park and Dragovic (2008)). However, the emphasis in this paper is on the use of dynamic simulations in consecutive phases of the software implementation process. The first step being rapid prototyping of core control algorithms, part of the TOS under development.

Rapid prototyping is a software development technique that emphasizes the creation of a working model of a system or application in a very short period. The purpose of rapid prototyping is to quickly test and evaluate the viability of an application and to refine the design to meet the needs of the target audience. Rapid prototyping can be applied to a wide range of software algorithms, including those for artificial intelligence, machine learning, but also control algorithms, as those two described here. In this paper, we focus on the use of simulation models in enabling rapid prototyping of the two key modules of our TOS: the "Scheduler" and the "Yard module". These modules can be easily recognized as essential modules for running container terminal operations (e.g. Kim, 2008; Mallidis and Dekker, 2009).

Simulation models are used to represent the real world by mimicking the behavior of a system or process. Simulation models are particularly useful in software development because they allow developers to experiment with the algorithms that will be used to control the behavior of the system. For example, an algorithm that is intended to control the behavior of an autonomous vehicle might be tested using a simulation model that mimics the driving environment and the behavior of other vehicles on the road. This allows developers to identify and fix problems with the algorithm before it is implemented in the real world, which can help to minimize the risk of unintended consequences. Another example is the analysis of how to handle situations with breakdowns of equipment or wrong information. How does the controlling algorithm react? What are the alternative flows (e.g., handling a damaged container that cannot be placed in the automated yard)?

Benefits of Simulation Based Rapid Prototyping

One of the benefits of using simulation models for rapid prototyping (e.g. Buchenrieder, 2000) of control algorithms is that they can be created and modified relatively quickly. This makes it possible for developers to rapidly test and evaluate different designs for the algorithm under different, controlled and reproducible conditions, and to refine the design as needed. For our purpose, we have used our validated container terminal library (also known as "Timesquare®") (Timesquare, 2017). This simulation library has been developed over the last 20 years and has been applied and validated in container terminals around the world. It covers both the physical world of cranes and vehicles as well as the software which controls terminals.

Another advantage of using simulation models for rapid prototyping is that they can be used to represent a wide range of real-world scenarios. For example, a simulation model can be used to represent the behavior of a system in different environments and circumstances, such as busy and less busy periods, or under various operational assumptions. This makes it possible to test the algorithm in a wide range of conditions, which helps to ensure that it will perform well in the real world.

Challenges with Simulation-Based Rapid Prototyping

One of the challenges of using simulation models for rapid prototyping of software algorithms is that they can be limited by the accuracy of the simulation model. The simulation model must accurately represent the real-world system or process to provide meaningful results. This can be difficult because real-world systems are often complex and difficult to model accurately. For example, a simulation model of a machine driven by an operator will only represent the "typical operator", and unlikely represent all the behaviors a human operator will show. This could lead to problems with the algorithm when it is tested in the real world. The same applies for operators of the control room software: they may make different decisions from the simulated operator in the model.

Another challenge of using simulation models for rapid prototyping is that they can be limited by the computing resources available. Simulation models can be computationally intensive and require significant processing power and memory to run. This can make it difficult to test algorithms quickly and efficiently, especially when the simulation models are complex or require a large amount of data. In our environment, we tested the algorithms inside the "Scheduler" and the "Yard module" over a period of up to 6 weeks – to reflect the high degree of variations and the impacts of strategies that would have their effects only over a longer period. The speed of the simulation was limited to approximately 6 times of the real-time (not to negatively affect the decision-making within the tested real software), which means that models would run at least for 1 week.

2.2 Testing Individual Modules by Means of Dynamic Simulation Models

Software testing is a critical stage in the software development process to ensure the quality and functionality of the end product. Over the years, various testing methods have been developed to evaluate the performance of software systems. One such method is dynamic simulation (see for instance Korn and Granino, 2007), which has gained popularity in recent times for its ability to mimic real-world scenarios and predict software behavior under different conditions.

Dynamic simulation is a testing technique that involves creating a virtual environment closely resembling the real-world conditions under which the software will operate. These days, simulation is often addressed as "digital twin", although digital twins search for an as detailed as possible representation of reality, whereas simulation searches for the right amount of detail (which in practice *simplifies* models). The simulation is designed to mimic the relevant behavior of the system components and their interactions, allowing the tester to evaluate the software's performance in a controlled environment. This testing method enables software developers to identify and fix bugs and performance issues before the software is released to the market.

Benefits of Dynamic Simulation

Dynamic simulation provides several benefits to software testing, including:

- Improved accuracy: Dynamic simulation reduces the need for physical testing, which is time-consuming and expensive. The virtual environment created during simulation allows the tester to evaluate the software's behavior under various conditions, providing a more accurate representation of its real-world performance.
- Faster and more efficient testing: Dynamic simulation enables the tester to conduct tests in a controlled environment, reducing the time and effort required to identify and fix bugs (see Fig. 2). This results in faster and more efficient testing, which can significantly reduce the overall development time of the software. In addition, the difference of the risk profile of implementing changes in a lab environment compared to production is large. The changes can be applied much earlier in the process, and the time to test a large series of test scenarios can be reduced compared to real-life testing.

- Increased flexibility: Dynamic simulation allows the tester to change the parameters of the simulation, such as inputs, outputs, and conditions, to evaluate the software's behavior under different scenarios. This increased flexibility provides a better understanding of the software's performance and behavior.
- Cost savings: Dynamic simulation eliminates the need for physical testing equipment, which can be expensive and requires frequent maintenance. The virtual environment created during simulation reduces the costs associated with physical testing and allows the tester to conduct tests at a lower cost.

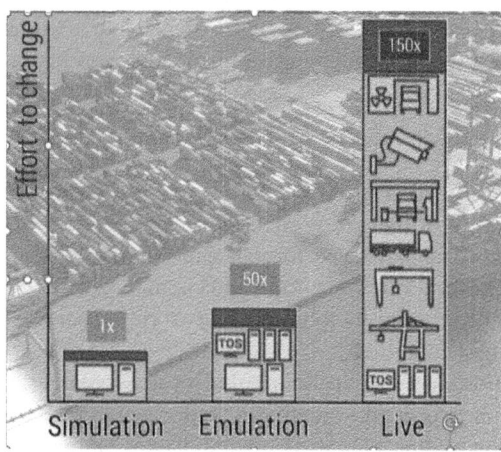

Fig. 2. Impact of using simulation on the cost of changing software (source: The National Institute of Standard Technology (NIST))

2.3   Testing the Entire System with a Virtual Terminal Emulation

Another technique which we use in our software development process is emulation (Boer and Saanen, 2012a, 2012b, 2014, 2016, 2017). Emulation involves creating a virtual environment that is a close representation of the target hardware or operating system. The software algorithm is run on this virtual environment, and its behavior and performance are observed. This technique is used to evaluate the software algorithm's compatibility with the target hardware and operating system.

Both simulation and emulation are useful for verifying the quality and performance of software algorithms (Boer and Saanen, 2016). They provide a way to test the software algorithm in a controlled environment, without the risk of damaging the real-world hardware or operating system. These techniques can also help identify potential bugs and performance issues, allowing developers to make necessary adjustments before releasing the software to the market.

3 Findings and Lessons Learnt from Our Test Cases

The afore-described stepwise approach using simulations enables control over the entire software development lifecycle and insight in the expected behaviors as early in the process as possible. However, it is not a solution for everything. Drawbacks and limitations are also found from our case study. This section will shed more lights on these findings, and the lessons learnt from our experience that can be possibly utilized to mitigate and overcome (part of) these issues.

Difference Between Simulation and Real-World

A simulation model is an approximation, not an exact representation of the real-world. To mimic reality, variation and randomness is typically built into the simulation model. Yet, it is impossible to exactly represent the reality. Therefore, it becomes essential to achieve the required right level of accuracy for the needs and the right level of simulation details compared to reality. This requires emphasis on multiple activities in the process already from model preparation in the start. However, still limitations and drawbacks were observed. These findings and lessons learnt from our experiences include:

- It is important to clearly define the objectives and scope. These form the underlying foundation for model development, configurations, and customization.
- Preparing the simulation model is an iterative and interactive process and requires continuous cross-measurements against the defined objectives.
- Garbage in is garbage out: Selecting the right scenarios and conditions and defining the right inputs and assumptions determines the accuracy, the reliability, and the value of simulation outcomes.
- Model verification and validation are critical to ensure that the model is developed and configured in the right way and has an accurate representation of the real system.
- Simulation is approximation. It is therefore important to always use common sense in assessing results, and never base any conclusions on one single simulation replication, as the results of an individual simulation differ depending on random events, and random durations of individual processes.
- Simulation scenarios, even when they are long and comprehensive, will typically start with a "perfect" situation (e.g., no database pollution, all positions of equipment and containers correct), whereas live operations never start like that, and instead often start with dealing with full history of events.
- Simulation tends to focus too much on happy day scenarios with little disturbances. Even when running long experiments (tests), the amount of "alternative flows" happening in a simulation environment is typically quite limited, compared to real production.
- Simulation experiments are typically unattended, without need for user input during the experiment. However, in reality, users are constantly interacting with the system, not per se interfering with the operation (e.g. planning of the next ship).

Being mindful about these points and limitations is therefore already advised even for the initial prototyping phase. However, there are more drawbacks and limitations observed in follow-up testing phase. These are elaborated in the remainder of this section, with advice also added from whatever lessons learnt from our real case study.

Testing of Standalone Modules

The simulation model from the prototyping phase could be used standalone, or in different combinations with modules from the actual software package. It could be run with just the "Yard module" (*responsible for decking containers*), just the "Scheduler" (*responsible for scheduling and dispatching of transport system and yard handling system*), or a combination of the two.

When running with one of the two modules only, the approach of using simulation as a testing tool for the actual software was really proving its value. Because only one component is exchanged from the simulated option to actual software, comparing performance indicators to the fully simulated environment is straightforward. Any differences (either positive or negative) that are observed must be due to the module that is being tested. Test scenarios could be fully catered to the needs of the team developing the module. Having this setup available before go-live allowed the teams to test new features, discover missing features, solve bugs, and tune the various parameters used in the modules.

An example of testing with simulation coupled with the "Yard module" is given in Fig. 3. Test runs simulating 30 days of operation were completed, both for a simulation only setup as benchmark, and for the setup with the "Yard module" coupled to the simulation to take care of decking requests both on landside and waterside. The "Yard module" performance could be compared to the benchmark operation, and additional functionality and tuning could be verified via KPIs obtained from the tests. Average truck handling time was worse than the simulation benchmark result at first, but further improvements and tuning brought it down to even lower values than the benchmark. Quay crane (QC) productivity was already better than the benchmark but was also further improved.

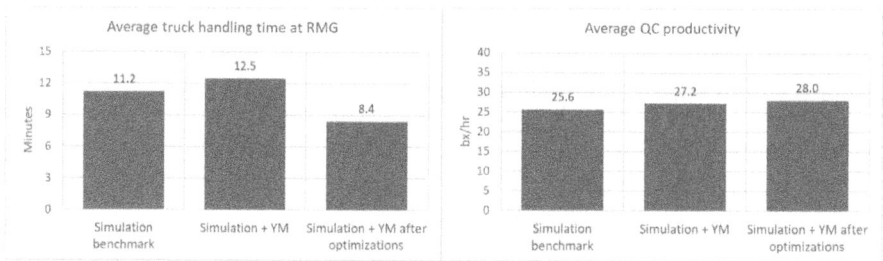

Fig. 3. Left: Average truck handling time at RMG (Rail-Mounted Gantry Crane) yard in simulation benchmark, versus different versions of the "Yard module". Right: Average QC (Quay Crane) productivity in simulation benchmark versus different versions of the "Yard module"

One drawback that could be observed is that with the tested modules in development, changes in the simulation were required between software versions as well, for example to support updates to the APIs (Application Programming Interface). To get and keep the simulation model running together with the tested modules required substantial effort. Still, the effort of fixing bugs later in live operation would be much larger.

Testing of Both Modules Combined

Besides tests for the individual modules, we also ran tests with both the "Yard module" and the "Scheduler" active at the same time. When doing so, some drawbacks started to show:

- First of all: to be able to test a specific integral functionality, it must be supported (i.e. implemented) in both of the software modules. Inevitably the development speed and supported operations by the two modules deviated, which limited what could be tested. In essence, this meant that the pace was set by the software module that was most behind, making tests less useful for the other team involved.
- Secondly: any differences between KPIs from the fully simulated environment versus the test done with two software modules cannot directly be attributed to one of the systems. This means that more effort was required to evaluate tests and find out reasons for deviations or lacking performance.
- Finally: when tests did not successfully complete, finding out the root cause was becoming much more difficult. There is a lot of information passing via different interfaces, and finding out what went wrong where is not always directly obvious, requiring time for all teams involved to solve.

To test performance of the "Yard module" and the "Scheduler", working together and benchmark the performance against simulation results is still important to understand where the software package stands in both performance and consistency. Figure 4 shows the performance level of the actual software components, versus the simulated benchmark. This concerned a fixed scenario applied to all experiments, for a peak operation on both waterside and landside for 8 consecutive hours. Average QC (Quay Crane) productivity is within 2% of simulation benchmark and could be reproduced in a second run. Also, by looking into more detail over the different hours, insights can be gained in when differences occur, and what simulation time to focus investigation on.

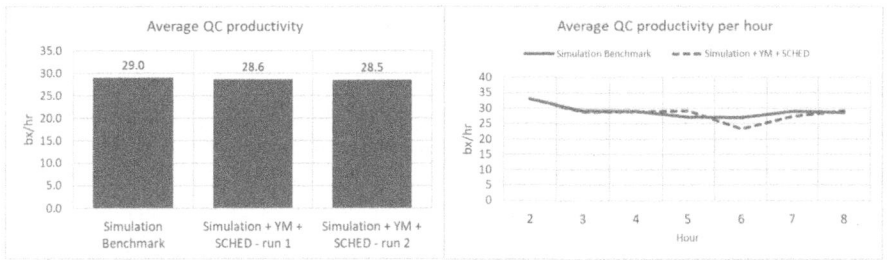

Fig. 4. Average QC (Quay Crane) productivity for simulation benchmark and two replications of a peak run with the same setup for the "Yard module (YM)" and the "Scheduler (SCHED)" software versions. Left shows the average QC productivity over the entire run and right shows the productivity per hour, helping to pinpoint specific periods to focus investigation on.

Timing is of the Essence
Figure 1 shows the different phases neatly separated. In reality, these can and will overlap to a high degree. When emulation tests are being executed, that might be with a limited scope, depending on what is supported by all involved modules at that time. Still, it will already require attention and effort that is taken away from standalone tests within a simulated environment. Even though there is still a huge benefit in doing those simulation tests, especially for the modules that are ahead on the roadmap.

These benefits will still be there even when the operational phase with live testing starts. But as Fig. 2 shows, the effort of fixing things in live operation will be substantial, leaving limited to no time for continued testing with the simulated environment.

The simulation model was coupled to the "Yard module" first, and later to the "Scheduler". The benefits from testing with a simulated environment were correspondingly much larger for the "Yard module" development than for the "Scheduler".

In short, the key lesson learnt is better to start early or be prepared to be overtaken by higher prioritized 'live' problems.

Happy days Versus Murphy's Law
The use of our simulation models in design and prototyping phases is largely focused on happy day operations. This means that some situations can be easily setup in the testing environment. An example could be busy vs. less busy periods, different amounts of equipment in operation, different operations at the waterside (e.g. twin lifting, balance between load and discharge). This makes it possible to test algorithms in a wide range of conditions, which helps to ensure that it will perform well in the real world.

The real world, however, does not consist of 365 happy days in a year. In fact, the large majority of use cases involve exception flows. Especially for integrated testing with both modules working together in the simulated environment, being able to test those exceptions could have been time saving for the live testing. In our test case, by the time the happy-day flows were working for the two modules, the emulation and live tests were already ongoing and became the main focus. It would have required additional work on the simulated environment, as that is designed mainly for uninterrupted operation.

Documentation is Key
The software architecture for the real system is based on microservices, relying heavily on information flow via APIs. The simulated environment has to support these APIs and information flow. Some of the interactions between the systems are highly dependent on timing. During our test case at times, documentation was lacking all required information – for example, the interface messages were clearly defined, but the exact timings and sequence in which they should be sent were not. The APIs for the modules were also still in development during testing, and information flows could change between software versions. If documentation is lagging behind those changes, there is a risk of having failed tests not due to issues in the tested Module, but due to errors in the simulated software APIs. Therefore, complete, accurate, and up to date documentation is key to making sure the simulated environment meets the need for testing the software modules. Without complete and accurate documentation, it might become impossible

using the software modules to test the simulated APIs, not fulfilling the full requirement of testing purpose.

Key Takeaways from the Test Case
To summarize the lessons learnt from the test case:

- Testing single modules with a simulated environment allows for extensive testing and tuning very early on in the process. This saves precious time, and improves the performance of the real software. A big benefit is that it can be done independently of other modules.
- Simulation benchmarking followed by testing of a single or multiple software module(s) gives insights into performance levels and allows for testing of software improvements and parameter tuning. It can provide insights into specific situations or periods of the test run that are interesting to zoom into for further optimizations.
- Testing multiple modules combined in a simulated environment requires synchronization in development speed. This challenge often makes testing harder to complete in time – which means before other project phases prevail - and to bring the same benefits observed for the single module.
- Live testing will inevitably come into the mix at some point. Given that this is very time-consuming, it will take away the focus from simulation testing. To make optimal use of the time before start of live testing becomes critical to reap the maximum benefits of the simulated environment.
- Testing using a simulated environment of complex modules should be started as soon as possible in the process.
- Complete, up to date and accurate documentation of APIs is a requirement.
- The biggest potential in saving time in live tests, is using the simulated testing for integrated operations (multiple modules working together), also supporting exception flows. As this typically comes after first single module testing followed by multiple modules during happy day operation, it has the greatest risk of being overtaken by live testing.

4   Conclusions

Developing a new process control system from scratch is a large undertaking. Such systems are subject to many dynamic and unpredictable interactions with users and other systems. They also need to manage a great variety of operational circumstances. As the operations the TOS controls are 24/7, downtime is undesired, and largely unavailable. For this purpose, the systems need to be reliable, easy to operate, and resilient to a large range of operational circumstances. The development and testing approach we have used, allows for testing in a comprehensive way, identifying issues in the software and specifically in complex control algorithms in an early phase of the development process. The approach also allows for verifying whether the targeted performance levels can be achieved by these control algorithms.

The stepwise approach from rapid prototyping to testing control algorithms in a simulation environment to testing the entire system against a virtual terminal ("digital twin") provides control over the development and insight in the expected behaviors as early in the process as possible. However, it is not a solution for everything. We found that the use cases in live operations are far greater than the use cases reflected in the model environments, meaning that the software and the control algorithms will be confronted with situations they have not been tested on. The extremes in operations are higher, the data provided are more abundant, and more polluted, and the impact of the user is hard to predict. Even when such scenarios have been identified – based on real-life cases, it is not always easy to get them implemented in the test environments. They require a series of events to coincide, which can be hard (complexity and effort wise) to realize.

Another finding is that although one may have provided an extensive set of assumptions about the future, these assumptions may not come true in reality. For example, we saw that the yard occupancy rate was expected never beyond 65%, but in reality it went quite soon well over this level. Also, the behavior of the trucking society was quite different from what was expected. Appointments for pick-ups were not made well in advance, but only made when already approaching the terminal. This makes yard preparation almost impossible. These are just a few examples of how a prognosed reality may turn out very different, impacting the way control algorithms should work.

Does this mean that an approach like this is in vain? Not at all. In general, the control algorithms which were developed following this approach did largely meet the criteria. They were robust to many of the even unexpected situations caused by the complexity of live operations (not all though!). Besides, a lot of errors were found, before the terminal was operational. Today, the test environments are still in use as basis for all (automated) regression tests, run against every new release. They provide insights, allow to minimize operational risk, and are also used for training purposes.

References

Agerschou, H., Dand, I., Sorensen, T., Ernst, T.: Planning and Design of Ports and Maritime Terminals, 2nd edn. Thomas Telford Ltd., London (2004)

Bish, E.K., Chen, F.Y., Leong, Y.T., Nelson, B.L., Ng, J.W.C., Simchi-Levi, D.: Dispatching vehicles in a mega container terminal. OR Spectrum **27**(4), 491–506 (2005)

Boer, C.A., Saanen, Y.A.: CONTROLS: Emulation to improve the performance of container terminals. In: Mason, S.J., Hill, R.R., Monch, L., Rose, O., Jefferson, T., Fowler, J.W. (eds.) Proceedings of the 2008 Winter Simulation Conference, pp. 1094–1102. Institute of Electrical and Electronics Engineers, Inc., Piscataway, New Jersey (2008)

Boer, C.A., Saanen, Y.A.: Improving container terminal efficiency through emulation. J. Simul. **6**(4), 267–278 (2012)

Boer, C.A., Saanen, Y.A.: Testing, tuning and training terminal operating systems. A modern approach. In: Gunther, H.O., Kim, K.H., Kopfer, H. (eds.) International Conference on Logistics and Maritime Systems (LOGMS), pp., 25–35. Bremen, Germany (2012b)

Boer C.A., Saanen, Y.A.: Plan validation for container terminals. In: Tolk, A., Diallo, S.D., Ryzhov, I.O., Yilmaz, L., Buckley, S., A. Miller, J. (eds.) Proceedings of the 2014 Winter Simulation Conference (2014)

Boer, C.A., Saanen, Y.A.: The journey of controls. Port Technol. Int. **30**(2), 30–32 (2016)

Boer C.A., Saanen Y.A.: Using simulation and emulation throughout the life cycle of a container terminal. In: Chan et al., W.K.V. (eds.) Proceedings of 2017 Wintersim Simulation Conference (2017)

Buchenrieder, K.: Design Automation for Embedded Systems, vol. 5, pp. 215–221. Kluwer Academic Publishers, Boston (2000)

Grifo, C.: The economic value of Terminal Operating Systems. Erasmus University Rotterdam, Master thesis (2008)

Kim, H.S., Nguyen, D.: Allocation model of container yard for ATC optimal operation in automated container terminal. J. Navig. Port Res. Int. Ed. **32**(9), 737–742 (2008)

Korn, G.A.: Advanced Dynamic-system Simulation: Model-replication Techniques and Monte Carlo Simulation. Wiley (2007)

Mallidis, I., et al: Yard management for improving efficiency of a container terminal. MIBES (2009)

Park, N.K., Dragović: Simulation study of container terminal throughput optimization. In: Proceedings of the 8th International Conference "Research and Development in Mechanical Industry" – RaDMI, 14–17.09.2008, Užice, Serbia, pp. 78–89 (2008)

Rusca, F.V., et al.: Simulation model of for maritime container terminal. Transp. Probl. **13**(4), 47–54 (2018)

Schütt, H.: Simulation technology in planning, implementation and operation of container terminals. In: Böse, J.W. (ed.) Handbook of Terminal Planning, pp. 103–116. Springer, New York (2011)

Zeng, Q., Yang, Z.: Integrating simulation and optimization to schedule loading operations in container terminals. Comput. Oper. Res. **36**, 1935–1944 (2009)

Timesquare ® Homepage, https://www.portwiseconsultancy.com/services/terminal-simulation-analysis/. Last accessed 17 Feb 2023

CONTROLS® Homepage. https://tba.group/en/software/emulation-software. Last accessed 17 Feb 2023

Xi, G.: A simulation based hybrid algorithm for yard crane dispatching in container terminals. In: Rossetti et al. (eds.) Proceedings of the 2009 Wintersim Simulation Conference (2009)

Application of Augmented reality in the Plaza Simón Bolivar in Manaure, Department of Cesar Colombia, as a Strengthening of Tourism and Culture

Paola-Patricia Ariza-Colpas[1,3]([✉]), Marlon-Alberto Piñeres-Melo[2,3],
Roberto-Cesar Morales-Ortega[1,4], Andres -Felipe Rodriguez-Bonilla[3],
Shariq Butt-Aziz[5], Maribel Romero Mestre[6], Combita-Niño Harold Arturo[4],
and Sumera Naz[7]

[1] Department of Computer Science and Electronics, Universidad de la Costa CUC, 080002 Barranquilla, Colombia
pariza1@cuc.edu.co, rmorales@certika.co
[2] Department of Systems Engineering, Universidad del Norte, 081001 Barranquilla, Colombia
pineresm@uninorte.edu.co
[3] Blazing Soft Company, 081001 Barranquilla, Colombia
andres.rodriguez@blazingsoft.com
[4] Certika Company, 081001 Barranquilla, Colombia
hcombita@certika.co
[5] Department of Computer Science, University of south Asia, Lahore, Pakistan
[6] Faculty of Engineering and Technology, Universidad Popular del Cesar, 200004 Valledupar, Cesar, Colombia
[7] Department of Mathematics, Division of Science and Technology, University of Education, Lahore, Pakistan
sumera.naz@ue.edu.pk

Abstract. Currently, one of the sectors most affected by the pandemic effect is the tourism sector, especially cultural tourism. The municipality of Manaure, located in the department of Cesar in the northeast of Colombia, has been characterized for having a high potential of the historical and cultural heritage of the department. However, there is a strong weakness concerning the dissemination, use, and appropriation of technology to support the processes of attracting and retaining local and foreign tourists. That is why the application "Enamorate del Cesar" has been developed as an application that combines the inclusion of augmented reality, applying the concept of time capsules and gamification to strengthen cultural tourism in both locals and foreigners. In the specific case of this department, the popular Plaza de Simón Bolívar in the municipality of Manaure has been taken as the epicenter for the use of new technologies applied to the tourism sector. To measure the impact of this application, which is the first one developed for the department of Cesar in Colombia, validation instruments have been designed to validate the use of the application with the community, which has resulted in progress in the processes of appropriation and improvement of the visibility of the cultural heritage. The objective of this article is to show the characteristics of the

© The Author(s), under exclusive license to Springer Nature Switzerland AG 2024
M. Mujica Mota and P. Scala (Eds.): EUROSIM 2023, CCIS 2033, pp. 84–97, 2024.
https://doi.org/10.1007/978-3-031-68438-8_7

application and the impact it has generated in the processes of social appropriation of knowledge and post-pandemic economic dynamization.

Keywords: Tangible and intangible heritage · cultural heritage · experiential experiences · augmented reality · Software application · Gamification

1 Introduction

Crises are the best time to reinvent yourself and adapt to new circumstances to make progress and improve. That is why the future of the tourism sector is marked by several technological trends that direct it toward a "smarter" and more sustainable model (environmental, territorial, and socioeconomic). More and more companies are using it to show their potential customers the destinations they can travel to without leaving their seats. Given the growing demand when choosing the place to go or the type of trip, augmented reality is perfect for showing the inside of a cruise cabin or the views you will have when hiring a hot air balloon ride. The possibilities are endless, and there are already companies that are offering their services around this technology, preparing virtual gymkhanas, stories, and mysteries to solve while visiting a city. Only in 2020, it became a business with a value of 120,000 million dollars a year. world level. And with the arrival of the metaverse, this trend will grow in a way never seen before.

The pandemic, beyond transforming the way trips are carried out, has also generated the need to develop and strengthen new activities and products. Some aspects that have become more relevant are the high value and social impact, the commitment to environmental sustainability, and, of course, the innovative offer framed in more conscious practices. An example of this type of product is Cultural Tourism, which is based on the traveler's motivation to learn, discover, and live cultural experiences in a tourist destination.

The land that today corresponds to the Municipality of Manaure, had in previous times a great indigenous predominance, due to this reason, it has a great historical and cultural importance among the municipalities of the Department of Cesar. Due to the importance of the cultural richness of this municipality, a tool has been developed that combines the use of cutting-edge technologies such as augmented reality and gamification called "Enamorate del Cesar", which allows to boost tourism in this region after the post-pandemic impact.

The selected location for the augmented reality application was The Simon Bolivar Square in the municipality of Manaure, located in the department of Cesar, is a central public space that has become a gathering place for the local community and visitors to the region. The square is surrounded by important buildings, including the Municipal Mayor's Office, the San Jose Parish, and the Public Library. In the center of the square is an imposing statue of Simon Bolivar, the Liberator of several South American countries, including Colombia, Venezuela, and Ecuador. The statue serves as a reminder of the influence and legacy of the Liberator in the region.

The square is decorated with beautiful gardens, benches, and fountains, making it a pleasant place to relax and enjoy the scenery. Additionally, during festivals and special events, the square becomes an ideal venue for cultural performances and other activities.

The Simon Bolivar Square in Manaure is an important historical and cultural landmark in the region. The square was built in the early 20th century and became an important public space for the local community. The square is named in honor of Simon Bolivar, who played a significant role in the independence of several Latin American countries, including Colombia.

The square is a place of great importance to the community of Manaure. Here, local citizens gather to discuss important issues and celebrate cultural events and festivals. Additionally, the square is also an important tourist attraction, especially for those interested in the history and culture of the region. The Simon Bolivar Square is a place of great importance to the municipality of Manaure and the department of Cesar in general. The square is a symbol of the history and culture of the region, and an important gathering point for the local community and visitors to the area.

The objective of this article is to provide the reader with a clear conceptualization of the components of augmented reality, to describe similar applications that have been developed in this line of application, and to show the details and characteristics of the application "Enamorate del Cesar".

2   Conceptual Information

2.1   Augmented Reality

Augmented reality is a very interesting technology that allows us to create a point of union between the real world and the digital world; what it does is, fundamentally, offer us a viewer through which we see our environment with visual information added or modified by a program. As can be intuited from the definition of "augmented reality" we have seen, this technology leaves the door open to many functions and ways of modifying the way we perceive the world around us. And since there is room for many different operating dynamics within this idea, there are different types of augmented reality [1].

Marker-Based Augmented Reality: This type of augmented reality, also known as marker-based augmented reality, is the simplest and easiest to develop. In this case, as the name suggests, "markers" or visual cues are used which, once physically placed in the location in space where we are interested in seeing changes across the screen. In other words, this type of augmented reality depends on an external visual element to know where to place the digital asset that will be seen by the viewer. It is therefore a work of substitution of a real element for another virtual element, a relatively simple process and one that results in a crude modification of what we see through the screen. Due to its limitations, this resource was used mainly in the early years of research and development of augmented reality systems and is now rarely used [2].

Augmented Reality Without Markers: This is the most technologically developed type of augmented reality since it is not necessary to place visual clues in the material environment that surrounds us; the software is already capable of identifying shapes and patterns in real time and places its virtual assets on these elements, having generated a three-dimensional map of the place in its memory system. In other words, in this case, the electronic device is not limited to placing virtual assets overlapping or substituting the track placed on the physical space, but rather "understanding" that physical environment and applying changes to it through digital editing in real-time, thanks to artificial intelligence. In turn, this type of augmented reality is divided into several subtypes:

Projection-Based Augmented Reality: In this case, the digital asset added by augmented reality remains immobile and tied to a particular place in a place. In this way it is possible, for example, to make a text remain on a wall in a room, as would happen with a projector [3].

Location-Based Augmented Reality: In this type of augmented reality, location technologies and detection of the user's movement are used to represent a zoom in or out of the digital asset, which appears in the person's field of vision if they are in the right time and place. This can be achieved by receiving information from the GPS system, using the compass, acceleration detectors, etc. [4].

Overlay-Based Augmented Reality: As its name indicates, this technology superimposes visual information on a physical element; for example, by making its different parts colored differently [5].

Contour-Based Augmented Reality: In this case, the contours of the elements captured by the camera are highlighted, so that the human eye has it easier to distinguish between different aspects of what surrounds it. It is widely used in cars with a screen, as it helps prevent traffic accidents.

2.2 Municipality of Manaure

Manaure is a municipality in the department of Cesar, located 32 km from distance from Valledupar near the border with the department of La Guajira. Due to this location used to be a place of transit for people of Vallenato and Guajiro origin; Some of these carried with them some type of musical instrument that they sang from the beginning of his journey to his point of arrival, demonstrating his talent by interpreting a walk, a son, a puya or merengue; so little by little Manaure stood out as a place of meeting, partying and exhibition of famous minstrels vallenatos. The strong guajiro-vallenata musical heritage in the municipality led to the creation of the Festival of Voices, Accordions and Guitar, which takes place on July 16. evidence of a strong social recognition of the territory.

The proximity of the municipality to the Serranía de los Motilones, at the foot of the Serranía del Perijá and its unique natural landscape led to the creation of a 150-hectare protected area known as Los Tananeos Natural Reserve, characterized by an ecosystem of tropical dry forest and the presence of a great diversity of endemic fauna and birds such as the Perijá Chamicero, the Perijá Hummingbird, the Mountain Sparrow, the Perijá Painted Parakeet and the Tapaculo from Perijá, among others.

Literature Review

Within the framework of the systematic review of the literature, a set of works focused on the application of augmented reality to cultural tourism can be evidenced. The most representative works of tourism application using augmented reality and Web applications are highlighted below (Table 1).

Table 1. Articles related to cultural tourism using technology.

Ref	Description	Type of augmented reality application
[6]	The authors present a software application that allows generating additional tourist information about iconic buildings in China to capture the attention of users to learn about the history related to these temples and oriental culture by creating virtual characters that show relevant information about the place in detail, generating missions. to comply that they can identify that users have managed to appropriate the socialized information of the tourist place.	Projection-Based Augmented Reality
[7]	The authors designed an augmented reality application that allows describing the cultural richness of the Roman Orthodox Church of Barcelona, applying time capsule processes where historical details about the characteristics of religious cults and rites that are performed in these settings are shown. and how they have strengthened the history of Barcelona.	Contour-based augmented reality
[8]	The authors made applications of Unity 3D and Vuforia SDK, to carry out a segmented mapping of different buildings in China, to offer interactive video game schemes that allow the user to interact with the history of these emblematic places.	Overlay-based augmented reality
[9]	The authors developed an application based on Mixed Reality using Web Ontologies to strengthen the meaning and exemplification of past customs in the Islamic context, through 3D modeling it is possible to access relevant information about the tourist site and to know the historical evolution of the different developments. that are carried out at present and its difference with the current scenarios.	Projection-Based Augmented Reality

(*continued*)

Table 1. (*continued*)

Ref	Description	Type of augmented reality application
[10]	The authors describe the Flaneur tool, which is constituted not only as an augmented reality application that shows historical information about a particular site, but also allows users to share information about characteristics of their experience on the site using the application, giving an important evaluation to the architectural component of the visited place.	Projection-Based Augmented Reality
[11]	The authors highlight the different forms of interaction of augmented reality in the inclusion and visualization of Balinese architecture allowing to work with visualization of architectural forms, aggregation of information in an augmented reality environment and how the user can connect with relevant information from the sites you visit.	Contour-based augmented reality
[12]	The authors carried out an implementation of augmented reality in the cave church of Matera, a medieval architectural building built in the Roman villa Mola di Mardi, where a reconstruction process of the place was carried out using augmented reality resources in order to carry out the process of reconstruction. Teaching how lighting works in this type of scenario in ancient times through the use of 3D modeling and photometry.	Overlay-based augmented reality
[13]	An augmented reality implementation is carried out in the old power plant in Piestany, with the purpose of explaining the basic conceptualization of factors that affect the quality and energy efficiency developed in this plant in its historical moment.	Contour-based augmented reality

(*continued*)

Table 1. (*continued*)

Ref	Description	Type of augmented reality application
[14]	The authors propose a system based on mixed reality that aims to visualize the reconstruction processes of archaeological sites in situ, for this an adaptive methodology is defined that allows a mixture between real and virtual scenes that the user can visualize the progress they have had. the transitions in the archaeo reconstruction processes	Projection-Based Augmented Reality
[15]	The purpose of this research was to validate the influence of augmented reality on the perception of the route of the Gothic building of the silk market in Valencia, which is considered a world heritage site, the idea was to be able to validate in a group of users how it had been the experience of knowing the historical place with or without the help of the augmented reality application and validating the perception of the visitors, which was very positive when using the application.	Augmented reality without markers

3 Methodology

The development of the software platform was carried out using an iterative and incremental approach based on the Agile methodology. The team was composed of three software engineers, a project manager, and a cultural tourism expert. The first step was to define the requirements and user stories, based on the needs of potential users, experts in cultural tourism, and local stakeholders. These requirements were prioritized and organized into a product backlog, which was continuously refined throughout the development process.

The platform was developed using Unity 3D, an engine widely used in the development of video games and augmented reality applications. The team designed and developed a series of markers and 3D models to represent different cultural attractions and landmarks in the department of Cesar. After the initial development of the software, a series of usability tests were conducted with potential users to identify and address any issues or bugs. Feedback from these tests was incorporated into subsequent iterations of the software, and new features and improvements were added based on user feedback.

The platform was also tested in the field, with a group of tourists and local stakeholders using the application to explore cultural sites in the department of Cesar. This allowed for further feedback and improvements to be made to the platform, ensuring that it met the needs of its intended users. The development of the platform based on augmented reality to support cultural tourism in the department of Cesar, Colombia, was a collaborative effort between software engineers, project managers, and cultural tourism experts. The Agile methodology, along with user-centered design principles, were used to ensure that the platform met the needs and expectations of its users, and provided a unique and engaging experience for cultural tourists visiting the department of Cesar.

4   Augmented reality software in the plaza Simon bolivar in Manaure, Cesar

4.1   Plaza Simón Bolivar

The Main Square, where it is possible to find the statue of Simón Bolívar, and the Parish of Our Lady of Mount Carmel. The square serves as a setting for various recreational, cultural, and musical activities such as Band rehearsals Municipal Martial. In its surroundings it is possible to find formal commerce such as the Brasa Manaurera restaurants, Casa Vieja restaurant, Lo Nuestro cafeteria, bakeries, among others; and informal trade such as street sales of wafers, breads, mango biche, among others; and even bathroom service (Fig. 1).

Fig. 1. Plaza Simón Bolivar of Municipality of Manaure.

4.2   Details of Software Development

This software application was developed using the agile SCRUM methodology, Unity 3D and Vuforia software, allowing the user to an interaction with the different cultural components of this historical heritage, for its achievement process a series of stages were developed. For the development of this application, a review of the historical and cultural context of the Simón Bolívar square in the municipality of Manaure was first carried out, allowing a process of characterization of the different historical and cultural components, which leads to capturing high-quality information. cultural relevance of this region of the country. Then, the audiovisual conceptualization of the different resources was achieved to generate a satisfactory user experience, where together with the artifacts

Fig. 2. Start of the application.

of software engineering, the achievement of an application that includes 360° videos, interactive games based on gamification is achieved, see Fig. 2.

The application first allows you to locate yourself after providing the credentials to view all the municipalities of the department of Cesar, when the municipality of Manaure is selected, it allows the user to select among the sites the "Plaza de Simón Bolívar", where he finds a series of icons where he can Interact with information about the history of the square and of course the history of the Nuestra Señora del Carmen church, identifying demographic and cultural aspects in two languages, English and Spanish, see Fig. 3.

Fig. 3. Nuestra Señora del Carmen church.

Within the application, when positioned on the statue of Simón Bolívar, the application allows access to different resources that show the history of the liberator Simon Bolívar and how this sculpture was made in his honor, which allows locals and foreigners to travel to the colonial era and learn about the history of Colombian independence interacting with videos and different resources based on gamification, see Fig. 4.

Fig. 4. Statue of Simon Bolivar.

5   User Experience Evaluation of the Enamorate del Cesar Application

There are various techniques for evaluating the user experience in software that can be used to measure user satisfaction and usability when interacting with a system or application. Some of these techniques include heuristic evaluation, user testing, surveys, metrics analysis, and user observation. Heuristic evaluation involves a group of experts evaluating the system's interface and design based on a list of established heuristics. User testing involves a group of real users performing specific tasks on the system while being observed and asked for feedback on their experience. Surveys are questionnaires sent to users to obtain information about their satisfaction and opinions about the application. Metrics analysis involves using tools to track and measure user behavior within the system. User observation involves observing users as they interact with the application to gather information about their experience and identify areas for improvement.

To validate the user experience of the application, a hybrid structured and unstructured validation instrument was developed. This approach combines elements of structured and unstructured evaluation to obtain a more complete understanding of the user's experience when interacting with the application. In the structured evaluation, predefined questionnaires and evaluation scales are used to measure the user's experience. For the specific case of this instrument, the validation scale was as follows: 1: Strongly agree, 2: Somewhat disagree, 3: Neither agree nor disagree, 4: Somewhat agree, 5: Strongly agree.

On the other hand, in the unstructured evaluation, a more open and flexible approach is used, allowing users to provide free comments and express their opinions and suggestions. The combination of these two techniques allows for the collection of both quantitative and qualitative data to better understand the user's experience in the augmented reality application. In this way, developers can identify areas for improvement and take measures to enhance usability and user satisfaction. The instrument considered the validation of different dimensions: Presentation, Individualization, Interactivity, Handling, Content, Help, Performance/Effectiveness, and Engagement. Regarding the presentation dimension, we analysed if the overall aesthetics of the application are adequate, if the colors used respect visual comfort, if the information on the screens is clear and sufficient to be understood, if the graphical elements used facilitate usability, and if the choice of options is done through accessible menus and buttons. The individualization dimension allowed assessing whether the program's level of adaptability to your needs is adequate and if the program adapts to different devices on which it has been executed.

Interactivity dimension analyses if the program allows for the exchange of information between the user and the application. As a user, you have found freedom in the choices made on different screens, and have found a sufficient number of options in each request. The dimension also considers if the program provides adequate help to the user when incorrect information is entered. Regarding the dimension of handling, it was possible to identify if the program was easy to use, intuitive, had the ability to navigate easily between options, if searches were fast and simple, allowing for easy retrieval and consultation of previously saved tasks.

When analyzing the contents, it is validated whether the objectives pursued in each option are easy to identify, verifying the technical quality of the videos if they can be easily understood and help to understand the cultural component associated with the historical place. In the help dimension, it is verified if it is clear, accessible, and easy to navigate, supporting in the resolution of doubts that arise in the use of the application, and if the application is able to detect handling errors and inform what should be done.

In the dimension of functionality/efficiency, it is validated whether there have been no errors that force to interrupt the action, verifying if the speed of access to information and switching between screens is appropriate, also checking that the program facilitates the user's work regarding the design and development of activities in the program. Finally, in the commitment dimension, it is validated if the program has met your expectations and if users would recommend the program to other users.

The instrument was applied to 200 users. The result of the software evaluation was very satisfactory, as it obtained high scores in all evaluated dimensions. In the Presentation dimension, the software obtained a score of 4.68, indicating that the general aesthetics and visual elements used are adequate and respect visual comfort. Regarding the Individualization dimension, the software obtained a score of 4.86, indicating that it adapts perfectly to the user's needs and can be executed on different devices.

The interactivity of the software was valued with a score of 4.86, indicating that the exchange of information between the user and the application is fluid, and the options offered are adequate and sufficient. The ease of handling the software obtained a score of 4.63, indicating that it is intuitive and easy to use, allowing the user to move easily between different options. Regarding the content of the software, it obtained a score of 4.59, indicating that the objectives pursued in each option are easy to identify, and the cultural component associated with the historical place is easily understandable. In the Help dimension, the software obtained a score of 4.55, indicating that the help offered is clear, accessible, and easy to navigate, supporting in resolving doubts that arise in the use of the application. In the Performance/Effectiveness dimension, the software obtained a score of 4.61, indicating that the speed of access to information and the passage from one screen to another is appropriate, and the program facilitates the user's work regarding the design and development of activities in the program. Finally, in the Engagement dimension, the software obtained a score of 4.79, indicating that it has met the user's expectations, and users would recommend the program to other users. In summary, the evaluated software has obtained very positive results in all evaluated dimensions, indicating that it is an effective and satisfactory tool for users, see figure 5.

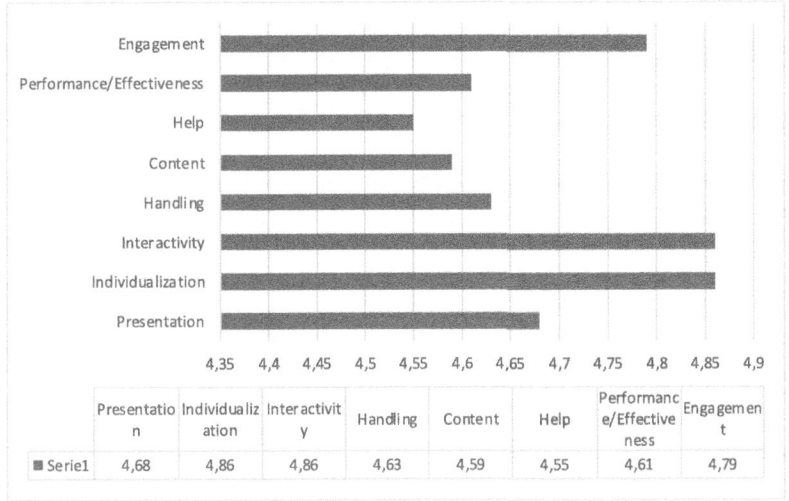

Fig. 5. Result of software validation.

6 Impact of the Enamorate del Cesar app in the municipality of Manaure, Cesar department

The mobile application 'Enamórate del Cesar' has had a significant impact in the municipality of Manaure, in the department of Cesar. Since its launch, this app has transformed the way residents and visitors interact with the municipality by providing useful information about tourist attractions, cultural events, local festivals, as well as public and private services.

In terms of tourism, the app has been fundamental in attracting visitors to the region by effectively and attractively promoting the natural and cultural beauties of Manaure. The app has helped improve the municipality's image and its ability to compete in the tourism market, and has fostered the development of small local businesses such as restaurants, souvenir shops, and tourist accommodations. Additionally, the app has had a positive impact on the local economy by promoting tourism and supporting local commerce. Residents have also found benefits in the app, as it allows them to access relevant information about public and private services such as hospitals, schools, transportation, and emergency services. It has also become a platform for promoting local events, such as fairs and festivals, which has helped strengthen the cultural and community identity of the municipality.

Another important benefit of the app is its ability to improve citizen security, as it provides real-time information on security incidents and emergencies. Residents can receive alerts about risky situations and take preventive measures to avoid security problems. The app has also been useful in helping local authorities maintain constant monitoring of the municipality, which has improved their capacity to respond and provide attention in case of emergencies. The 'Enamórate del Cesar' app has had a significant impact in the municipality of Manaure in terms of tourism, local economy, citizen security, and cultural promotion. The app has proven to be a powerful tool for fostering economic and social development in the municipality, and has improved the quality of life of residents by providing them with useful and relevant information about local services and events.

7   Conclusions

The development of this application constitutes the first approach to technologies based on augmented reality in the municipality of Manaure in the department of Cesar, it has made it possible to rescue both cultural and historical components that allow native and foreign people to appropriate traditions and culture of the region. It has revitalized the economic and social component and has allowed the attraction of foreign tourists, strengthening the tourism sector in the department of Cesar and positioning it as a leader in the implementation of technologies for tourism promotion.

References

1. Boulahrouz, M.: Salidas escolares, geolocalización y realidad aumentada en Educación Superior. Una revisión sistemática de la literatura. EDMETIC **12**(1), 5 (2023)
2. Robles, B.F., Pérez, S.M.: Experiencia formativa sobre el uso de realidad aumentada con estudiantes del grado de Pedagogía. Revista Tecnología, Ciencia y Educación **24**, 119–140 (2023)
3. Piñeres-Melo, M.A., Ariza-Colpas, P.P., Nieto-Bernal, W., Morales-Ortega, R.: SSwWS: structural model of information architecture. In: Advances in Swarm Intelligence: 10th International Conference, ICSI 2019, Chiang Mai, Thailand, July 26–30, 2019, Proceedings, Part II, pp. 400–410. Springer International Publishing, Cham (2019). https://doi.org/10.1007/978-3-030-26354-6_40
4. Martínez, N.M.M., Mariscal, A.J.F.: Posibilidades didácticas de la herramienta de realidad aumentada ZapWorks en la enseñanza de las ciencias. Una experiencia con estudiantes de un Máster en Profesorado. Revista Tecnología, Ciencia y Educación **24**, 91–118 (2023)
5. Ariza-Colpas, P.P., Piñeres-Melo, M.A., Nieto-Bernal, W., Morales-Ortega, R.: WSIA: web ontological search engine based on smart agents applied to scientific articles. In: Tan, Y., Shi, Y., Niu, B. (eds.) Advances in Swarm Intelligence: 10th International Conference, ICSI 2019, Chiang Mai, Thailand, July 26–30, 2019, Proceedings, Part II, pp. 338–347. Springer International Publishing, Cham (2019). https://doi.org/10.1007/978-3-030-26354-6_34
6. Kalloori, S., Chalumattu, R., Chalet, F., Zimper, M., Klingler, S., Gross, M.: Talking houses: transforming touristic buildings into intelligent characters in augmented reality. In: Information and Communication Technologies in Tourism 2023: Proceedings of the ENTER 2023 eTourism Conference, January 18–20, 2023, pp. 334–339. Springer Nature Switzerland, Cham (2023)
7. Lluis, G.M., Isidro, N.D., Galdric, S.R., Ernest, R.D.: The experience of using ICT in the development of a singular architectural project: the Romanian Orthodox Church in Barcelona. In: 2014 9th Iberian Conference on Information Systems and Technologies (CISTI), pp. 1–7. IEEE (2014)
8. Kalarat, K.: Applying relief mapping on augmented reality. In: 2015 12th International Joint Conference on Computer Science and Software Engineering (JCSSE), pp. 315–318. IEEE (2015)
9. Elrawi, O.M.: The use of mixed-realities techniques for the representation of Islamic cultural heritage. In: 2017 International Conference on Machine Vision and Information Technology (CMVIT), pp. 58–63. IEEE (2017)

10. Ioannidi, A., Gavalas, D., Kasapakis, V.: Flaneur: Augmented exploration of the architectural urbanscape. In 2017 IEEE Symposium on Computers and Communications (ISCC), pp. 529-533. IEEE (2017)
11. Indraprastha, A.: An interactive augmented reality architectural design model: a prototype for digital heritage preservation. In: 2019 International Conference on Advanced Computer Science and information Systems (ICACSIS), pp. 83–88. IEEE (2019)
12. Lassandro, P., Fioriello, C.S., Lepore, M., Zonno, M.: Analysing, modelling and promoting tangible and intangible values of building heritage with historic flame lighting system. J. Cultural Heritage **47**, 166–179 (2021)
13. Hain, V., Löffler, R., Zajíček, V.: Interdisciplinary cooperation in the virtual presentation of industrial heritage development. Procedia Eng. **161**, 2030–2035 (2016)
14. Magalhaes, L.G., et al.: Proposal of an information system for an adaptive mixed reality system for archaeological sites. Procedia Technol. **16**, 499–507 (2014)
15. Puyuelo, M., Higón, J.L., Merino, L., Contero, M.: Experiencing augmented reality as an accessibility resource in the UNESCO heritage site called "La Lonja", Valencia. Procedia Comput. Sci. **25**, 171–178 (2013)

Simulation Model to the Online Course Advising

Alejandro Felipe Zárate Pérez[✉] and Idalia Flores de la Mota

Universidad Nacional Autónoma de México, Av. Universidad 3004, Col Copilco Universidad,
04510 Coyoacán, Ciudad de México, Mexico
alejandro.zarate@unam.mx

Abstract. The processes involved in online teaching have been inherited from traditional face-to-face courses, in almost every case we transfer the face-to-face to the virtual without making an analysis about how the things work in virtuality. In this paper we propose a simulation model for the course taught process, and from the model we will review a point that we assume to be definitive: the quantity of hours that an advisor dedicates to grading a course. Even the size of the student's group assigned to an advisor is defined by administrative policy but not necessarily by the needs of the training program, which one can be different according to the project's aim.

Keywords: simulation · optimization · teaching analytics · e-learning analytics

1 Introduction

In the last thirty years, technology has been advancing faster and this has also been reflected in education, an example of this is the modality called "Distance Education", which has gained more strength, as it is more accessible thanks to innovative technologies. This modality has the purpose to give education access to diverse sectors that have not been able to be attended, due to situations such as geographic, employee, time, among others.

Nowadays, with the incorporation of ICT (Information and Communication Technologies), it is possible to glimpse the scope that these represent for distance education, thus playing an essential role, because of the application of these recent technologies to the educational and training field, what is called "e-learning".

E-learning is a way of using ICT as a means of distribution for educational materials and other services, in which there is also an interrelation between teachers and students. Thus, in this new teaching-learning environment, web technology is used through the Internet.

Within education we find two types of education: face to face and continuing education. In this paper we will focus only on the second: continuing education.

There are a lot of definitions about continuing education, some of which vary according to the country to which we refer. However, for this paper we will take the UNAM definition of continuing education:

© The Author(s), under exclusive license to Springer Nature Switzerland AG 2024
M. Mujica Mota and P. Scala (Eds.): EUROSIM 2023, CCIS 2033, pp. 98–109, 2024.
https://doi.org/10.1007/978-3-031-68438-8_8

"It is an educational modality designed, organized, systematized, and programmed that complements the curricular formation and deepens and broadens knowledge in all fields of knowledge; it trains and updates professionally and is aimed at the university community and the public."

The Dirección General de Cómputo y de Tecnologías de Información y Comunicación (DGTIC) part of Universidad Nacional Autónoma de México, offers online continuing education courses, mainly in computing, through the Coordination of Continuing Distance Training, which are aimed to the public, the university community and institutions and companies that request them (Fig. 1).

Fig. 1. Online course website

2 State of Art

For each of the courses taught, the course materials and activities are developed according to a didactic planning made for the needs of each course. This implies that all groups in the same course see the same materials and perform the same activities during the course (Fig. 2).

By using a model of this type, the online courses that are taught do not require the figure of a teacher who oversees transmitting the knowledge, since this is planned through the materials and activities that are developed. For this reason, for each group that is opened, an Advisor is assigned, who is only in charge of resolving doubts and evaluating the necessary activities.

It is also necessary to mention that the activities of these courses are planned on a weekly basis, i.e., the materials and activities must meet the learning aims each week.

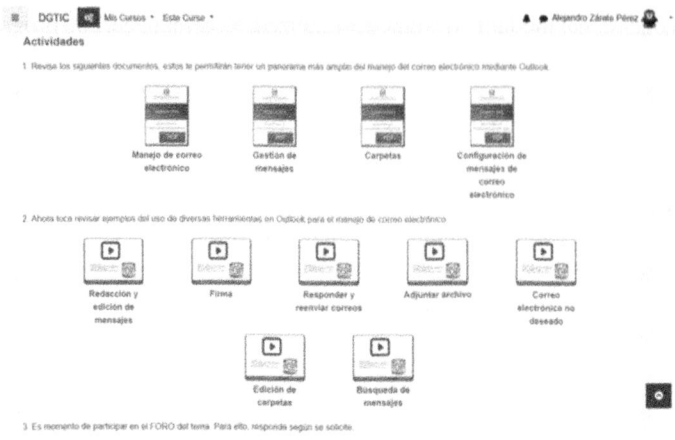

Fig. 2. Example of online course materials.

Although in some cases this advisor may take part in the development of the materials and activities, in general, the advisors assigned to the courses are not involved in the development of the course materials and activities and are limited only to the resolution of doubts and evaluation of the activities (Fig. 3).

Fig. 3. Topic evaluation example

Under this advisors' model we can define the process as follows:

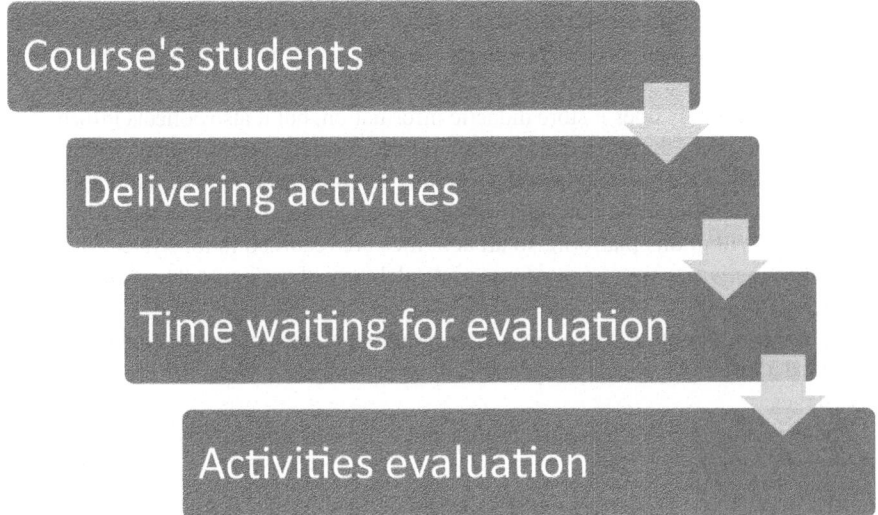

Each of these parts are described as follows:

Course's Students. We refer to students who actively perform the activities showed before submitting an activity that requires evaluation. Such activities are materials consultation, forums participation, etc.

- **Delivering activities.** This part corresponds exclusively to the delivery by the student of the activity that will be evaluated for the corresponding week. As mentioned before, the activities are planned by week, so the submitting deadline for the evaluation will be at the end of the week, but it does not necessarily happens this way, since students submit activities from the beginning of the week until the end of the week.
- **Time waiting for evaluation.** In this waiting time we refer to the time elapsed between a student's submission and the time when an advisor starts grading; this is because students can submit activities at any time but must wait for the advisor to log in to the course site to grade. In this case, since advisors log in daily, the maximum time for an advisor to begin grading an activity is 24 h.
- **Activities evaluation.** Here, as its name shows, the advisor evaluates the activities delivered by the student, the time it takes for an assessor to evaluate an activity depends on the delivery made by the student.

3 Modeling Problem

At a technical level, we can affirm that all the systems necessary to teach online courses use a database for their correct operation, in which the information of the lessons, activities and even the participants' grades are stored.

However, not only does it store didactic information, but it also collects information on the interactions of the participants within the platform, this information can be useful because through it we can find how the teaching-learning process is developing.

The operation of online courses is done through a Moodle LMS, the entire course is conducted within this platform, so all activities are recorded in the system database, among these activities are review of materials, delivery of assignments, participation in forums, etc.

From this information we can make observations of each of the parts of the process mentioned above, with which it is possible to statistically model the behavior of each of these stages.

Therefore, in general, the simulation model will be as follows (Fig. 4):

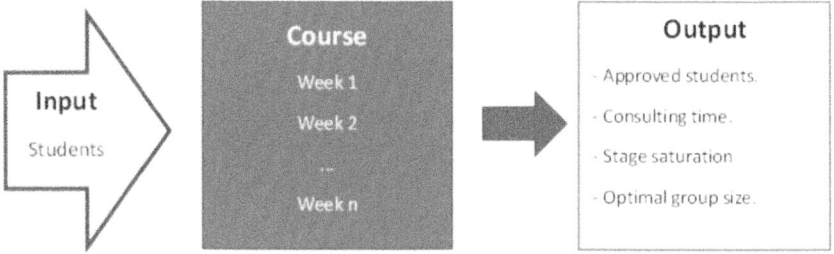

Fig. 4. Simulation model

This implies that the simulation model takes a certain number of students, who enter the course, perform their weekly activities and at the end we could answer questions such as: How many students pass? how long was the counseling time? in which phase is the model most saturated? what is the optimum group size?

However, as we can see within the course modeling students must pass the activities for each week, within each week the simulation model is as follows (Fig. 5):

Therefore, the simulation model of the course is composed of the individual simulations of each week until the end of the course.

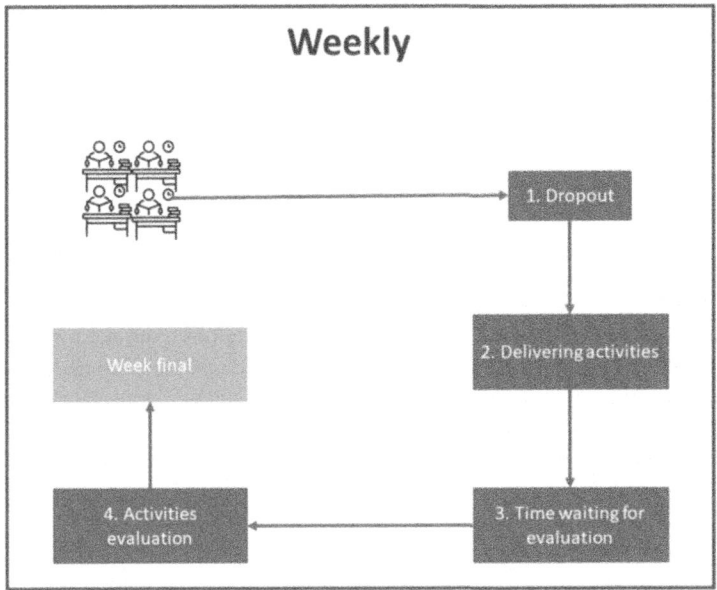

Fig. 5. Weekly model

4 Example

To exemplify this model, we are going to calculate the counseling time employed, this is important because up to now each advisor has been given a time of 20 h. This time is decided by the policies of the institution, but it has not been proven that this time is sufficient.

The data that we will use in each phase are described below. These data were taken during a training program given to an institution where 4-week courses were given with an assigned counseling time of 20 h for each course.

1. **Dropout (D)**. The observed dropout rate was 5%, which is extremely low compared to the rates seen in courses open to the public.
2. **Delivering activities**. For this phase, we will use the following frequency table, which shows how the activities were delivered according to the day of the week (Table 1).

Table 1. Frequency table by day

Day	Frecuency
Monday	1.02%
Tuesday	3.06%
Wednesday	7.33%

(*continued*)

Table 1. (*continued*)

Day	Frecuency
Thursday	13.92%
Friday	26.03%
Saturday	48.59%
Sunday	0.05%

3. **Waiting time for evaluation (TE).** According to the observed data we will use an exponential distribution with $\lambda = 12$. This occurs because students make the activity submissions at the end of each day and the assigned advisor generally reviews these submissions the next morning.
4. **Grading (TG).** As mentioned before, this phase refers to the time it takes for an assessor to rate an activity, for this part we will use a normal distribution with $\mu = 25$ and $\sigma = 15$.

We also used groups of 20 students, so the result of a simulation is shown below (Table 2):

Table 2. Simulations results per week

Week		Monday	Tuesday	Wensday	Thursday	Friday	Saturday	Sunday
1	Students	1	1	0	5	5	8	0
	TE	7	19	0	8	16	11	0
	TG	36	4	0	78	114	153	0
2	Students	0	0	3	4	3	9	1
	TE	0	0	20	6	8	8	15
	TG	0	0	94	170	89	155	69
3	Students	0	0	1	0	5	13	0
	TE	0	0	4	0	25	13	0
	TG	0	0	16	0	103	351	0
4	Students	0	0	2	1	5	11	0
	TE	0	0	4	11	7	15	0
	TG	0	0	84	29	165	254	0

In summary, we have the following (Table 3):

Table 3. Simulation results for course

	Week 1	Week 2	Week 3	Week 4
Final students	20	20	19	19
TE' average	8.79	8.05	5.98	5.19
TG's sum	383.98	577.28	470.42	532.05

This shows that at the end of the course **19 students finished** the course, the average **waiting time** for grading was **7 h** and the time used by the **advisor to grade** the activities was **32.73 h**.

We run this simulation process 20 times, the results are presented in the following Table 4.

Table 4. Simulations results

Simulation	Final students	TE	TG
1	19	7.00	32.73
2	18	9.83	33.91
3	18	7.83	34.97
4	18	7.55	33.01
5	19	7.02	35.14
6	19	5.85	37.70
7	18	7.35	31.96
8	17	6.82	31.93
9	16	6.71	25.06
10	19	7.50	30.57
11	17	8.84	30.39
12	19	7.74	30.38
13	19	8.69	29.11
14	18	7.86	31.57
15	19	8.02	31.01
16	19	6.73	31.17
17	15	6.59	26.99
18	19	7.32	31.45
19	18	7.04	27.74
20	19	8.28	31.96

The results of these simulations show that the average waiting time was 7.53 h and the average time to grade was 31.44 h. In this case we can see that the minimum time to grade was 25 h when the student dropout rate was the highest.

However, the 20 runs were only to exemplify the results because they are no representative. So, the question here will be, how many runs we need to consider stable the simulations results?

To answer this question, we had to run 10 simulations, then 20, 30, up to 1,000. In all simulations set we calculate the average final students, waiting time and grading time. With this data we obtain the next Table 5:

Table 5. Simulations results

Simulations	Average Final students	Average TE	Average TG
10	16.1	7.07	28.87
20	16.4	7.28	28.32
30	16.83	7.42	30.55
40	16.63	7.28	29.57
50	16.2	7.31	29.41
...	.	.	.
1000	16.25	7.3	29.36

The next step is normalizing the data, to do it we obtain the difference between the simulations n less the simulations n-1 and the calculate the absolute value. After this process we have 999 rows.

Finally, we calculate the moving average of 5 with the differences, we do this to smooth the data and then the graph will look a clearly tendence (Fig. 6).

We observe that over 800 simulations the differences in the 3 cases maintain under 0.2, this is small value, and we will consider acceptable. Therefore, to the next scenarios analysis we will use exactly 800 simulations to calculate the values.

The results after **800 simulations** are **16 final students**, **7.32 h to average waiting time** and **29.39 h to average grading time**. We have to remember the initial parameter: 20 students at the beginning, a Poisson distribution with lambda 12 for the waiting time for evaluation and a Normal distribution with media 25 and sigma 15 for the grading time; also, we assume that an advisor have assigned 20 h for grade students.

But the obtained results shows that the advisor needs 29 h for grade, i.e., 50% more time that assigned originally. What should we do to reach 20 h or less?

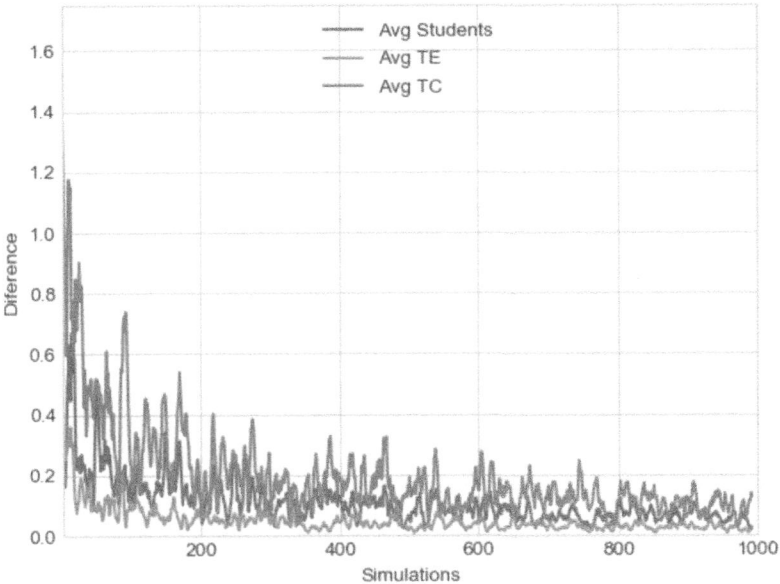

Fig. 6. Resulting graph from 1 to 1,000 simulations.

4.1 Scenarios

We are going to find the optimal values without exceeding the 20 h for grading time in the course. For this case we have two options: decrease the size group or decrease the grading time per activity.

Size Group
In this case we are going to move the size group from 10 to 20 and analyze its behavior with the grading time hours. The results are shown in the next Fig. 7.

According with these results we can observe that the optimal group size is between 13 and 14 students, the total grading time between 19.07 and 20.52 h. Therefore, if the group size is reduced between 30% and 35% its possible do not exceed the time grading assigned.

Grading Time Per Activity
For this scenario we will decrease the grading time, we will decrease the parameters for the Normal distribution, the values used, and the results are in the next Table 6.

Therefore, the optimal value is between Normal (16, 11) and Normal (17,12) with grading time between 18.89 and 20 h.

However, this case is the most difficult because the change in the Grading time per activity implies that its necessary.

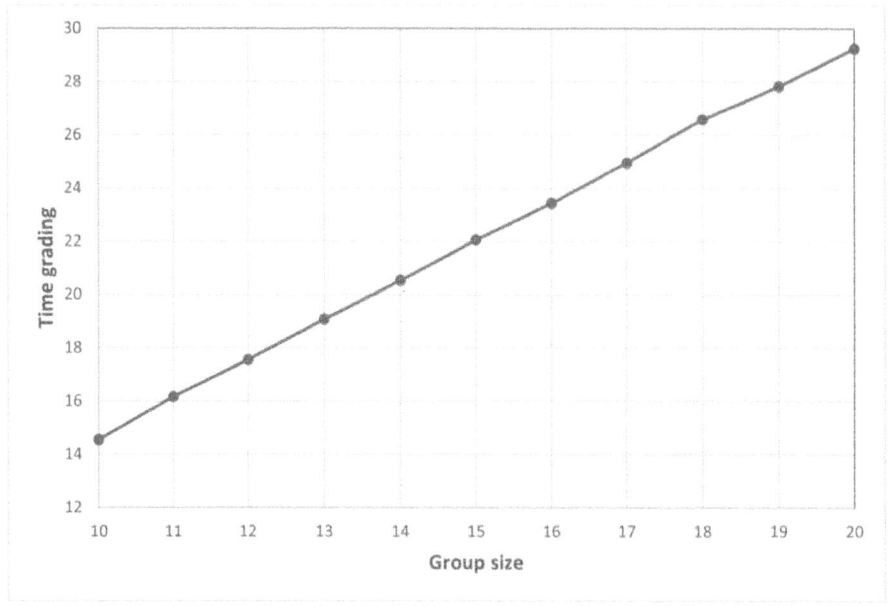

Fig. 7. Total grading time by size group

Table 6. Results moving the Normal distribution parameters.

Media	Sigma	Grading time
25	15	29.34
20	10	23.41
18	13	21.13
17	12	20.00
16	11	18.89
15	10	17.54
10	5	11.76

5   Conclusions

The results obtained during the simulations have allowed us to find the hours that an advisor needs to the grade students activities, which differs from the official hours allocated by an additional 50%. This is relevant because the assessor is exceeding by far the working time, which must be readjusted according to the resulting data.

In this case, more advisor time could be given, or the student group size could be reduced. For the latter case, the simulation model can be used in such a way that instead of fixing the number of students, what would be fixed would be the maximum number of hours to run the necessary simulations to find the student optimal group necessary to use only 20 h of advisor time.

However, this is not the only case where this simulation model can be used, as it can be used to set up before the start of a training project the necessary resources to complete without problems: advisor's quantity, time's advisor, group size, among others.

Likewise, the model described in this work can be used during the course or training program, with the purpose of predicting upcoming events by feeding the model with those generated in real time, i.e., using historical and current data to identify behaviors and make the necessary decisions if they will prevent the achievement of the established objectives.

One of the disadvantages of the model lies in the observed data, because it is necessary to make an adequate collection of them, to make the corresponding processing with them and to find the probabilistic models that best adapt to them. This is so that the simulation results are as close as possible to the real situation.

On the other hand, the greatest advantage of the model is the flexibility it gives us to answer questions such as: What is the optimal group size? How much time is dedicated to grading? How many advisors do we need for N students? among others. These questions can be solved using exact methods (queuing theory, dynamic programming), but these need assumptions that cannot be changed quickly, and it would be necessary to develop a model for each specific situation or training program.

Finally, the model allows us to answer questions quickly with only minor adjustments to the variables and to reuse it in different courses and/or training programs that behave differently.

References

1. Ndukwe, I.G., Daniel, B.K.: Teaching analytics, value and tools for teacher data literacy: a systematic and tripartite approach. Int. J. Educ. Technol. High. Educ. **17**, 22 (2020). https://doi.org/10.1186/s41239-020-00201-6
2. Robinson, S.: A tutorial on simulation conceptual modeling. In: 2017 Winter Simulation Conference (WSC), pp. 565–579. Las Vegas, NV, USA (2017). https://doi.org/10.1109/WSC.2017.8247815
3. Lynch, C.J., Gore, R., Collins, A.J., Cotter, T. S., Grigoryan, G., Leathrum, J.F.: Increased need for data analytics education in support of verification and validation. In: 2021 Winter Simulation Conference (WSC), pp. 1–12. Phoenix, AZ, USA (2021). https://doi.org/10.1109/WSC52266.2021.9715485.
4. Bell, M.G.H.: Stochastic user equilibrium assignment in networks with queues. Transport. Res. Part B: Methodological **29**(2), 125–137 (1995). https://doi.org/10.1016/0191-2615(94)00030-4
5. Abad, R.C.: Introduccion a la simulacion y a la teoria de colas. Netbiblo (2002). https://doi.org/10.4272/84-9745-017-5
6. Soon, W.M., Ang, K.C.: Introducing queuing theory through simulations. Electron. J. Math. Technol. **9**(2), 152–165 (2015)
7. Fuller, D., Arruda, E., Filho, M.F., Jose, V.: Learning-agent-based simulation for queue network systems. J. Operat. Res. Soc. (2019). https://doi.org/10.1080/01605682.2019.1633232

Architecture Based on Cloud Services to Boost Cultural and Heritage Tourism in the Department of Cesar, Colombia

Ariza-Colpas Paola Patricia[1,3]([✉]), Piñeres -Melo Marlon Alberto[2,3],
Morales-Ortega Roberto-Cesar[1,6], Rodriguez -Bonilla Andres Felipe[3],
But-Aziz Shariq[4], Leidys del Carmen Contreras Chinchilla[5],
Maribel Romero Mestre[5], Ronald Alexander Vacca Ascanio[5],
and José Caicedo-Ortiz[1,3]

[1] Department of Computer Science and Electronics, Universidad de la Costa CUC, 080002
Barranquilla, Colombia
{pariza1,rmorales1,jcaicedo1}@cuc.edu.co
[2] Department of Systems Engineering, Universidad del Norte, 081001 Barranquilla, Colombia
pineresm@uninorte.edu.co
[3] Blazing Soft Company, 081001 Barranquilla, Colombia
andres.rodriguez@blazingsoft.com
[4] Department of Computer Science, University of South Asia, Lahore 44000, Pakistan
[5] Faculty of Engineering and Technology, Universidad Popular del Cesar, 200004 Valledupar,
Cesar, Colombia
{leidyscontreras,maribelromero,
ronaldalexandervacca}@unicesar.edu.co
[6] Certika Company, 081001 Barranquilla, Colombia
rmorales@certika.co

Abstract. Cultural and heritage tourism is an important source of income for many regions around the world, including Colombia. However, it often faces a series of challenges that hinder its development and success. In the Cesar department of Colombia, the lack of updated information and effective management of tourism infrastructure have limited the potential of cultural and heritage tourism in the region. Tourists often do not have access to detailed information about places of interest, such as the history behind them and available activities. This makes travel planning difficult and reduces the quality of the tourist experience. To address these issues, the implementation of a cloud-based service architecture has been proposed to boost cultural and heritage tourism in the Cesar department. This architecture focuses on enhancing the tourist experience by providing updated and personalized information about tourist sites, allowing tour reservations, and facilitating the management of tourism infrastructure. The application called Enamorate del Cesar allows centralized and real-time management of tourist sites, which improves the quality of the tourist experience and reduces waiting times. Additionally, the architecture provides access to detailed and personalized information about tourist sites, which facilitates travel planning and improves tourist satisfaction.

© The Author(s), under exclusive license to Springer Nature Switzerland AG 2024
M. Mujica Mota and P. Scala (Eds.): EUROSIM 2023, CCIS 2033, pp. 110–121, 2024.
https://doi.org/10.1007/978-3-031-68438-8_9

Keywords: Tangible and intangible heritage · cultural heritage · economic reactivation · cultural tourism · experiential experiences · history · augmented reality

1 Introduction

Tourism is a significant source of revenue for many regions around the world. In Colombia, cultural and heritage tourism is a crucial part of the country's economy, providing jobs and driving local development. However, despite the potential benefits of cultural and heritage tourism, it often faces a series of challenges that limit its development and success. In the Cesar department of Colombia, the lack of updated information and effective management of tourism infrastructure have hampered the growth of cultural and heritage tourism in the region.

To address these issues, a proposed solution is the implementation of a cloud-based service architecture to boost cultural and heritage tourism in the Cesar department. This architecture is designed to enhance the tourist experience by providing updated and personalized information about tourist sites, allowing tour reservations, and facilitating the management of tourism infrastructure. The Enamorate del Cesar application is a central component of this architecture and provides tourists with a comprehensive and interactive platform to plan and enjoy their trip to the region.

This article aims to explore the challenges facing cultural and heritage tourism in the Cesar department of Colombia and how a cloud-based service architecture can help overcome these obstacles. The article will analyze the Enamorate del Cesar application, its features, and how it can improve the tourist experience in the region. Additionally, the article will examine the potential impact of the cloud-based service architecture on the local economy and how it can contribute to the sustainable development of cultural and heritage tourism in the region.

The following sections will delve deeper into the challenges facing cultural and heritage tourism in the Cesar department, the proposed solution, and the Enamorate del Cesar application. Finally, the article will conclude with a discussion of the potential impact of the cloud-based service architecture on the local economy and the future of cultural and heritage tourism in the region.

2 Historical Evolution of AR in Tourism

The evolution of augmented reality (AR) in tourism has been a fascinating journey, with numerous advancements over the years. AR technology has revolutionized the way tourists explore and experience new places, providing them with interactive and engaging experiences that make their travels more memorable [1]. In the early days, AR in tourism was mainly limited to basic mobile apps that displayed information about tourist attractions, such as maps, descriptions, and pictures. However, with the advent of more sophisticated technology, AR has become much more advanced and interactive.

One of the earliest applications of AR in tourism was the development of virtual tours. This technology allowed tourists to take virtual tours of tourist attractions, such as museums and historical sites, without having too physically be there. Virtual tours were also used in the hospitality industry to provide virtual views of hotel rooms and resort facilities [2]. As AR technology improved, the focus shifted to creating more immersive experiences for tourists. This led to the development of applications that combined AR with virtual reality (VR) to create more realistic experiences. For example, tourists can now use AR and VR to virtually visit distant places, such as national parks or even space.

Another major advancement in AR technology in tourism is the ability to customize the tourist experience. By using AR, tourists can now personalize their tours and select attractions that are of interest to them. AR also enables tourists to access real-time information about their destinations, including weather updates, flight information, and hotel reservations. Looking ahead, the future of AR in tourism looks promising. As technology continues to improve, we can expect to see even more interactive and engaging experiences for tourists [3]. AR will continue to provide new ways for tourists to explore and experience new places, creating unforgettable memories and enhancing the overall tourism experience.

3 AR Architectures in Tourism

Augmented Reality (AR) has revolutionized the tourism industry by providing immersive and interactive experiences for tourists. Various types of AR architectures have been developed and implemented in tourism to enhance the travel experience. These include marker-based AR, markerless AR, location-based AR, and sensor-based AR [4].

Marker-based AR is one of the most common types of AR architecture used in tourism. It uses physical markers, such as QR codes or images, to trigger the AR experience. Once the marker is detected by the device's camera, the AR content is superimposed on top of the marker. This technology has been widely used in museums, galleries, and historical sites to provide visitors with additional information about exhibits and artifacts. Markerless AR, on the other hand, uses image recognition technology to detect objects in the environment without the need for physical markers [5, 6]. This allows for a more seamless AR experience, as users do not need to scan a marker to trigger the AR content. This technology has been used in various tourism applications, such as providing information about landmarks or tourist attractions.

Location-based AR uses GPS to provide location-specific information to users. This technology is particularly useful for guiding tourists around a city, providing them with information about nearby attractions and landmarks. Location-based AR has also been used to create virtual tours of cities, providing tourists with an immersive and interactive way to explore a new destination. Sensor-based AR uses data from the device's sensors, such as accelerometer and gyroscope, to provide contextual information about the environment [7]. This technology can be used to provide tourists with information about their surroundings, such as the history or cultural significance of a particular area.

Each architecture has its own advantages and limitations, and the choice of architecture depends on the specific needs of the application. For example, marker-based AR is ideal for providing detailed information about specific objects, while location-based AR is more suitable for guiding tourists around a city [8]. By understanding the different types of AR architectures available, tourism professionals can select the most appropriate technology to enhance the travel experience for their customers.

4 Applications of AR in Cultural and Heritage Tourism

Tourism is a vital industry for many regions worldwide, and cultural and heritage tourism is a significant source of revenue for many destinations, including Colombia. However, it faces several challenges that hinder its development and success. In the department of Cesar, Colombia, the lack of updated information and effective management of tourism infrastructure has limited the potential of cultural and heritage tourism in the region. To address these issues, various software architectures have been developed to improve the tourist experience and make travel planning more accessible. In this context, this state-of-the-art article explores different software architectures that have been applied in the context of cultural and heritage tourism, with a specific focus on cloud-based service architectures for the promotion of cultural and heritage tourism in the department of Cesar, Colombia.

TripLens [9] is an augmented reality application that allows tourists to explore a city and find the best places to visit in real-time. By using the camera of the smartphone, users can point at any monument, restaurant or point of interest and the application will display detailed information about the place. Users can browse through different categories such as food, culture, and nightlife to find the best places to visit in a specific city. Additionally, users can also read reviews and ratings from other tourists to make informed decisions about the places they want to visit. The Time Travel application [10] uses augmented reality technology to show tourists how a city looked in different historical periods. The application allows users to select different historical periods and see how the city has changed over time. By using the camera of the smartphone, users can point at any place in the city and see how it was in the past. The application provides detailed information about each historical period, allowing tourists to have an immersive and educational experience. This application is ideal for those who want to learn more about the history and evolution of a specific city.

SkyView [11] is an AR application that helps tourists explore the night sky by identifying stars, planets, and constellations through their smartphone camera. Using the phone's camera, users can simply point it at the sky and the app will overlay information about the objects in view, such as their names, distances, and other interesting facts. This app is perfect for astronomy enthusiasts or anyone who wants to learn more about the wonders of the night sky. Thyssen-Bornemisza Museum AR [12] is an AR application that offers a guided tour of the Thyssen-Bornemisza Museum in Madrid. Using the camera of the smartphone, the app shows information and details about the works of art on display in real-time. The app provides a more interactive experience for visitors, allowing them to learn more about the art pieces and their history. This app is ideal for art lovers who want to enhance their museum experience.

Hieroglyph [13] is an augmented reality application that allows tourists to leave virtual messages and notes at tourist attractions for other tourists to see in the future. Using the smartphone camera, users can leave notes, drawings, or images in a specific location, which are saved to the specific location through geolocation technology. In this way, future visitors can see and read what other tourists have left in that place, which can add an additional layer of information and context for their visit. Wikitude [14] is an augmented reality application that allows tourists to access information about tourist attractions through the smartphone camera. By pointing the camera at a place of interest, the application shows detailed information about the place, such as reviews, photos, and videos. In addition, the application uses geolocation technology to provide real-time information about nearby places and their opening hours. Tourists can explore different categories, such as restaurants, hotels, museums, and attractions, to find the best places to visit in a particular city. The application also allows users to create and share their own personalized tour guides.

Gamar [15] is an augmented reality platform designed to allow tourists to explore cities and tourist sites in an educational and entertaining way. The application uses interactive games and challenges to engage users in the experience and teach them about the history and culture of the places they visit. Tourists can discover new places and complete missions, earning points and prizes along the way. AR City [16] is an augmented reality application that allows tourists to explore cities around the world through the camera of their smartphones. The application displays information about monuments, restaurants, and other points of interest nearby as the tourist moves around the city. Users can customize their experience and filter information to find what they're looking for, such as activities for children, souvenir shops, or museums. Additionally, the application offers tourist guides and detailed maps to help tourists plan their itinerary and explore more efficiently.

National Geographic World Atlas [17] is an augmented reality application that allows tourists to explore the world through the camera of their smartphone, providing detailed information about countries and regions. By pointing the camera at a specific location, users can obtain relevant information about the place, such as its history, culture, climate, geography, and much more. The application also offers detailed maps to help tourists navigate and plan their trip.

Landmarker [18] is another augmented reality application designed to help tourists discover places of historical and cultural interest. By pointing the smartphone camera at a specific location, the application shows detailed information about the place, including its history, cultural significance, and other relevant details. The application can also provide recommendations for nearby places that may be of interest to tourists. In summary, both applications can be useful tools for tourists looking to learn more about the world and the places they visit.

5   Architecture of the Proposal Solution

Augmented reality is a technology that combines virtual and physical elements to create an immersive real-time experience. To achieve this, it is necessary to have a software architecture that allows for the capture, processing, and presentation of the information necessary to create the augmented reality experience. This architecture is composed of several components, each of which plays a fundamental role in creating the augmented reality experience. In this section, we will describe in detail each of the components that are part of an augmented reality software architecture, from display devices to content servers, through computer vision libraries, rendering engines, and more, see Fig. 1.

Display Devices: Display devices are a key component of augmented reality, as they allow users to see computer-generated objects and graphics overlaid on the real world [3, 19]. The developed application allows the use of smartphones, tablets, and wearable

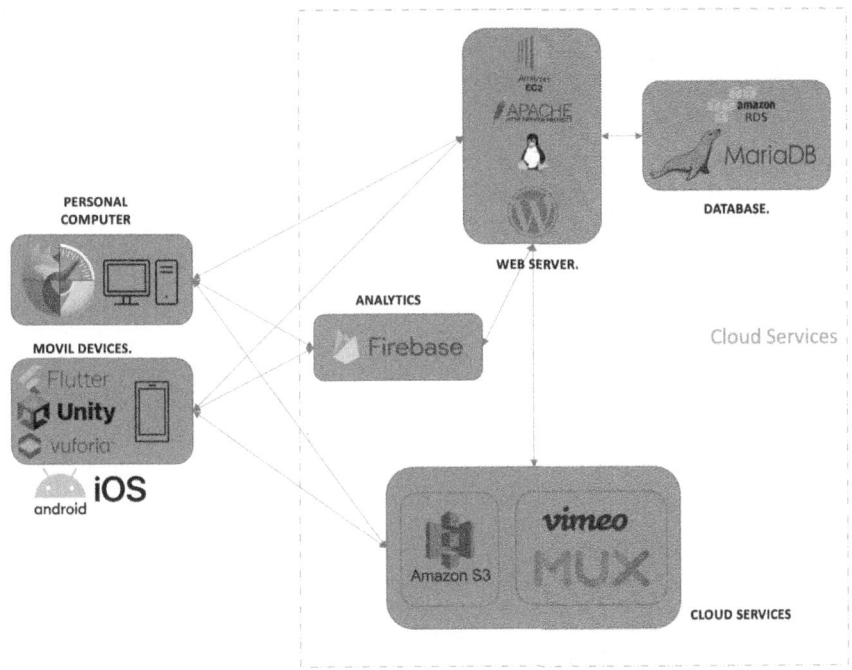

Fig. 1. Arquitecture of "Enamorate del Cesar" app.

devices, which supported by the development of a responsive web application, can generate a rewarding user experience with the application.

Sensors: Sensors are critical components in augmented reality, as they capture real-world data and use it to generate the augmented reality experience [20]. In the specific case of the "Enamorate del Cesar application", interaction with tourist attractions is done using smartphone cameras that allow for the capture of the user's position, movement, lighting, and orientation.

Computer Vision Libraries: Computer vision libraries are a collection of algorithms and tools used to process the data captured by sensors [21]. These libraries are used to interpret the information captured by sensors, such as images and videos, and to create 3D models of the real world. The algorithms are also used to detect objects and patterns in the real world.

Rendering Engines: Rendering engines are responsible for combining real-world data and computer-generated graphics to create the augmented reality experience [22]. Rendering engines use image and graphics processing techniques to create a composite image of the augmented reality that is displayed on the display device.

Content Databases: Content databases contain information about real-world objects and locations that will be used in augmented reality, such as images, 3D models, videos, and audio [23]. The data in the database is used to overlay computer-generated objects onto the real world and to provide additional information to users. Each of the emblematic sites in the Department of Cesar was modeled considering historical information provided during the information gathering process and added to a database of audiovisual resources.

Application Development Platform: The application development platform is a set of tools and APIs used to create augmented reality applications [24]. The platform provides an interface for accessing sensors, computer vision libraries, and rendering engines, and for developing applications that use augmented reality. In the specific case of the developed application, Vuforia and Unity3D software were used for the development of the platform.

Data Management Systems: Data management systems are used to store and manage user and application data. These systems can include user databases, tracking information, and behavioral data [25]. Data management systems are also used to personalize the augmented reality experience for users. In the case of the developed application, it also includes a Responsive Web Interface that allows storing information and recent user search history.

Content Servers Content servers are responsible for transmitting augmented reality data through the cloud [26]. Content servers are used to provide content and applications to users in real-time, and to ensure a smooth augmented reality experience. These servers can also collect user data and send it to data management systems. For the developed application, Amazon server was used to manage the sending and receiving of information.

6 "Enamorate del Cesar" application

The Department of Cesar is in the northeastern part of Colombia, bordered by the Caribbean Sea to the north, and the departments of La Guajira, Magdalena, Bolivar, Santander, Norte de Santander, and Arauca. The department is known for its rich cultural heritage, historical sites, diverse ecosystems, and agricultural production.

Cesar has 25 municipalities, each with its unique attractions and characteristics. Valledupar, the department's capital city, is famous for its music, literature, and folklore. It is the birthplace of vallenato, a genre of music that blends European, Indigenous, and African rhythms. Other important municipalities include Aguachica, known for its coffee production and nature reserves, and La Paz, which is home to the Cienaga de Zapatosa, a freshwater lagoon with an ecosystem that supports numerous bird species, see Fig. 2 and 3.

The process of selecting sites for the implementation of augmented reality involves several steps. First, it is necessary to identify the potential sites that are relevant to the project's goals and objectives. This can involve research into historical, cultural, or natural landmarks in the region. Once a list of potential sites has been compiled, the next step is to assess their suitability for the implementation of augmented reality.

Factors that may be considered in this assessment include the availability of sensor data, such as GPS or image recognition, the quality of the data that can be collected from

Fig. 2. The location of the Department of Cesar in Colombia

Fig. 3. Distribution of the 25 municipalities in the department of Cesar

the site, and the potential for user engagement with the site. Other considerations may include the level of infrastructure required to support the implementation of augmented reality at the site, such as the availability of Wi-Fi or cellular data networks.

After evaluating the potential sites, a shortlist of the most suitable locations for implementation can be created. These sites can then be further evaluated based on criteria such as their accessibility to users, the level of interest and engagement that they generate, and the potential for generating revenue through sponsorship or advertising. Ultimately, the sites that are selected for implementation will be those that best meet the project's goals and objectives, as well as the needs of its users. The following are the selected sites from the department of Cesar to be included in the application, see table 1.

The proposed augmented reality application for cultural and historical sites in the Cesar department works by using the camera and of the user's device to capture real-world images and overlay computer-generated objects onto them. The application uses computer vision libraries to detect and interpret the real-world objects and features, and then uses rendering engines to generate and overlay virtual objects on top of them, see Figs. 4 and 5.

The application will be pre-loaded with a database of content that includes images, 3D models, videos, and audio files for each selected cultural and historical site. When the user points their device's camera at a specific site, the application uses the device's GPS and computer vision to identify the site and retrieve the relevant content from the database.

The application will then superimpose the relevant virtual objects and information onto the real-world view of the site, allowing users to interact with the virtual content as if it were part of the real-world site. For example, users can view historical photos and videos of the site, listen to audio guides, and even see 3D models of historical buildings and structures.

Table 1. The selected places in the Department of Cesar visible in the application.

Municipality	Place	Municipality	Place
Manaure	Plaza Simón Bolívar	La Paz	Parque de la Almojabanera
	Rio Manaure		Plaza Olaya Herrera
San Diego	Plaza principal	Agustín Codazzi	Parque de la Guitarra
	Plaza principal Los Tupes		Plaza Alfonso Avila Quintero
Pueblo bello	Cascada La Tranquilidad	Bosconia	Parque árbol de la paz
	Parque La Pista		Estación de tren Bosconia
El Copey	Balneario Salto del Chimila	Becerril	Balneario "El 5"
	Parque Central El Copey		Monumento a Rafael Orozco
Valledupar	Balneario Hurtado Río Guatapurí	Pelaya	Cienaga del Sahaya
	Plaza Alfonso López		Parque Central Simón Bolívar
La Gloria	Malecón y Río Magdalena	Aguachica	Plaza de San Roque
	Iglesia San José		Bosque del aguil
Gamarra	Plaza Principal	Rio de Oro	Parque principal
	Ciénaga Baquero		Cerro de la Virgen
Gonzalez	Parque principal	San Martín	Parque Simón Bolívar
	Mirador de Cristo Rey		Parque junto a la carretera vieja
San Alberto	Parque principal	Astrea	Brazo del río Cesar (balneareo)
	Parque lineal "Malecón"		Parque central de Astrea
El Paso	Calle de los Durán	Chimichagua	Plaza Santander
	Plaza principal		Muelle turístico
Tamalameque	Malecón Puerto Bocas	Curumani	Plaza de La Santísima Trinidad
	Parque Simón Bolívar		Vía al Trigre
Chiriguana	Puerto de Chiriguaná	La jagua de ibirico	Plaza principal
	Plaza principal		

The application will also include features that allow users to save and share their experiences with others, such as the ability to take photos and videos of the augmented

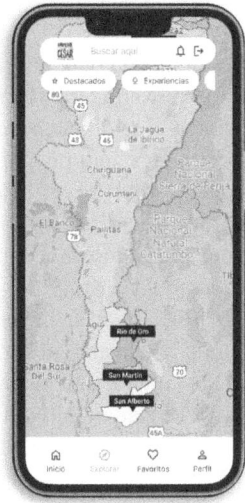

Fig. 4. Exploring the municipalities of the department of Cesar in the application

Fig. 5. Main page for exploring tourist sites

reality content and share them on social media. The user interface will be designed to be intuitive and easy to use, allowing users of all ages and backgrounds to enjoy and learn from the cultural and historical sites of the Cesar department, see Figs. 6 and 7.

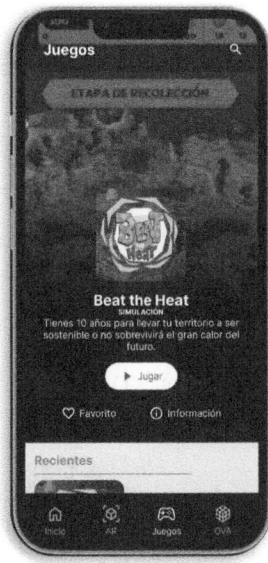

Fig. 6. Gamification-based games to discover tourist sites

Fig. 7. Virtual objects to enhance the cultural learning of the region

7 Conclusions

From a technological perspective, the implementation of augmented reality software in the Department of Cesar is a significant advancement in terms of incorporating cutting-edge technology in the tourism sector. The use of augmented reality can improve the experience of tourists and enhance their learning of the region's cultural and historical sites. Economically, the implementation of augmented reality software can help promote tourism in the region, attracting more visitors and generating revenue for local businesses. Additionally, the development of the software can also create employment opportunities in the technology sector, contributing to the region's economic growth.

From a social perspective, the augmented reality software can help raise awareness and appreciation of the Department of Cesar's cultural and historical heritage. The software can also enhance the accessibility of the region's cultural and historical sites, making them more inclusive for people with disabilities or limited mobility. In terms of cultural education, the augmented reality software provides an engaging and interactive way for visitors to learn about the region's cultural and historical sites. It can also help preserve the cultural heritage of the Department of Cesar by providing a platform for the dissemination of information to future generations. Overall, the implementation of augmented reality software in the Department of Cesar has the potential to positively impact the region's technological advancement, economic growth, social inclusion, and cultural preservation.

References

1. Jayawardena, N.S., Thaichon, P., Quach, S., Razzaq, A., Behl, A.: The persuasion effects of virtual reality (VR) and augmented reality (AR) video advertisements: a conceptual review. J. Bus. Res. **160**, 113739 (2023)
2. Yin, Y., Zheng, P., Li, C., Wang, L.: A state-of-the-art survey on Augmented Reality-assisted Digital Twin for futuristic human-centric industry transformation. Robot. Comput.-Integr. Manuf. **81**, 102515 (2023)
3. Fang, W., Chen, L., Zhang, T., Chen, C., Teng, Z., Wang, L.: Head-mounted display augmented reality in manufacturing: a systematic review. Robot. Comput.-Integr. Manuf. **83**, 102567 (2023)
4. Kozinets, R.V.: Immersive netnography: a novel method for service experience research in virtual reality, augmented reality and metaverse contexts. J. Serv. Manag. **34**(1), 100–125 (2023)
5. Kumar, H., Gupta, P., Chauhan, S.: Meta-analysis of augmented reality marketing. Mark. Intell. Plan. **41**(1), 110–123 (2023)
6. Ariza-Colpas, P.P., Piñeres-Melo, M.A., Nieto-Bernal, W., Morales-Ortega, R.: WSIA: web ontological search engine based on smart agents applied to scientific articles. In: Tan, Y., Shi, Y., Niu, B. (eds.) ICSI 2019. LNCS, vol. 11656, pp. 338–347. Springer, Cham (2019). https://doi.org/10.1007/978-3-030-26354-6_34
7. Najmi, A.H., Alhalafawy, W.S., Zaki, M.Z.T.: Developing a sustainable environment based on augmented reality to educate adolescents about the dangers of electronic gaming addiction. Sustainability **15**(4), 3185 (2023)
8. Özeren, S., Top, E.: The effects of Augmented Reality applications on the academic achievement and motivation of secondary school students. Malays. Online J. Educ. Technol. **11**(1), 25–40 (2023)

9. Triplens Application. https://apps.apple.com/cr/app/ajusta-y-traduce-triplens/id1450813744
10. Time Travel Application. https://apps.apple.com/us/app/venice-travel-guide-and-map/id3 32120643
11. SkyView application. https://play.google.com/store/apps/details?id=com.t11.skyviewfree& hl=en_US
12. Thyssen-Bornemisza Museo Nacional. https://www.museothyssen.org/thyssenmultimedia/ visitas-virtuales
13. Hieroglyph Application. https://apps.apple.com/us/app/hieroglyph-pro/id1505115635
14. Wikitude app. https://play.google.com/store/apps/details?id=com.wikitude&hl=es&gl=US
15. Gamar Applicarion. https://play.google.com/store/apps/details?id=com.gamar.platform&hl= es_419&gl=US
16. AR City Application. https://play.google.com/store/apps/details?id=com.beColumbus.ARC ity&hl=es_419&gl=US
17. National Geographic World Atlas Application. https://apps.apple.com/us/app/the-world-hd/ id364904503
18. Landmarker Application. https://apps.apple.com/us/app/landmarker-social-travel-app/id1 441002314
19. Colpas, P.P.A., Tapias, B.A.H., Comas, A.G.S., Melo, M.A.P., Royert, J.M.: Aula touch game: digital tablets and their incidence in the development of citizen competences of middle education students in the district of Barranquilla-Colombia. In: Tan, Y., Shi, Y., Tuba, M. (eds.) ICSI 2020. LNCS, vol. 12145, pp. 537–546. Springer, Cham (2020). https://doi.org/10.1007/ 978-3-030-53956-6_49
20. Fernandes, J., Brandão, T., Almeida, S.M., Santana, P.: An educational game to teach children about air quality using augmented reality and tangible interaction with sensors. Int. J. Environ. Res. Public Health **20**(5), 3814 (2023)
21. Coronado, E., Itadera, S., Ramirez-Alpizar, I.G.: Integrating virtual, mixed, and augmented reality to human-robot interaction applications using game engines: a brief review of accessible software tools and frameworks. Appl. Sci. **13**(3), 1292 (2023)
22. Xinzhen, D., Zhifen, R.: Building English Audio-Visual and Oral Mobile Teaching System Based on Virtual Augmented Reality Technology. Wireless Communications and Mobile Computing (2023)
23. Villagran-Vizcarra, D.C., Luviano-Cruz, D., Pérez-Domínguez, L.A., Méndez-González, L.C., Garcia-Luna, F.: Applications analyses, challenges and development of augmented reality in education, industry, marketing, medicine, and entertainment. Appl. Sci. **13**(5), 2766 (2023)
24. Cao, J., Lam, K.Y., Lee, L.H., Liu, X., Hui, P., Su, X.: Mobile augmented reality: user interfaces, frameworks, and intelligence. ACM Comput. Surv. **55**(9), 1–36 (2023)
25. Sadhu, A., Peplinski, J.E., Mohammadkhorasani, A., Moreu, F.: A review of data management and visualization techniques for structural health monitoring using BIM and virtual or augmented reality. J. Struct. Eng. **149**(1), 03122006 (2023)
26. Zhang, H., Mao, S., Du Niyato, Z., Han,: Location-dependent Augmented Reality Services in Wireless Edge-enabled Metaverse Systems. IEEE Open J. Commun. Soc. **4**, 171–183 (2023). https://doi.org/10.1109/OJCOMS.2023.3234254

Logistics and Transportation Systems

Modeling of Logistics Networks with Labeled Property Graphs for Simulation in Digital Twins

Alexander Wuttke$^{(\boxtimes)}$, Joachim Hunker, Anne Antonia Scheidler, and Markus Rabe

Department IT in Production and Logistics, TU Dortmund University, Leonhard-Euler-Strasse 5, 44227 Dortmund, Germany
`alexander2.wuttke@tu-dortmund.de`

Abstract. Digital twins have gained increasing attention in research and practice. This includes the use in logistics networks, which are the parts of supply chains focusing on implementing flows of goods. Digital twins are the digital representation of physical objects and feature a wide array of data. The tasks of digital twins in the domain of logistics include simulation, optimization, and monitoring. To fulfill these tasks, data models are required as a base. Using a common data model for all tasks is a promising approach, but there is a lack of modeling rules and guidelines on how to create appropriate models. In this paper, a modeling framework is proposed and discussed to fill this gap for logistics networks. It is based on labeled property graphs, which are a special graph structure. Also, the proposed modeling framework is applied exemplarily on a real-world use case scenario in the domain of city logistics.

Keywords: Logistics Networks · Simulation · Digital Twins · Modeling · Labeled Property Graphs

1 Introduction

Graphs and graph theory are not a new concept per se and have been around for several centuries, but gained a lot of interest in research and practice over the last years [17]. This development was driven in particular by advances in database technology, especially graph databases in the context of databases that do not follow a relational structure (NoSQL) [19]. Behind these developments is the realization that data are of high value, especially when they are highly interconnected. While the mathematical foundation and the tools are available, theory and application in the domain of logistics systems and simulation are lacking. Typically, logistics systems are understood as networks consisting of a multitude of different interconnected elements [21]. Therefore, they are also referred to as logistics networks (LNs). To analyze such complex networks, digital twins (DTs) can be used. A DT is a virtual representation of a physical product or

© The Author(s), under exclusive license to Springer Nature Switzerland AG 2024
M. Mujica Mota and P. Scala (Eds.): EUROSIM 2023, CCIS 2033, pp. 125–139, 2024.
https://doi.org/10.1007/978-3-031-68438-8_10

artifact, for example a LN, and a bi-directional data connection between physical and virtual product [31]. Following [8] and [6], simulation is the main application and at the core of a DT. Simulation is a well-established method that uses an experimentable model of a system, called simulation model, to run experiments and gain insights on the system under investigation [32]. Simulation has been proven beneficial and generated notable results in the domain of logistics, e.g., in decision support [22], routing and scheduling [15], and knowledge discovery [23]. Several reviews have analyzed the ties between simulation and DTs, for example, in [30].

To create a simulation model in general and for a DT in particular, different successive formalization steps are necessary (e.g., conceptual model, formal model, executable model), usually supported by a procedure model to guide the process of creating a model [24]. For a target-oriented modeling, a modeling framework (MF) is necessary [4]. However, most approaches assume different means of description in the steps of model creation, which often depend on the simulation tool used. As a result, it is often necessary to translate from one means of description to the next one. The authors of this paper emphasize the importance of consistently formalized modeling in the domain of LNs and propose a novel approach to a MF in simulation for LNs that makes use of labeled property graphs (LPGs). This is considered as a valuable addition to the portfolio for both researchers and practitioners in the domain of simulation of LNs.

Therefore, the goal of this research is to answer the following research question: How can graph theory be used to realize a stringent and beneficial modeling approach for simulation in DTs of LNs?

Section 2 introduces related work and highlights the research gap that is addressed. Section 3 introduces the developed MF and Sect. 4 presents a use case on which the framework is tested and findings are discussed. The paper closes in Sect. 5 with a brief summary and discussion of limitations as well as an outlook on possible further research.

2   Related Work

The theoretical background for the remainder of this work is presented in the following sections. First, DTs in the context of LNs are briefly introduced. Based on this, modeling and simulation of LNs as well as relevant fundamentals of graph theory are discussed. This section closes with a look at the research gaps.

2.1   Digital Twins in Logistics Networks

The concept of a DT was coined by Michael Grieves, who presented a first idea in 2002 [10]. Originating in this idea, a common definition describes three parts of a DT: physical products in a real space, virtual products in a virtual space, and a bidirectional connection between the former ones, allowing data to flow between real and virtual space [10]. A similar understanding is given by [29], which also includes data and services as two additional dimensions. They categorize DT-driven services in six categories: simulation, control, evaluation, monitoring, prediction, and optimization. Another widely used definition is presented by [8], describing a DT as a multi-physics, multi-scale, and probabilistic simulation of a system. It is based on best available models and data sources like sensors. For the remainder of the paper, a DT is understood according to the definition given by [31, p. 157]: "A Digital Twin is a virtual representation of a physical artifact that contains a bidirectional data flow between the virtual and digital representation. It provides solutions for further data handling, i.e., simulation, optimization, or monitoring".

Focusing on the simulation part of a DT, [26] identify DTs as one of the main research trends in simulation. DTs are a well-aligned and consistent environment concerning models and data. The concept of a DT is an optimal concept for the use of simulation. As working principle, the data collected in the real world can be used to instantiate and execute the simulation. Therefore, they enable, for example, simulation-based decision support. Reviewing the scientific literature, the dominant application field of a DT is manufacturing, whereas the concept of DT has been transferred to other research domains, for example, aviation and space [8]. A comprehensive analysis of the application domains of a DT is given by [6].

Another popular field of application for DTs are LNs. The physical assets in the real space, on which data for the DTs are obtained, include, e.g., goods, transportation vehicles, and physical infrastructure. DTs can be used to model the various nodes and flows of the network, including suppliers, transportation, warehouses, and customers. By utilizing DTs, it becomes possible to optimize the flow of goods, reduce inventory levels, and improve the overall efficiency of the LN. The findings made in the DTs are then applied to the physical assets, which concludes the bidirectional flow of data. However, based on a literature review, [31] conclude that DTs in the domain of logistics are still lacking focus compared to other domains such as production. Moreover, white spots in the research of DTs in logistics are identified. An example is the missing integration of several tasks, like simulation and monitoring, in one DT. Also, the need for research of design principles in the domain of DTs in logistics is pointed out. In interviews with several companies operating a DT, [33] identified eight practical requirements of DTs in logistics. Summarized, these requirements can be divided into the categories data handling, data policies, and DT services, which should be taken into account when developing new approaches. A concept for a DT in supply chains is introduced by [3]. The authors argue that simulation, optimization, and data analytics are the core technologies enabling DTs of supply chains, which coincides with the understanding in the established literature. The strategic planning is conducted by the means of simulation and optimization. Typical applications in analytics for LNs include, e.g., trend analysis, demand prediction, or customer behavior analysis and are commonly used based on high-volume data sets. An overview of relevant research activities is given by [28].

2.2   Modeling and Simulation of Logistics Networks

Nowadays, logistics systems are fairly complex. The terminology of a logistics system is essentially congruent with the term supply chain [21]. A logistics system consists of a network of interrelated elements for the spatiotemporal transformations of goods as well as the associated processes, for example, transportation and storing [21]. For this purpose, independent actors, such as suppliers and customers, interact in such a system. Those systems are typically understood as networks, consisting of interconnected nodes and edges. The concept of a logistics system as a network emphasizes the inherent emergent nature of the elements present in the system [21]. Logistics systems are, therefore, referred to as LNs. According to [18], there are different ways of examining such a LN, for example, by experiments on the real network. However, since investigations on the real network are often not possible or not reasonable due to cost reasons, a formalization of the network as a model is necessary. The understanding of the term model is complex and differs depending on the research domain. In the research domain of simulation of LNs, the complex interrelationships are typically represented using specific models, called simulation models. These are executable models that can be studied with the help of simulation [16]. Following [18], simulation models can be differentiated along three dimensions which are the time-dependent behavior (static/dynamic), the usage of randomness (deterministic/stochastic), and the behavior of changing states (continuous/discrete time). In this paper, the definition as proposed by [32, p. 3] is adopted: "Representation of a system with its dynamic processes in an experimentable model to reach findings which are transferable to reality; in particular, the processes are developed over time."

To conduct a simulation study in a purposeful and structured way, procedure models are used. The authors propose the procedure model for simulation from Rabe et al. (2008), which is well established in the literature and also part of a guideline [32, p. 20]. Figure 1 illustrates the procedure model. According to this procedure model, the model formalization consists of three phases.

- System analysis, resulting in a conceptual model (e.g., a whiteboard sketch)
- Model formalization, resulting in a formal model (e.g., a UML2 diagram)
- Implementation, resulting in a executable model (e.g., a simulation model developed in a general programming language)

Relevant in these phases is a modeling concept, which is also called conceptual framework [1] or MF [4]. Typically, these can be found in particular with regard to the simulation tool used. According to [11], established MFs are:

- Building-block-oriented MFs (application-oriented, most widely used in production and logistics). These concepts use libraries offering building blocks, containing states and state transitions as well as parameterizable flow logic.
- Object-oriented modeling concepts following the object-oriented paradigm from software development. Real-world objects are modeled by objects as instances of a class, which contain properties and methods.

- Theoretical MFs. These are, for example, graph-theoretical concepts (e.g., petri nets), queueing theory, or automata-theoretical concepts.
- Language frameworks. This includes in particular higher programming languages or specific languages with regard to simulation tools.

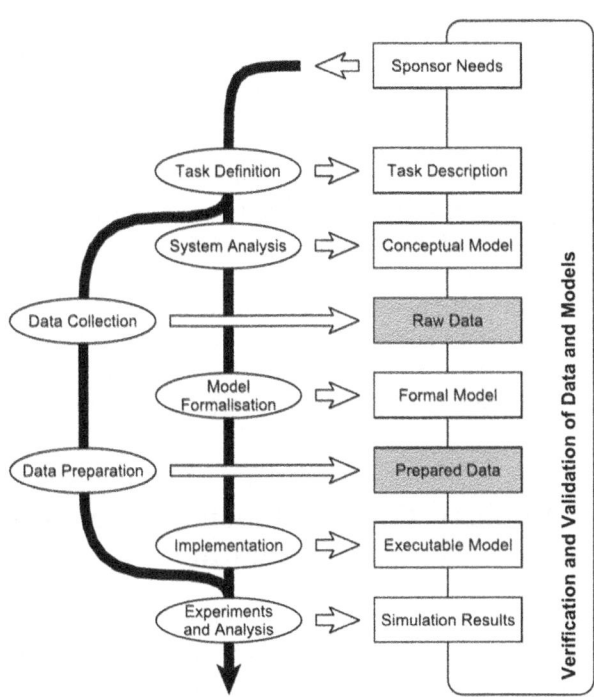

Fig. 1. Simulation procedure model according to [32, p. 20]

It should be noted that the modeling concepts overlap and often hybrid forms exist, especially in simulation tools. For example, a building block-oriented concept, in which building-blocks are linked together, can in the final result be interpreted as a graph, and, therefore, a theoretical MF. Of particular interest in this research are graph-theoretic modeling concepts, since logistic systems represent complex, highly interconnected networks. This must be considered conceptually in the way of modeling. For this, graph theory can be used for a corresponding formalization.

2.3 Graph Theory and Graph Algorithms

Graph theory is a branch of discrete mathematics. A Graph consists of a pair $G = (V, E)$ of a finite set of vertices, denoted by V, and a finite set of edges, denoted by E [9]. It holds for each edge e that it is a subset u, v of V, where u

and v are called the end-vertices of e. In the logistics context in particular, the term node has become synonymous with vertice. Therefore, for the remaining paper, vertices will be referred to as nodes. It is noted that only the mathematical definition, e.g., via an adjacency matrix, of a certain graph is relevant, since graph visualizations and diagrams are subjective and can vary. An edge always connects two nodes, which can be different or the same node, the latter being a so-called loop. Graphs are distinguished in particular based on their structural properties. A graph with undirected edges and without loops is called a simple graph. If the edges of a graph are directed, the graph is called a directed graph or digraph. This creates an order in the graph, in which nodes can be distinguished as predecessors and successors. Furthermore, one-sided or two-sided relationships can be described. Because systems in reality are more complex than they can be represented exclusively with nodes and directed edges, graphs can be enriched with labels and properties [25]. This type of graph is referred to as a LPG. Following [19], both nodes and edges can contain properties in form of an attribute-value pair (e.g., name: Supplier A) and have a label (e.g., supplier), making it possible to group nodes. A LPG is a commonly encountered graph in research and practice. For example, the well-known graph database Neo4j [20] uses an LPG as the data model to persistently store data in a native graph format.

3 Labeled Property Graph Modeling Framework for Logistics Networks for Digital Twins

The tasks of DTs in the domain of LNs, as described in Sect. 2, are simulation, optimization, and monitoring. These three tasks have in common that they require suitable data and a model to be used on. Handling as well as storing data and models is the task of the IT infrastructure, which is considered to be one of the key challenges of DTs [7]. Another important challenge of DTs according to [7] is the introduction of standardized ways of modeling. In this section, a design proposal is discussed that provides a common data model for enabling DTs for the aforementioned range of tasks and to tackle the shown challenges.

When analyzing the system of a LN for modeling, there is often a whiteboard sketch at the beginning. Likely, the sketch consists of several clusters of information that are interconnected with each other. Basically, such a sketch is a graph structure. In the next step, this sketch is formalized, so simulation, optimization, or monitoring can be used on an implementation of the formalized model. The form of a formalized model varies and is dependent on the intended later tasks. It is common to use multiple means of description for a formalized model, which are not necessarily a graph structure (see Sect. 2). For DTs, this paper encourages the formalization in a single graph structure, which makes it easier to formalize initial ideas and offers advantages in the downstream processes beyond. At first, graph-based models as a common data model in DTs are discussed in Sect. 3.1 and later a MF for LNs is given in Sect. 3.2.

3.1 Graph-Based Model as a Common Data Model for Digital Twins

In Sect. 2 it was shown that DTs of LNs should be able to perform the tasks of simulation, optimization, and monitoring on models. To reduce redundancies in creating, managing, and storing models, using a common model is a promising approach. Such a common model as a base of operation should be tailored towards the needs of those tasks above. Specifically, the authors of this paper propose to use a labeled property graph model (LPGM) with certain rules. The rules will be introduced in the next section. Despite being designed as a common data model, for multiple purposes, the MF could also be used for sole purpose of simulation, optimization, monitoring, or any combination of them. Besides the advantages of a common data model, there is also a drawback as the model probably contains more data and is more complex than needed for just one single task. To deal with that circumstance, adapters for each task have to be used, which only provide the specifically relevant parts of the model.

To enable DTs for optimization and monitoring, analytic studies of the underlying data models are needed. Considering analytic operations, research has shown that utilizing graph algorithms on graph structures yields promising results, especially on highly-connected data [25]. This is also true for graph mining applications like in basket analysis, where frequent patterns are searched for [14]. It is important that the model has a high information content that can possibly be extracted. For this purpose, LPG are particularly suitable as their properties offer the possibility to create semantically rich graphs. Though promising results with some graph-based analytic methods could be achieved, there is a strong need for further research in this area to provide a wider choice of methods and to exploit the potential of graphs. The data must not only have high information content, their quantity must also be extensive. In the field of LNs, this is usually the case (see Sect. 2). One challenge that arises from such large data sets is data management. Fortunately, data management of large data sets is a strength of graph databases, in which LPGs are usually stored. The suitability of graph databases as a starting point for analytic operations has already been discussed in [12] and was found to be promising.

For the optimization of LNs, approaches from the field of operations research are used among others. As LNs feature network structures, the approaches often base on mathematical graph representations of the networks at hand. Exemplary applications are tour planning [34] and scheduling of operations [5]. The required mathematical graph representations for those tasks are not LPGs in particular, but LPGs could be easily derived from them.

LPGMs can also hold the required information to be used as simulation models. To be of use for simulation, the LPGMs must be interpretable by a simulation tool. For this purpose, a special form is needed that is defined in a MF. At the beginning of a simulation, the model has to be loaded into the internal structure of the simulation tool's engine. Most simulation tools are developed in universal object-oriented programming languages. For these tools, a concept called object graph mapping resembles a possibility to copy a LPGM effectively to the internal structure of the program. This feature is supported by several graph databases like [20]. Using object graph mapping ensures a direct means of transformation with minimal overhead. It is most useful, when the internal structure of the simulation tools is also graph-oriented. It is also beneficial for those simulations that make use of graph algorithms like in [2] to solve vehicle routing problems. Another promising use case for a LPGM has been discussed by [13]. According to them, graph-based approaches favor the use of data farming, which is a method for simulation-based data generation that can be used to gain detailed insights of simulated objects [27]. Data farming generates large amounts of data, which are preferably stored in graph structures and databases. For further information on this topic and the implementation of data mining on farmed data in a farming-for-mining framework, please refer to [13].

3.2   Modeling Framework

The MF presented in this section makes use of LPGs and introduces a set of labels, rules for edges, as well as general modeling rules, which are used to model characteristics of LNs. The MF only concerns the nodes and their edges, but not their properties. This is due to the fact that properties entirely depend on the utilized tools as well as the available data and, therefore, vary greatly. Although a common set of properties could possibly be identified, this step is not considered by the authors to be purposeful for the MF at hand. However, exemplary properties of nodes will be selected to give a better idea on how a specific model might look like. Before the MF is presented it is noted that the process of modeling is highly subjective, and there are a large number of different objectives for models, which makes it nearly impossible to propose a one-fits-all solution. Therefore, users are encouraged to customize or extend the MF to their individual needs.

First, the set of labels for nodes is described. There is no dedicated set of labels for edges. Labels of a LPG categorize nodes to associate common traits. In terms of LNs, the authors propose to introduce eight labels. One of these labels must be assigned to each node used in the model. The eight labels are *agent, material, information, resource, behavior, environment, mechanic,* and *report*.

As shown previously, LNs consist of physical locations belonging to stakeholders. Even though they work together to a certain extent within the LNs, they can be considered as autonomous entities often referred to as agents. To reflect this in the model, an agent-centered approach is used, in which the various locations are considered as the agents. In the LPGM, each agent is represented with a node labeled as *agent*. The properties of these nodes include basic information about the location, e.g., the geographic location, the stakeholders company's name, or working hours. The *agent* nodes are the central building blocks of the LPGM. It should be noted that simulation tools that make use of the MF are not limited to agent-based simulation.

As shown in Sect. 2, LNs implement the flow of material and information. This implies the presence of material and information objects that can be transferred between stakeholders. When such objects should be modeled in the

LPGM, the *material* or respectively *information* label must be used. Material objects might refer to stock keeping units (SKUs) or load carriers. An example for an information object are orders that can communicate demand upstream.

Other objects in LNs, such as transport vehicles or loading docks, are utilized by the agents and can be referred to as their resources. Nodes resembling such objects are associated with the *resource* label. Exemplary properties are, for example, the capacity of vehicles or the number of bays in a loading dock.

To describe the behavior of agents, resources, material, and information objects, additional nodes are used. They are connected to the respective nodes by edges and are labeled as *behavior*. Examples for a behavior are the way of stock keeping or how deliveries are planned. Properties could describe rates of occurrences of certain actions or the probabilities to trigger specific behavior. To be able to represent more complex behavior, multiple behavior nodes can be cascaded. The definition of the LN's behavior in the model is needed for the purpose of simulation.

As shown earlier, agents operate in an environment. In the case of LNs, this might be information about entry restrictions to city centers or about holidays at the locations. To store this information in the model, nodes with the *environment* label are used. A further component are nodes with the *mechanic* label. They describe parts of the model that are related to the scope of functions of the simulation tool and do not fit to any other label. For example, a chosen stochastic distribution could be a node with its type and parameters as properties. The last component of the MF are nodes with the *report* label. These nodes represent data collected during the operation of the real LN or its simulation.

It should be noted that a model must not make use of all types of labels. Also, not every node needs to be utilized by each task in a DT. For example, the label *report* is not likely to be used for simulation models (although it is used for the results of simulation), but it is a crucial part of monitoring.

Edges according to the MF are always unidirectional, except for edges between *agent* nodes, which are bidirectional. The edges represent a hierarchy where the data from the nodes where the edges end are used by the nodes where the edges start. As agents act on an equal level of hierarchy, bidirectional edges are used. Apart from that, agents can serve as start nodes for edges to any other node except those labeled with *mechanic*. Nodes with the *behavior* label are required to be referenced by *agent* nodes, as they always detail those. They themselves can be starting points for edges that reference any other labeled nodes. Those nodes with the *material, information, resource, environment, mechanic,* and *report* label can never be used as starting nodes for an edge and can only be used as detailing data. Besides the labels and rules how to structure nodes with edges, there are general rules that must be followed. Overall, the MF consists of a total of four rules. For a model to be considered valid with respect to the MF, none of the rules may be violated.

1. Each node must have one of the labels defined in this section
2. There must be at least two nodes with *agent* labels

3. Each node with the *agent* label must have at least one edge to another node with the *agent* label
4. Models must only have edges starting and ending with nodes according to the rules defined in this section (e.g., behavior must start at an *agent* node)

As described at the beginning of this section, properties of edges will not be discussed in detail. However, one should keep in mind that they are probably useful, for example, for storing distances between agents when no geo-coordinate-based approach of distance calculation is used.

4   Exemplary Use Case

This section applies the previously introduced MF to a real-world use case as an example. The model is targeted for simulation and, therefore, not necessarily includes the characteristics for monitoring and optimization purposes. However, the model could be easily extended.

The use case is based on a simulation study conducted by [22]. It evaluates the consolidation of distribution flows in the metropolitan area of Athens (Greece). For this purpose, the utilization of consolidation centers was examined. Suppliers that want to deliver goods to the urban center of Athens deliver their goods to the consolidation center instead. From there, smaller vehicles carry out the transports to the urban centers. This can be modeled as a two-echelon LN, in which the first echelon resembles transports to the consolidation center and the second one to the end customers. [22] examined two scenarios. In the first one, they consider one consolidation center and in the second one, two centers were assumed. The following use case is based on the first scenario. [22] created a simulation model, consisting of five suppliers, one consolidation center and 8,537 customers. For demonstration purposes, only a part of the LN is modeled in this section. More specifically, one supplier, one consolidation center, and one customer have been modeled. Figure 2 shows a possible simple whiteboard sketch.

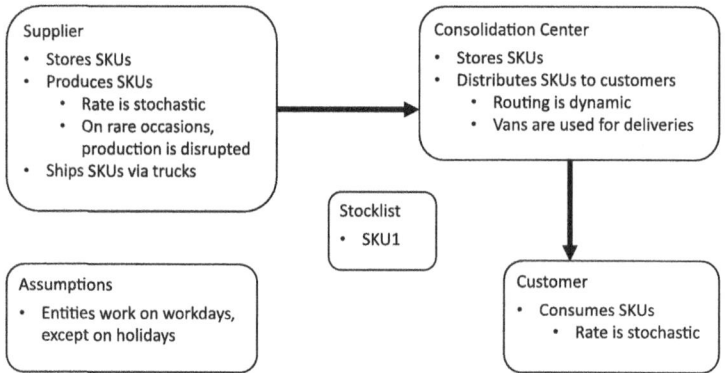

Fig. 2. Whiteboard sketch for the exemplary LPGM

Such a sketch is a typical result of a system analysis and created in the beginning. It outlines some of the key characteristics for the LN at hand. As described in Sect. 2, formalization steps are applied to obtain a model. In Sect. 4.2, the introduced MF is used for formalization. In the first place, the topology of an exemplary LPGM is discussed. Figure 3 shows the exemplary model.

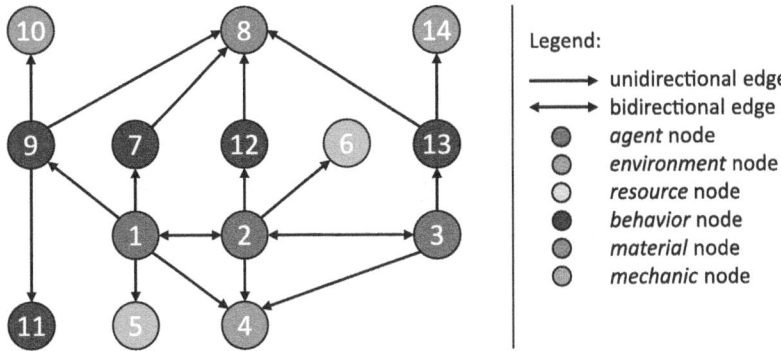

Fig. 3. Graph structure of the exemplary LPGM

Figure 3 includes a total of 14 nodes. Nodes labeled as *agent* are used for each of the three participants in the LN under consideration. Node 1 is the agent node of the supplier, Node 2 resembles the consolidation center, and Node 3 the customer. Between the supplier and the consolidation center as well as between the customer and the consolidation center are bidirectional edges, allowing the flow of material and information. In the scenario at hand, the information transmitted is limited to orders of SKUs and not explicitly modeled. All three *agent* nodes are adjacent to Node 4, which holds the dates of national holidays in Greece. This node is labeled as an *environment* node. In the scenario, two different types of vehicles are utilized for delivery. To model the vehicles in the LPGM, nodes with the label *resource* are used. The larger vehicle is represented by Node 5 and adjacent to Node 1, the smaller vehicle is represented by Node 6 and is connected to Node 2.

The behavior of the supplier is described by two nodes with *behavior* labels. Node 7 is intended to describe the supplier's ability to store SKUs. Which SKUs are storable is defined by adjacent nodes resembling SKUs. In Fig. 3, the node used for the SKU is Node 8. It is annotated with the label *material*. Since the task of the supplier is to deliver SKUs, it must source these beforehand. In the example, the sourcing behavior is specified in Node 9. Just like the third node, the sourcing must have a reference to the SKU it should source. To specify the amount of sourced SKUs, a stochastic approach is used. In the LPGM, Node 10 with the *mechanic* label holds the required information about the stochastic distribution. It is also possible to modify the behavior of Node 9 by an additional behavior like malfunctions. Therefore, another behavior node is used adjacent to the production node. The behavior of the consolidation center is specified by

Node 12. Equivalently to Node 7, the node describes the storage of its agent and is adjacent to the SKU node it should store. For the behavior of the customer, Node 13 resembling the consumption is used. The rate of consumption is stochastically described by the adjacent Node 14 with the *mechanic* label, which holds information about the distribution used. Node 13 is also adjacent to Node 8 to specify which SKU is consumed.

As described in the approach, the required properties are dependent on the used simulation tool. Nevertheless, an example of properties for all nodes is presented in Table 1. Exempted are Nodes 9 and 13, as they have no properties in the given example. It should be noted that *Parameter 1* and *Parameter 2* of nodes 10 and 14 are used to parameterize the described distributions, e.g., a normal distribution is parameterized by values for the mean (*Parameter 1*) and the variance (*Parameter 2*). With the properties in Table 1, it is conceivable that a simple simulation can be run. For sophisticated real-world applications, a larger number of properties to describe complex problems would be needed.

Table 1. Properties for the exemplary LPGM

Node	Property	Data Type
1, 2, 3	Name	String
	Coordinate	Point
	Working Hours	Float[7]
4	Holidays	List<Date>
5, 6	Type	String
	Capacity	Integer
	Speed	Float
7, 12	Capacity	Integer
	Initial Fill	Float
	Sourcing Strategy	String
8	Name	String
10, 14	Distribution	String
	Parameter 1	Float
	Parameter 2	Float
11	Probability	Float
	Rate	Float

The modeling presented in this section has shown the intuitive and versatile applicability of the MF. It has proven to be easy to use and fast to set up while ensuring rich information content. The transition from a whiteboard sketch to a LPGM could be easily accomplished. Since a real use case was considered, it can be concluded that the MF can be applied to those scenarios.

However, the discussed use case has limitations and is not sufficient to assess the performance of the system comprehensively as it is not put to test on a simulation, optimization, or monitoring task. Regarding simulation, the versatility of the approach is limited to the scope of functions of the simulation tool, as it has to interpret the function and meaning of each node, relationship, and property. Due to the lack of such a simulation tool, no assessment could be given. Although established simulation tools that are not specifically designed for the use of graph models could have been used with an additional transformation process, this process was not considered to be expedient, as the main idea of using a graph model would not have been investigatable.

5   Summary and Outlook

This paper introduced a novel MF for LNs. It is tailored to be used in a DT to act as a common data model for simulation, optimization, and monitoring. The approach is based on LPGs, which are a type of graph structure featuring nodes and edges as well as properties to describe their details. The MF includes a set of labels for each node, rules on how to structure a graph, and a general set of modeling rules. Moreover, to demonstrate the use of the MF, an exemplary model for a real use case was discussed.

The MF contributes to the introduction of common data models for multiple purposes in the domain of LNs in DTs. This research shows that graph theory can be applied for beneficial and value adding modeling of LNs. Thus, the authors encourage the scientific community to use DTs for LNs and common data models in them, especially those based on graph theory.

However, the research is subject to limitations. Though it was shown, how a model can be created with the MF, it was never put to test to actually carry out simulation, optimization, or monitoring. Therefore, no quantifiable results can be provided. Also, there is a lack of a comprehensive toolbox to build on models according to the MF. Another important missing part is a simulation tool, which can interpret and work with a simulation model based on LPGMs.

Future research and implementation work should focus on a simulation tool, specifically designed for the use of LPGM and to take advantage of the underlying graph structure. Projects at the authors' department have shown promising results. Future research will focus on developing a suitable simulation tool based on the results presented in this work. Moreover, the MF should be used in a DT for a real-world scenario and tested on the tasks of simulation, optimization, and monitoring.

References

1. Balci, O.: Guidelines for successful simulation studies. In: Balci, O., Sadowski, R.P., Nance, R.E. (eds.) 1990 Winter Simulation Conference Proceedings, pp. 25–32. IEEE, Piscataway (1990). https://doi.org/10.1109/WSC.1990.129482
2. Barceló, J., Grzybowska, H., Pardo, S.: Vehicle routing and scheduling models, simulation and city logistics. In: Zeimpekis, V., Tarantilis, C.D., Giaglis, G.M.,

Minis, I. (eds.) Dynamic Fleet Management. Operations Research/Computer Science Interfaces Series, vol. 38, pp. 163–195. Springer, Boston (2007). https://doi.org/10.1007/978-0-387-71722-7_8

3. Barykin, S.Y., Bochkarev, A.A., Kalinina, O.V., Yadykin, V.K.: Concept for a supply chain digital twin. Int. J. Math. Eng. Manag. Sci. **5**(6), 1498–1515 (2020). https://doi.org/10.33889/IJMEMS.2020.5.6.111 https://doi.org/10.33889/IJMEMS.2020.5.6.111

4. Cassandras, C.G., Lafortune, S.: Introduction to Discrete Event Systems. Springer, Cham (2021). https://doi.org/10.1007/978-3-030-72274-6

5. Dolgui, A., Ivanov, D., Sethi, S.P., Sokolov, B.: Scheduling in production, supply chain and industry 4.0 systems by optimal control: fundamentals, state-of-the-art and applications. Int. J. Prod. Res. **57**(2), 411–432 (2019). https://doi.org/10.1080/00207543.2018.1442948

6. Enders, M.R., Hoßbach, N.: Dimensions of digital twin applications. A literature review. In: Proceedings of the 25th Americas Conference on Information Systems, pp. 1–10. Association for Information Systems, Cancun (2019)

7. Fuller, A., Fan, Z., Day, C., Barlow, C.: Digital twin: enabling technologies, challenges and open research. IEEE Access **8**, 108952–108971 (2020). https://doi.org/10.1109/ACCESS.2020.2998358

8. Glaessgen, E., Stargel, D.: The digital twin paradigm for future NASA and U.S. air force vehicles. In: Structures, Structural Dynamics, and Materials and Co-located Conferences. American Institute of Aeronautics and Astronautics, Reston (2012). https://doi.org/10.2514/6.2012-1818

9. Golumbic, M.C.: Algorithmic Graph Theory and Perfect Graphs, vol. 57, 2nd edn. Elsevier, Amsterdam (2004)

10. Grieves, M., Vickers, J.: Digital twin: mitigating unpredictable, undesirable emergent behavior in complex systems. In: Kahlen, F.-J., Flumerfelt, S., Alves, A. (eds.) Transdisciplinary Perspectives on Complex Systems, pp. 85–113. Springer, Cham (2017). https://doi.org/10.1007/978-3-319-38756-7_4

11. Gutenschwager, K., Rabe, M., Spieckermann, S., Wenzel, S.: Simulation in Produktion und Logistik. Springer, Heidelberg (2017). https://doi.org/10.1007/978-3-662-55745-7

12. Hunker, J., Scheidler, A.A., Rabe, M.: A systematic classification of database solutions for data mining to support tasks in supply chains. In: Kersten, W., Blecker, T., Ringle, C. (eds.) Data Science and Innovation in Supply Chain Management: How Data Transforms the Value Chain, pp. 395–425. epubli, Berlin (2020). https://doi.org/10.15480/882.3121

13. Hunker, J., Wuttke, A., Scheidler, A.A., Rabe, M.: A farming-for-mining-framework to gain knowledge in supply chains. In: Kim, S., et al. (eds.) Proceedings of the 2021 Winter Simulation Conference. IEEE, Piscataway (2021)

14. Inokuchi, A., Washio, T., Motoda, H.: Complete mining of frequent patterns from graphs: mining graph data. Mach. Learn. **50**(3), 321–354 (2003). https://doi.org/10.1023/A:1021726221443

15. Juan, A.A., Rabe, M.: Combining simulation with heuristics to solve stochastic routing and scheduling problems. In: Dangelmaier, W., Laroque, C., Klaas, A. (eds.) Proceedings of the 15th ASIM Conference on Simulation in Production and Logistics, pp. 641–649. Heinz-Nixdorf-Institut, Paderborn (2013)

16. Kleijnen, J.P.: Design and Analysis of Simulation Experiments. Springer, Cham (2015). https://doi.org/10.1007/978-3-319-18087-8

17. Koessler Gosnell, D., Broecheler, M.: The Practitioner's Guide to Graph Data. O'Reilly, Sebastopol (2020)

18. Law, A.M.: Simulation Modeling and Analysis, 5th edn. McGraw-Hill Education, New York (2015)
19. Meier, A., Kaufmann, M.: SQL & NoSQL Databases: Models, Languages, Consistency Options and Architectures for Big Data Management. Springer, Wiesbaden (2019). https://doi.org/10.1007/978-3-658-24549-8
20. Neo4j: Neo4j (2022). https://neo4j.com/. Accessed 22 Dec 2022
21. Pfohl, H.C.: Logistics Systems: Business Fundamentals. Springer, Heidelberg (2022). https://doi.org/10.1007/978-3-662-64349-5
22. Rabe, M., Klueter, A., Wuttke, A.: Evaluating the consolidation of distribution flows using a discrete event supply chain simulation tool: Application to a case study in Greece. In: Rabe, M., Juan, A.A., Mustafee, A., Skoogh, S.J., Johansson, B. (eds.) Proceedings of the 2018 Winter Simulation Conference, pp. 2815–2826. IEEE, Piscataway (2018). https://doi.org/10.1109/WSC.2018.8632266
23. Rabe, M., Scheidler, A.A.: An approach for increasing the level of accuracy in supply chain simulation by using patterns on input data. In: Tolk, A., Diallo, S.Y., Ryzhov, I.O., Yilmaz, L., Buckley, S., Miller, J.A. (eds.) Proceedings of the 2014 Winter Simulation Conference, pp. 1897–1906. IEEE, Piscataway (2014)
24. Rabe, M., Spieckermann, S., Wenzel, S.: A new procedure model for verification and validation in production and logistics simulation. In: Mason, S.J., Hill, R.R., Mönch, L., Rose, O., Jefferson, T., Fowler, J.W. (eds.) Proceedings of the 2008 Winter Simulation Conference (WSC), pp. 1717–1726. IEEE, Piscataway (2008)
25. Robinson, I., Webber, J., Eifrem, E.: Graph Databases: New Opportunities for Connected Data, 2nd edn. O'Reilly, Sebastopol (2015)
26. Rosen, R., von Wichert, G., Lo, G., Bettenhausen, K.D.: About the importance of autonomy and digital twins for the future of manufacturing. IFAC-PapersOnLine 48(3), 567–572 (2015). https://doi.org/10.1016/j.ifacol.2015.06.141
27. Sanchez, S.M.: Simulation experiments: Better data, not just big data. In: Tolk, A., Diallo, S.Y., Ryzhov, I.O., Yilmaz, L., Buckley, S., Miller, J.A. (eds.) Proceedings of the 2014 Winter Simulation Conference, pp. 805–816. IEEE, Piscataway (2014). https://doi.org/10.1109/WSC.2014.7019942
28. Seyedan, M., Mafakheri, F.: Predictive big data analytics for supply chain demand forecasting: methods, applications, and research opportunities. J. Big Data 7(1) (2020). https://doi.org/10.1186/s40537-020-00329-2
29. Tao, F., Cheng, J., Qi, Q., Zhang, M., Zhang, H., Sui, F.: Digital twin-driven product design, manufacturing and service with big data. Int. J. Adv. Manuf. Technol. 94(9–12), 3563–3576 (2018). https://doi.org/10.1007/s00170-017-0233-1
30. van der Valk, H., Hunker, J., Rabe, M., Otto, B.: Digital twins in simulative applications: a taxonomy. In: Bae, K.H., Feng, B., Kim, S., Lazarova-Molnar, S., Zheng, Z., Roeder, T.R.T. (eds.) Proceedings of the 2020 Winter Simulation Conference (WSC), pp. 2695–2706. IEEE, Piscataway (2020). https://doi.org/10.1109/WSC48552.2020.9384051
31. van der Valk, H., Strobel, G., Winkelmann, S., Hunker, J., Tomczyk, M.: Supply chains in the era of digital twins - a review. Procedia Comput. Sci. 204, 156–163 (2022). https://doi.org/10.1016/j.procs.2022.08.019
32. Verein Deutscher Ingenieure: VDI 3633 - Simulation of Systems in Materials Handling, Logistics and Production: Fundamentals. Beuth, Berlin (2014)
33. Winkelmann, S., van der Valk, H.: Openness of digital twins in logistics – a review. In: Herberger, D., Hübner, M. (eds.) Proceedings of the 2022 Conference on Production Systems and Logistics. publish-Ing., Hannover (2022)
34. Zhao, L., Han, G., Li, Z., Shu, L.: Intelligent digital twin-based software-defined vehicular networks. IEEE Netw. 34(5), 178–184 (2020)

Simulation-Optimisation-Based Decision Support System for Managing Airport Security Resources

Geoffrey Scozzaro[1]([⊠]), Miguel Mujica Mota[2], Daniel Delahaye[1], and Catherine Mancel[1]

[1] ENAC, Université de Toulouse, 31400 Toulouse, France
geoffrey.scozzaro@enac.fr
[2] Amsterdam University of Applied Sciences, Amsterdam, The Netherlands
m.mujica.mota@hva.nl

Abstract. Airport access mode disruptions have a significant impact on passenger arrival flow. Such events can lead to a wave of late outbound passengers, that can congest airport facilities such as check-in counters or security systems. In this context, efficient handling of airport resources is crucial to ensure passengers' reliable door-to-door journeys and maintain a high airport Level-of Service. Through a simulation-optimisation-based decision support system, we investigate the relevance of a dynamic opening of fast security line facilities for delayed passengers. An optimisation model solved through simulated annealing is proposed to improve security resource handling at a tactical level. Airport's Level of Service is evaluated before and after optimisation through a microscopic passenger flow simulation model. The methodology is tested on a study case based on one of the terminal buildings of Mexico-City airport. Results indicate that using dedicated security lines for passengers in a hurry due to disruptions reduces the number of stranded passengers at the end of the day; out of the three scenarios evaluated the one that uses a dynamic management of resources provides the best results; however this task should be governed by a DSS that adapts to the particular case under study.

Keywords: Airport · Optimisation · Simulation · Disruption Management

1 Introduction

Airport security screening systems are key components of passenger processing time at the airport. A poor handling of security screening resources could lead to excessive waiting times and, thus, threaten the whole passenger journey. This could be especially the case when actual passenger flows deviate from the forecast ones, e.g. when a disruption occurs on a ground transportation mode than

© The Author(s), under exclusive license to Springer Nature Switzerland AG 2024
M. Mujica Mota and P. Scala (Eds.): EUROSIM 2023, CCIS 2033, pp. 140–155, 2024.
https://doi.org/10.1007/978-3-031-68438-8_11

enables passengers to access the airport. In this context, reactivity and adaptability of airport operators are essential to keep an efficient security process and ensure passengers a reliable and smooth door-to-door journey. In the so-called passenger trajectory [21] within airport terminals, the security checkpoints are locations that are necessary and fundamental for keeping the aviation operations protected from unlawful actions that might put at risk the lives of people or the value of goods. The objective of those checkpoints is to detect threats as objects/liquids/aerosols/gels that can potentially impact the safety of the transport operation or of the users. In the case of Aviation, the International Civil Aviation Organisation describes the requisites and suggestions on security in the Annex 17 of the Chicago convention [10]. The capacity of these systems is determined by the amount of passengers they can process in a specific period of time (i.e. the throughput). They are designed based on the expected demand of passengers and the processing capacity of passengers. This one can be affected by diverse factors such as the experience of personnel as some of the screening is done manually, the efficiency of the automatic systems like body or baggage scanners, how rigorous the screen is performed, the tighter control, etc. The slower performance is achieved in most of the situations, reducing the system's capacity as discussed by [4]. In many cases, due to the rigidness of the scan procedure, the throughput is reduced causing delays and also stress to the passengers which, in the worst situations, can even make them miss their flights. There have been attempts and studies to make the processing of passengers more efficient [16,17], they range from dedicated lines to new categories of passengers. They demonstrated that using a one-size-fits-all approach for processing passengers is not the most efficient way to improve the capacity or efficiency. They studied the impact of adjusting the system to the categories of passengers giving good results in most of the cases.

In the current paper, we explore the combination of optimisation with simulation for developing a decision support tool that allows coordinating the management of the security resources in a dynamic fashion while adjusting the system to different categories of passengers. In particular, we put emphasis on the situation where passengers are under pressure with a tight schedule to reach their gate on time. We propose (as in previous studies) to use different categories of passengers and adapt the system to them and to provide dedicated resources when needed such in a time of disruption. Furthermore, we consider that the airport has limited resources (workforce) to operate the security points making the situation even more challenging. To consider other elements that are out of the scope of a traditional mathematical model, we develop a simulation model of the complete terminal building to identify the effects of the allocation in the different systems that take place in the passenger trajectory in comparison with focusing only on the security area like most of the studies do. We combine the two techniques to develop the principles of a Decision Support System (DSS) that allows managing in a dynamic fashion the opening and closing of security points; in contrast with the traditional inflexible approach where some points are open during the day and kept like that, just wasting some capacity that might be better used in critical times. We illustrate the approach with the case of one of the terminals of the main airport in Mexico City which is well-known for its limited capacity. A real data profile is used for the analysis of the system. The approach can provide benefits to passengers with a better service level but also to the other stakeholders of the airport system. With the use of a dynamic approach, the airport can save costs on personnel and operations and airlines will gain in punctuality as the schedule becomes more robust. The approach illustrates how the managing of the resources of a complex system like an airport can be improved with the simultaneous use of simulation and optimisation techniques.

2   Literature Review

The idea that business passengers tend to be quicker on the security lines while families and handicapped passengers take longer is validated by [11]. Furthermore, [1] showed that the quantity of clothing of a passenger directly influences the passengers' queuing time. Unfairness and long waiting times might as well be incremented due to social bias of passengers from certain countries or ethnicity at security checkpoints [3]. Long waiting times can contribute to passenger's frustration or to the experience of bad service as a recent survey showed that security plays an important role in the service perceived by passengers at airports [12,19].

Regarding resource optimisation handling, queuing theory is a tool that can be used to manage airport operations as explained by [5]. Such an approach can provide analytical optimal solutions. However, restrictive assumptions on the demand (i.e. passenger flow entrance in an airport security context) are required to apply the results of such theory. Since, in the general case, passenger arrival flow does not follow standard distribution such as Poisson law, most of the analytical results are not applicable. On the other side, simulation approaches have been vastly employed to assess airport capacity [14], or even optimise resource allocation at check-in or security screening system for large airports. [13] used an evolutionary algorithm combined with a simulation model to solve the check-in allocation problem. [15] studied the impact of new procedures linked to Covid-19 restrictions on security checkpoint capacity. [16] proposed for the first time passenger segregation depending on their characteristics to improve airport security screening Level of Service. [17] studied the impact of a specific type of passengers, who, in exchange for sharing personal information, could reduce their airport processing time by using a new type of facility. [18] used a simulation model for dynamic resource allocation of security screening at airports. Passenger queue length or waiting time are minimised under staff resource constraints.

In the current paper, we propose optimising the security line opening policy through a combined simulation-optimisation framework. The optimisation model is based on a deterministic optimisation model like in [22], with the additional consideration of opening a priority line for a subset of passengers. Moreover, a detailed airport simulation model, which captures effects of other airport processes on passengers, is used in combination with the optimisation approach to evaluate the benefits of such optimisation. Such an approach enables finding a good compromise between deterministic approaches, which generally overestimate solution quality (since uncertainties can drastically reduce airport Level of Service as explained in [9]), and simulation approaches, which tend to take excessive computational time to find optimised solutions. The next section provides a detailed description of the problem considered in this study.

3   Problem Description

The airport security screening system is composed of a set of Security Checkpoints (SC). An illustration of one SC is provided in Fig. 1. Each SC is composed of a set of security points (SP), a common queuing line, one fast security point, and its associated queuing line. An SP is a facility where hold-baggage and passengers are screened by security teams. An SP or a fast SP can be operated by a security team, composed of five or six agents. The day of operations is divided into 30-min time slots whether each SP is operated or not. No distinction is made between the different SP. The First-In First-Out (FIFO) scheme is considered for each queuing line. Each passenger entering the security system is characterised by: an entering time at the SC, a status, and a boarding time. Two categories of passengers are considered: regular and priority. A constant service rate is assumed for each SP and another one for fast-SP (higher than the common service rate). When a fast SP is operated, priority passengers go through the fast queuing system to be served. Otherwise, they are processed through the common security system. Passenger boarding times are used to infer, for each passenger, the maximum waiting time at security after which he/she would miss the flight.

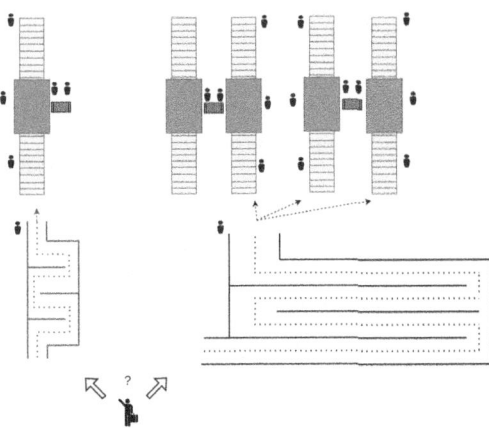

Fig. 1. Scheme of a security checkpoint. The fast queuing line is displayed in red. (Color figure online)

The optimisation problem considered relies upon deciding, for each time slot, how many SP (including fast-SP) should be operated on each SC. A resource pooling constraint is considered, i.e. a maximum number of SP and fast-SP can be open during each time slot across the airport. The proposed reallocation aims at improving the airport security system efficiency. Different Key Performance Indicators (KPIs) can be considered to evaluate security system efficiency such as average passenger queuing time, maximum passenger waiting time, passenger share waiting less than 10 min, maximum waiting queue length, among other. However, in this study, we retain a bi-criteria objective function to minimise. The first criterion is the total number of passengers missing their flight. A passenger is assumed to miss his flight if his arrival time at the boarding area is later than the closure boarding time of its flight. The second criterion is related to the maximum passenger waiting time experienced across all passengers. It relies in a penalty activated if a passenger experiences a queuing time higher than a waiting threshold noted W_{max}. This penalty is implemented to favour fairness solutions by avoiding large passenger waiting times. In the following, the value of W_{max} is set to 30 min.

3.1   Methodology

To design an optimal line opening policy, a mathematical model and a resolution approach based on a Simulated Annealing are proposed. The detailed mathematical model is presented in Sect. 5. Solving a mathematical model which does not consider all the details of the real system can lead to overestimation of real benefits of the proposed solution. Therefore, we consider in the loop the use of a simulation framework developed with a tool called Simio [23] allowing a post-solution evaluation. This tool is described in Sect. 4. The simulation framework is able to consider other characteristics such as arriving time, boarding time, category, flight gate among others. This information is used as input for the mathematical model. Then, the mathematical model resolution provides an opening line strategy that is evaluated through several replications of the simulation model.

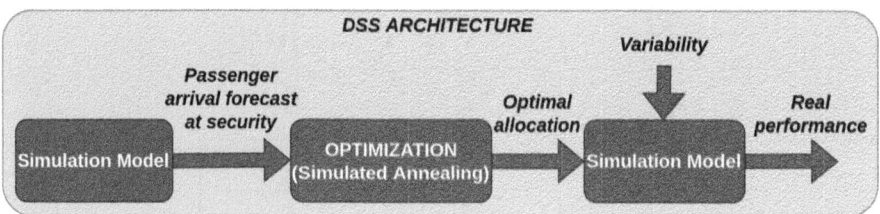

Fig. 2. Architecture of the approach

Figure 2 illustrates the integration of the mathematical model with the simulation framework.

4   Simulation

Simulation techniques have high descriptive power in comparison with other ones like mathematical programming or machine learning which are purely based on observational data. One of the advantages of simulation is that the causal relationships are transparent to the analyst if the model is descriptive enough. For that reason, we developed a detailed model of an airport terminal and combined it with optimisation for developing a tool that allows reacting to disruptions on a timely basis; while at the same time making transparent the consequences of making decisions on one element of the system (airport). In that way, we can foresee the overall effect on the passengers (in this example) when modifying some capacity in one area with the ultimate goal of minimising the passengers missing their flights at the end of the day. For that purpose, we developed a complete model of one of the terminal buildings of an airport that encompasses the different processes undertaken by any passenger namely check-in, waiting in departure halls, boarding pass control, security, and boarding gates. The descriptive level is important as the geometry of the airport, as well as the physical dimensions, play an important role in determining what is the time consumed in between processes. The following subsections describe the simulation part of the developed architecture.

4.1   Mexico City Case Study

This work focuses on Terminal 2 of Mexico City Airport. The model encompasses the complete terminal building under study (Terminal 2). We focus on the security checkpoints of that terminal as security in general is the main hurdle to overcome when passengers arrive late due to disruptions in their surface access to the airport. The security screening system associated with this terminal is composed of two security checkpoints each having five security lines. We considered the actual demand for the airport based on real data, and we simulated the demand under disrupted circumstances. The model is parameterised considering the actual capacity and expected performance of the different facilities. It is verified and partially validated using real historical data from 2019. Then the model is used to evaluate three different situations; one in which the system opens only regular points and no priority lines, one in which a priority line is open for each checkpoint (and a regular one close) and one in which the security team allocation is made on dynamic fashion using the results of the mathematical model. The main objective is to minimise the number of stranded passengers. By doing this implementation, we illustrate the benefit of having a DSS that can provide support when critical situations appear where human capacities are set up to the limit. Illustrations of the terminal model and the checkpoints under study are provided in Figs. 3 and 4 respectively.

The Terminal 2 model has two security points that are circled in red. Passengers enter the departures area of the terminal through three different entrances and they go to check in, if they have time spend some time in the departure hall, then go to security and then boarding gates.

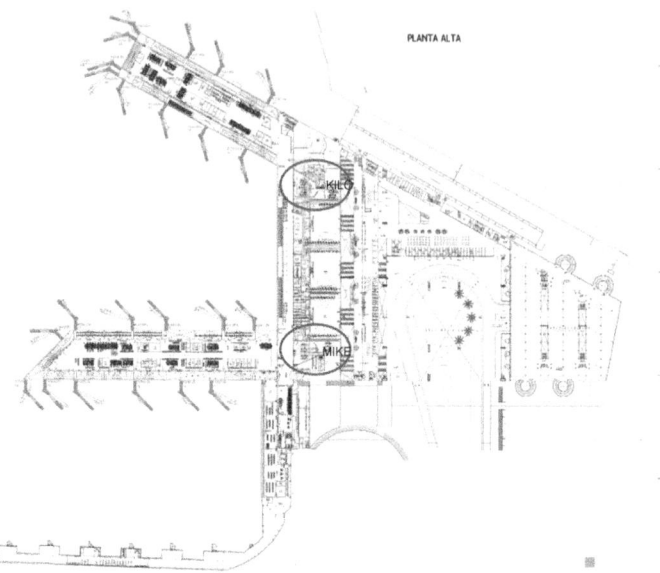

Fig. 3. Illustration of Mexico city airport. Security checkpoints are circled in red. (Color figure online)

4.2 System Description

In this section we present a description of the different elements that can be considered and/or play a relevant role in the problem:

– regular passengers: most common passengers found in airport terminals;
– elderly passengers: passengers characterized by slower motion than the regular ones and who sometimes need special attention or support;
– priority passengers: passengers that will be in a hurry within the airport due to a disruption and who are likely to miss their flight;

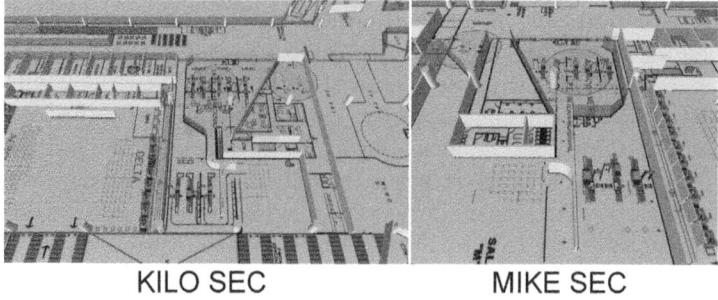

KILO SEC   MIKE SEC

Fig. 4. Illustration of security checkpoints in Mexico city airport (displayed in yellow). (Color figure online)

- Common queuing line: security queuing line for all the passengers which follows the First-In First-Out scheme; i.e. where the first passenger who entered the queuing system will be the first one served. No distinction is made between passengers. In the line, the three categories are mixed which causes slower passengers to obstruct the flow of the other ones;
- Fast queuing line: a segregated-queuing line which is considered in the alternative scenario; this one will be used by the passengers that have priority status. These could either represent passengers severely impacted by the ground side, business, or Passengers with Reduced Mobility;
- Physical dimensions and layout of the real system: we considered the geometrical configuration and scale of the system as it might play a relevant role.

Each passenger is characterised by a service time at security screening that can vary (fast for priority, slow for elderly or common passengers) and the walking speed. Other elements that play a role in the model might be the following ones:

- Check-In areas; the three areas for check-in are considered in the model
- Departure Hall; area where passengers and their companions spent time after check-in. Priority passengers skip those areas as they have very limited time
- Security points; these are the areas within the terminal where the passengers are screened for metals or liquids
- Boarding gates; these are the rooms where the passengers spend their final time before boarding the aircraft
- Physical dimensions; as the passengers spent time in the terminal while they move around it, the physical dimensions of the terminal should be considered for the model.

4.3   Conceptual Modelling

The flow of passengers followed a standard path as depicted in Fig. 5. In our approach, the passenger enters with a companion that stays in the departure

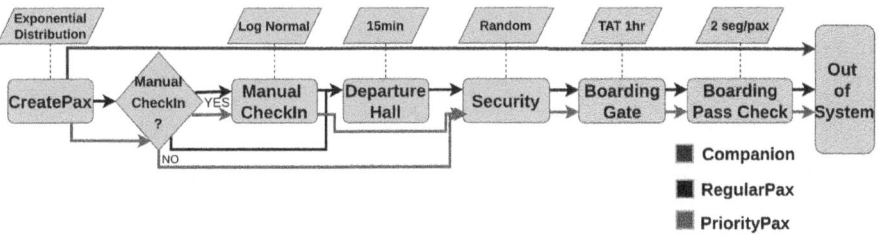

Fig. 5. Standard path followed by passengers within the airport. A distinction is made between companions (in blue), regular passengers (in black) and priority ones (in red). (Color figure online)

hall while the passenger performs his check-in. Once he finishes and if there is still time, they spend some time together and after some minutes the companion leaves the system and the passenger goes to security. In the case of priority passengers, they are walking faster and skip the departure hall. In the alternative scenarios when priority passengers noticed that a priority line is open they take it to speed up the process. From the system description, we included the most relevant elements. We developed a base case scenario that has the characteristics presented in Table 1.

Table 1. Modelling Assumptions

Parameter	Value	Comment
Simulation Run	7 h	assumption
Peak Hour	1 hr (3rd hr)	assumption
Regular Pax Speed	1.1 m/s	assumption
Elderly Pax Speed	0.5 m/s	assumption
Priority Pax Speed	1.4 m/s	assumption
Distribution of arrivals	discrete distribution	assumption for priority and regular
Processing time	35 s + variability	regular pax
Processing time	20 s + variability	priority pax
Security Lines	10 lines and 2 optional for priority	finite number of personnel

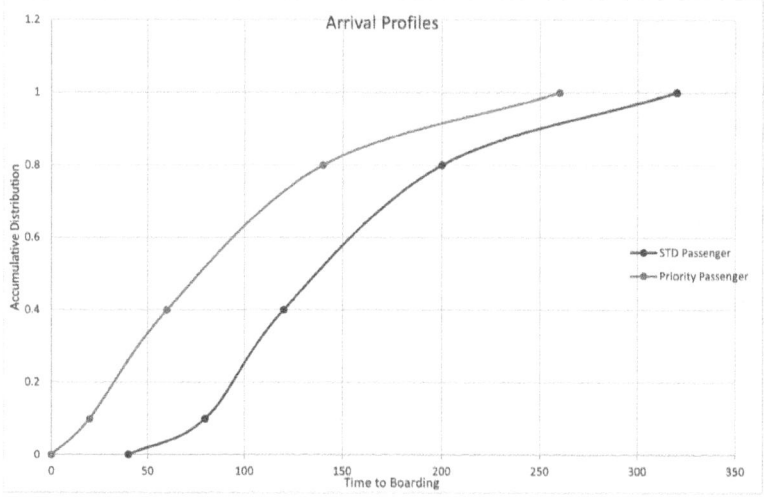

Fig. 6. Arrival profiles of passengers with and without disruption (in orange and blue respectively). (Color figure online)

To generate the nominal passenger flow, a simulation representing a 7-hour duration is run. One peak hour is considered and the complete passenger trajectory from entering the building until reaching the gate is simulated. From previous studies, we considered realistic processing time and variability sources, mainly on arrival times and processing times. For evaluating the approach we considered two scenarios alternatives to the base case. In all the scenarios, several passengers arrive with very limited time in advance due to a disruption. The airport can decide to open a priority security checkpoint to handle passengers having a priority status; the final scenarios deal with the modes of managing this alternative.

Figure 6 presents the arrival profiles considered for all the passengers; in the graph, it is depicted how the arrival profile of passengers is affected by disruption and how the new arrival profile is considered in the simulation model. Finally, the share of elderly passengers is fixed at 7%. This number is fixed according to the 2015 survey led in France on air passengers [7], this number should be adapted to the specific case under study for more realistic analysis.

5   Optimisation Approach

First, a short description of the mathematical model is proposed in the following. Data, decision variables, constraints and the objective function are detailed. Then, the resolution approach based on simulated annealing is detailed. Notations related to the input data for the mathematical model are summarised below.

Data

$\mathcal{S}$	set of security checkpoints
$\mathcal{T} = \{1, 2, ..., \|\mathcal{T}\|\}$	indice set of discrete times
$\mathcal{H} = \{1, 2, ..., \|\mathcal{H}\|\}$	indice set of interval times where lines opening are decided
$s \in \mathcal{S},\ t \in \mathcal{T},\ d_{s,t}^{\mathrm{R/P}}$	number of regular/priority passengers expected to arrive at t on security system s
θ_C/θ_F	service rate per time step for one common/fast security line
$h \in \mathcal{H},\ L_h$	maximum number of security lines that can be open during interval time h
$s \in \mathcal{S},\ L_s^C$	maximum number of common/fast security lines that can be operated on security system s
$t \in \mathcal{T},\ h_t$	interval time included in $\mathcal{H}$ that covers t
$s \in \mathcal{S},\ Q_s^{C/F}$	initial number of passengers on common/fast queuing line of security checkpoint s
W_{max}	maximum acceptable waiting time
$s \in \mathcal{S},\ t \in \mathcal{T},\ g_{s,t}^R(w)$	function that returns for a waiting time w the number of stranded regular passengers arrived at t on s.
$s \in \mathcal{S},\ t \in \mathcal{T},\ g_{s,t}^P(w)$	function that returns for an average waiting time w the number of stranded priority passengers arrived at t on s.

Main and auxiliary decision variables are presented below.

Decision Variables

Main decision variables

$s \in \mathcal{S}, h \in \mathcal{H},\ x^C_{s,h}$	number of common security lines open on security checkpoint s during hour h
$s \in \mathcal{S}, h \in \mathcal{H},\ x^F_{s,h}$	binary variable equal to 1 if a fast security line is open on security checkpoint s during hour h , 0 else

Auxiliary decision variables

$s \in \mathcal{S},\ t \in \mathcal{T},\ y^C_{s,t}/y^F_{s,t}$	number of passengers served from common/fast queuing system of s at t / between t and t+1
$s \in \mathcal{S},\ t \in \mathcal{T} \cup \{0\},\ d^C_{s,t}/d^F_{s,t}$	number of passengers that arrive on common/fast queuing line at t
$s \in \mathcal{S},\ t \in \mathcal{T} \cup \{0\},\ q^C_{s,t}/q^F_{s,t}$	number of passengers that are still waiting in common/fast queuing line after passengers served at t
$s \in \mathcal{S},\ t \in \mathcal{T},\ w^C_{s,t}/w^F_{s,t}$	average waiting time experienced by passengers entered in common/fast queuing system of s at t
$f^{C/F}\left(\boldsymbol{y}^{C/F}_s, d_{st}, q_{st}, t'\right)$	function that returns the number of passengers arrived at t on common/fast queuing system s and served at t'
$\bar{w}$	maximum passenger waiting time
$\bar{t}^{C/F}_{s,t}$	time when the last passenger arrived at t on common/fast queuing line of s is served
$\bar{w}$	maximum passenger waiting time

Constraints

$$x^C_{s,h} \le L^C_s \qquad\qquad\qquad h \in \mathcal{H},\ s \in \mathcal{S}\ (1)$$

$$\sum_{s \in \mathcal{S}} \left(x^C_{s,h} + x^F_{s,h}\right) \le L_h \qquad\qquad\qquad h \in \mathcal{H}\ (2)$$

$$d^C_{s,t} = d^R_{s,t} + \left(1 - x^F_{s,t}\right).d^P_{s,t} \qquad\qquad\qquad s \in \mathcal{S}\ (3)$$

$$d^F_{s,t} = x^F_{s,t}.d^P_{s,t} \qquad\qquad\qquad s \in \mathcal{S}\ (4)$$

$$q^{C/F}_{s,0} = Q^{C/F}_s \qquad\qquad\qquad s \in \mathcal{S}\ (5)$$

$$q^C_{s,t} = d^C_{s,t} + q^C_{s,t-1} - y^C_{s,t} \qquad\qquad\qquad s \in \mathcal{S},\ t \in \mathcal{T}\ (6)$$

$$q^F_{s,t} = x^F_{s,t}.d^P_{s,t} + q^F_{s,t-1} - y^F_{s,t} \qquad\qquad\qquad s \in \mathcal{S},\ t \in \mathcal{T}\ (7)$$

$$y_{s,t}^{C} = \min\left(\theta_C.x_{s,h_t}^{C},\, d_{s,t}^{C} + q_{s,t-1}^{C}\right) \qquad\qquad s \in \mathcal{S},\, t \in \mathcal{T} \quad (8)$$

$$y_{s,t}^{F} = \min\left(\theta_F,\, d_{s,t}^{F} + q_{s,t-1}^{F}\right) \qquad\qquad\qquad s \in \mathcal{S},\, t \in \mathcal{T} \quad (9)$$

$$w_{s,t}^{C/F} = \frac{1}{\max\left(1,\, d_{s,t}^{C/F}\right)} \cdot \sum_{t'=t}^{|\mathcal{T}|} (t'-t).f^{C/F}\left(\boldsymbol{y}_s^{C/F}, d_{st}^{C/F}, q_{st}^{C/F}, t'\right) \quad s \in \mathcal{S},\, t \in \mathcal{T}$$

$$(10)$$

$$\bar{t}_{s,t}^{C/F} = \min\left\{t' \in \mathcal{T}\mid d_{st}^{C/F} + q_{st}^{C/F} \leq \sum_{i=t}^{t'} y_{si}^{C/F}\right\} \qquad\qquad s \in \mathcal{S},\, t \in \mathcal{T} \quad (11)$$

$$\bar{w} = \max_{s \in \mathcal{S}, t \in \mathcal{T}} \bar{t}_{s,t}^{C/F} \qquad\qquad\qquad\qquad\qquad\qquad\qquad (12)$$

Constraint (1) refers to the number of common/priority lines that can be operated on one security checkpoint. The maximum number of lines that can be open during interval time h across the airport is defined through constraint (2). (3) and (4) compute the common and fast demand. Initial common and priority services are set thanks to equation (5). Common and fast queue lengths are computed through (6) and (7) respectively. (8) and (9) compute the number of passengers served at the common and fast queuing lines of s at t. Waiting time experienced on fast and common queuing lines are set through equation (10). Finally, (11) and (12) fix the maximum passenger waiting time.

Objective Function

The objective function relies in minimising the number of stranded passengers:

$$\min \sum_{s \in \mathcal{S}} \sum_{t \in \mathcal{T}} \left(g_{s,t}^{R}\left(w_{s,t}^{C}\right) + g_{s,t}^{P}\left(x_{s,t}^{F}.w_{s,t}^{F} + \left(1 - x_{s,t}^{F}\right)\right).w_{s,t}^{C}\right) + K.\bar{w}.\mathbb{1}_{[W_{\max};+\infty]}(\bar{w}),$$

$$(13)$$

where K is a coefficient such as a penalty $K.\bar{w}$ is activated when the maximum waiting time $\bar{w}$ exceeds $W_{\max}$.

5.1   Simulated Annealing

The mathematical model presented here is non-linear and a multi-criteria objective function is considered. Therefore, we propose a resolution of the problem through Simulated Annealing (SA). Such a method has been effective for optimising airport handling resources [2,8,20]. A short description of SA is presented below. For a complete description of the SA, the reader can refer to [6].

SA is a global optimisation technique based on an analogy with a cooling process in metallurgy. This metaheuristic is a single-solution method, contrary to population methods such as genetic algorithms. Starting from an initial solution, a new solution is selected in the neighbourhood of the current one. The new solution is kept if it improves the objective value. Otherwise it can be kept depending on an acceptance probability that is a function of a temperature parameter. The method is initialised with a high-temperature parameter to have a high acceptance rate, authorising the degradation of the current objective value, to favour the exploration of the solution search space. The temperature parameter is steadily decreasing along algorithm iterations to favour intensification and, at the end of the search, only accept solutions that improve the objective value.

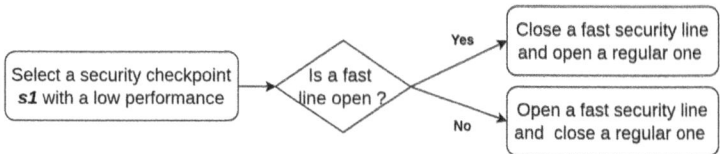

Fig. 7. Neighbour generation process

To implement this method, an initial solution and a neighbour selection process need to be defined. The initial solution is the initial allocation that would have been operated if no airport access mode disruption happened. For each security checkpoint, five regular security lines are initially open. The neighbourhood generation process is detailed in Fig. 7. A neighbour of the solution is generated by closing a fast or a regular security line on a security checkpoint and by opening a new one. The selection of the lines that are open and close is based on a performance computed for each security checkpoint. The lower the performance, the higher the chance a new security line will be open and vice versa. The process is randomised to favour the exploration of new solutions.

6   Results

We run three scenarios to evaluate the impact of the different policies to manage the disruption for the passengers. For those scenarios we used different performance indicators (PI) like queue waiting time, and queue length among others; however, for the objective of this article we only presented the ultimate one that determines if the system is able to perform its main function which allows passenger take their aircraft on time; in the results graph, we present the total of stranded passengers making a distinction between the ones that suffered a disruption (priority) from those that not.

– Standard Scenario; in this scenario, the passengers follow the trajectory already presented and use the full capacity of the system. No priority line is open for priority passengers.
– Priority Line Scenario; in this scenario, we assume the airport operator opens a priority line during the peak hour of the airport. We considered also finite resources (i.e. a regular point is close when a priority one is open)

– Dynamic opening of lines; in this case, we use the proposed SIM-OPT approach to open dynamically priority lines in time according to the expected amount of passengers at both security checkpoints during the day. Finite resources are considered and lines can be open/close in blocks of 30 min.

Figure 8 presents the average number of stranded passengers obtained after running the different scenarios. It is noticeable that the number of stranded passengers is reduced progressively revealing the potential of the actions taken by the airport operator. First, with the addition of a priority line during peak hours, the number of stranded passengers is reduced in comparison with the base case; the number of regular passengers that are affected increases in comparison with the base case, but this is compensated with the priority ones and the total amount is reduced. When we combine the optimisation with the simulation in a dynamic fashion, the number of stranded passengers in total is reduced in comparison with the previous scenarios. The results show that segregating passengers flow at the right moment is beneficial for both regular and priority passengers; however, this action can not be easily managed by the airport operator on a day-day basis being the reason why a DSS based on our approach is required.

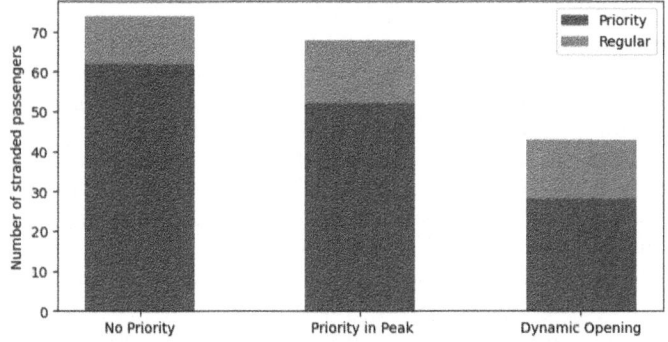

Fig. 8. Average number of stranded passengers depending on the security allocation

7 Conclusions

This work highlights the importance of airport reactivity when disturbances in passenger flows occur. We propose a simulation-optimisation framework to improve the security line opening policy at a tactical level. Results demonstrate that the dynamic allocation of resources adapted to the demand is the best alternative for managing disruptions without increasing the number of resources, which in the case study is limited. The dynamic segregation of passenger flows and the prioritisation of the delayed ones succeeded in reducing the number of passengers missing their flights. The next steps should focus on improving the models by considering the contractual restrictions that are in place for the security team allocation and their transfer time between security checkpoints among other ones. Finally, an implementation of the proposed opening line policy can be evaluated in a field experiment at an airport.

Acknowledgements. The authors would like to thank IGAMT for their help in data provided and also would like to thank the Dutch Benelux Simulation Society (www.dutchBSS.org) and EUROSIM for disseminating the article.

References

1. Bullock, D.M., Haseman, R., Wasson, J.S., Spitler, R.: Automated measurement of wait times at airport security: deployment at Indianapolis international airport, Indiana. Transp. Res. Rec. **2177**(1), 60–68 (2010)
2. Candalino, T.J., Jr., Kobza, J.E., Jacobson, S.H.: Designing optimal aviation baggage screening strategies using simulated annealing. Comput. Oper. Res. **31**(10), 1753–1767 (2004)
3. Carr, A., Biswas, T., Wheeler, J.V.: Airport operations and security screening: an examination of social justice. J. Air Transp. Manage. (2020)
4. De Barros, A.G., Tomber, D.D.: Quantitative analysis of passenger and baggage security screening at airports. J. Adv. Transp. (2007)
5. De Neufville, R., Odoni, A.R., Belobaba, P.P., Reynolds, T.G.: Airport Systems: Planning, Design, and Management. McGraw-Hill Education (2013)
6. Delahaye, D., Chaimatanan, S., Mongeau, M.: Simulated annealing: from basics to applications. In: Gendreau, M., Potvin, J.Y. (eds.) Handbook of Metaheuristics. International Series in Operations Research & Management Science, vol. 272, pp. 1–35. Springer, Cham (2019). https://doi.org/10.1007/978-3-319-91086-4_1
7. DGAC: Enquête nationale des passagers aériens (2015). https://www.ecologie.gouv.fr/sites/default/files/ENPA_2015_2016.pdf
8. Drexl, A., Nikulin, Y.: Multicriteria airport gate assignment and pareto simulated annealing. IIE Trans. **40**(4), 385–397 (2008)
9. Harchol-Balter, M.: Performance Modeling and Design of Computer Systems: Queueing Theory in Action. Cambridge University Press, Cambridge (2013)
10. ICAO: Annex 17-aviation security. https://store.icao.int/en/annexes/annex-17
11. Janssen, S., van der Sommen, R., Dilweg, A., Sharpanskykh, A.: Data-driven analysis of airport security checkpoint operations. Aerospace **7**(6), 69 (2020)
12. Maliwat, J.D.: Five typical emotional reactions to airport security screening: a case study. Psychology **8**(12), 594–602 (2018)
13. Mujica Mota, M.: Check-in allocation improvements through the use of a simulation-optimization approach. Transp. Res. Part A: Policy Pract. **77**, 320–335 (2015)
14. Mujica Mota, M., Flores, I.: Revisiting the flaws and pitfalls using simulation in the analysis of aviation capacity problems. Case Stud. Transp. Policy (2020)
15. Mujica Mota, M., Scala, P., Di Bernardi, A., Orozco, A.: Restart: analysis of post-COVID 19 capacity in security checkpoints at Mexico City airport. Transp. Res. Procedia **59**, 234–243 (2021)
16. Mujica Mota, M., Scala, P., Murrieta-Mendoza, A., Orozco, A., Di Bernardi, A.: Analysis of security lines policies for improving capacity in airports: Mexico City case. Case Stud. Transp. Policy **9**(4), 1476–1494 (2021)

17. Mujica Mota, M., Scala, P., Schultz, M., Lubig, D., Luo, M., Perez, E.J.: The rise of the smart passenger I: analysis of impact on departing passenger flow in airports. In: SESAR Innovation Days 2021 (2021)
18. Pérez, E., Taunton, L., Sefair, J.A.: A simulation-optimization approach to improve the allocation of security screening resources in airport terminal checkpoints. In: 2021 Winter Simulation Conference (WSC), pp. 1–11. IEEE (2021)
19. Sakano, R., Obeng, K., Fuller, K.: Airport security and screening satisfaction: a case study of us. J. Air Transp. Manag. **55**, 129–138 (2016)
20. Scala, P., Mujica Mota, M., Wu, C.L., Delahaye, D.: SIM-OPT in the loop: algorithmic framework for solving airport capacity problems. In: 2018 Winter Simulation Conference (WSC), pp. 2261–2272. IEEE (2018)
21. Schultz, M., Fricke, H.: Managing passenger handling at airport terminals. In: 9th Air Traffic Management Research and Development Seminars (2011)
22. Scozzaro, G., Mancel, C., Delahaye, D., Feron, E.: Optimising security screening resources during airport access mode disruptions. In: SESAR Innovation Days (2022)
23. Simio: Airport simulation software. https://www.simio.com/applications/airport-simulation-software/index.php

Simulation-Based Allocation and Routing Optimization: A Case Study of Single Versus Team Driving

Sajjad Amrollahi Biyouki[1]([✉]), Chinedu Ufodike[2], Xueping Li[1] [iD],
Hoon Hwangbo[1], and John E. Bell[2]

[1] Department of Industrial and Systems Engineering, Tickle College of Engineering,
The University of Tennessee, Knoxville, USA
`samrolla@vols.utk.edu`
[2] Department of Supply Chain Management, Haslam College of Business,
The University of Tennessee, Knoxville, USA

Abstract. Planning vehicle routes in supply chain management is a challenging problem since all demands need to be met at the minimum cost with various limitations on available resources. This study tries to find the optimal routes for multi-type vehicles based on information provided by a real-world supply chain dataset, which includes the delivery information of eleven vehicles with both single-type and team-type drivers. To achieve this, this paper also simulates routes and re-allocates delivery locations to different vehicles by the particle swarm optimization (PSO) algorithm and optimizes the routes within the newly allocated delivery locations. Experimental results show that this simulation-based approach improves the routes in terms of overall cost and distance for the deliveries in comparison with the original routes that have been operated. This study also evaluates the necessity of having additional drivers in a cost-effective manner. The optimal newly-assigned routes could help minimize the logistical costs of the company and efficiently manage its supply delivery network.

Keywords: Vehicle routing · supply chain management · particle swarm optimization · simulation-based allocation · meta-heuristics

1 Introduction

Mobile technology and the rise in e-commerce have placed the logistic industries at the front end of consumers' purchase journeys, and how they should rethink their supply chain concept in retail distribution [5, 33]. The growth in e-commerce sales and widespread use of the internet and smartphones brings about increased distribution competitiveness, especially for the logistic companies to have enough vehicles available to meet the demands of their customers while gaining their satisfaction [3, 9, 22] and more importantly, to be able to improve the efficiency of their delivery operations. This development has helped retail companies reach

© The Author(s), under exclusive license to Springer Nature Switzerland AG 2024
M. Mujica Mota and P. Scala (Eds.): EUROSIM 2023, CCIS 2033, pp. 156–170, 2024.
https://doi.org/10.1007/978-3-031-68438-8_12

more consumers by allowing the consumer to experience seamless online shopping twenty-four hours a day and expanding their market [1,9]. This expansion of the market and growth in e-commerce deliveries create a steady increase in the shipment volumes to be handled on the last mile for corresponding customers [8]. Additionally, it may be necessary to operate a task with two or more drivers due to distance and time requirements. Hence, the logistics companies must decide whether they require to hire a seasonal driver to be able to deal with the high-volume delivery challenge [23]. In addition to the aforementioned challenges, the fastest delivery time tends to be considered as a constraint for companies in such a way that they necessarily have to meet it in order to maintain in the competitive market. Hence, given these determined delivery requirements and last mile challenges, the logistics companies seek to minimize their costs in terms of the number of required drivers and delivery paths.

To clarify the necessity of multiple drivers, we explain a situation where hiring seasonal drivers has been used by the logistic companies to deal with these challenges. During the COVID-19 pandemic in 2020, as one of the most extensive shipping years, billions of packages were delivered to the customers. To deal with the increased number of package deliveries in a cost-effective manner, UPS had hired a significant number of seasonal drivers and raised their payment, in addition to the existing drivers during the pandemic while millions of employees lost their jobs. Hence, they were able to reduce overtime work for their drivers by hiring seasonal drivers since working overtime increases direct labor costs. Additionally, truck drivers have been reported to be assigned unrealistic delivery schedules and are likely to violate the driving time regulations. In the United States, it has been estimated by the U.S. Department of Transportation that up to 28% percent of truck drivers could experience sleep apnea while driving due to their fatigue, which could result in roadway accidents. Hence, a delivery driver must comply with the required hour of service [25], which could include taking multiple breaks within a certain driving time to reduce the negative impact on motor carrier safety performance [26,27]. Given the required regulations, the logistics companies have to either ease unreasonably tight time windows or employ an assistant driver to meet the delivery time demand of the consumers.

This paper studies the use of single-type and team-type drivers in the dispatching problem of a multinational pizza restaurant in order to obtain insights into how to provide the restaurant with a better servicing cost (if using a team driver is cost-effective or not) and improve the total distance of the routes by re-allocating the delivery locations and optimizing the delivery path for each truck. The particle swarm optimization (PSO) technique [20], as one of the well-known and popular meta-heuristic algorithms in the vehicle routing problem [14], is used to re-allocate the delivery locations and subsequently optimize the delivery path by using the mathematical formulation of the vehicle routing problem.

The remainder of the paper is organized as follows. Section 2 outlines the literature related to the vehicle routing problem, including the simulation-based as well as heuristic-based optimization works in supply chain management. In Sect. 3, the real-world problem with its assumptions and dataset structure is described. Section 4 briefly explains the particle swarm optimization technique as well as the mathematical formulation of the problem. In Sect. 5, we present the optimized delivery path models and discuss the results. Finally, Sect. 6 concludes the major remarks of the paper with potential future works of this study.

2 Literature Review

The study of understanding the phenomenon of whether it is cost-effective to use a single or team of two drivers being assigned to a truck and obtain their optimal delivery path can be informed by literature in the last mile logistics, drivers helper/team decisions, optimal as well as near-optimal methods [2,6,28], and vehicle routing research streams. The growth of e-commerce has significantly influenced the distribution strategies of logistics companies as they are competing for a faster delivery service to their customers [18]. Boyer et al. [7] concluded that last-mile delivery is adjusted based on factors like the consumer density and duration of the delivery window in such a way that a greater customer density and longer delivery windows impact the deliveries significantly. There has been some research that has experimented with on-time delivery delays to decrease customer loyalty levels since faster delivery (within 24 h) increases customer satisfaction [35]. According to an exploratory factor studied by Xing et al. [34], they found that distribution punctuality in e-commerce is highly valued by the customer when options like delivery date and the ability to deliver orders quickly are included. Another framework by Castillo et al. [10] introduced the idea of crowdsourcing delivery as a new possibility to increase the responsiveness in the last mile of the supply chain.

The literature considering whether specific routes shall be conducted by a single driver or team driver vehicle is scarce. The vehicle routing problem with time windows (VRPTW) is regarded when the deliveries must occur within some specified service time frame [32], and it helps in the avoidance of unnecessary long routes [12]. The issues of route planning and the schedule of the vehicle fleet in order to satisfy the demand of retail stores have been a challenging problem for distribution centers [17]. In their study, they modeled a time-dependent vehicle routing problem (TDVRP) by partitioning the day into three speed zones, where the free flow of speed was used as a factor to check the speed differences at different times of the day [17]. Additionally, vehicle route departure time has been treated as a decision variable in some works, where each of the service nodes is allowed to have a delayed departure [11]. Hence, a heuristic method was proposed with an efficient technique to select the optimal departure times.

Regarding the demands from different retail stores, shipment companies have been faced with logistic challenges during certain peak periods when they attempt to reach out to their customers who might be further away from their distribution centers in a cost-effective and timely manner. This issue of finding ways to improve the efficiency of the delivery operations is not just related to the logistic companies but also to manufacturing companies [23]. These real-world challenges could be addressed by the application of a driver helper in a situation where the customer destinations might be further away from the distribution center and by avoiding letting a single-type driver works overtime, which could cause unpleasant employment conditions [23]. The case of the vehicle routing problem, in which some vehicles are operated by single-type and team-type drivers subject to European Union regulations, was studied by Goel and Kok [16]. They presented an algorithm to schedule the drivers working and driving hours to avoid exceeding standard driving time limits. Other studies have used simulation to obtain insights into how driver helpers can be utilized in the most effective way to minimize cost [23]. Similarly, Rhodes et al. [30] developed criteria for UPS driver helpers intending to identify which routes would need a dependent helper, how the helpers can be deployed, and what locations need an independent helper for delivery during the peak season.

Simulation-based optimization has been thoroughly reviewed in the literature [4,15]. They have presented the taxonomy of various simulation-based optimization methods in terms of the methodology, the model structure, and their applications. The simulation optimization is applied in broad areas of supply chain management, including inventory management [31], risk [24], resilience [36], and vehicle routing [19]. Furthermore, meta-heuristic algorithms have been applied to solve varying vehicle routing problems [14,21]. Particle swarm optimization (PSO) [20] is one of the most widely used meta-heuristic techniques to solve the vehicle routing problem in the literature. Hence, this paper would leverage the PSO method to simulate and allocate the most applicable route for each vehicle and then optimize the route given the newly-assigned delivery locations.

3   Real-World Urban-Based Distribution Problem

3.1   Problem Description: Case Study

The scope of this paper is primarily inspired by the perspective of a multinational pizza restaurant chain located in the state of Minnesota. In the given city of Eagan, Minnesota, the chain owns a central distribution center (DC) and a group of retail stores located in various surrounding cities. While this pizza restaurant wants to avoid additional costs related to hiring a seasonal driver, a truck that is operated by a single-type driver can only go within a two hundred mile radius of the DC. Team-type drivers, on the other hand, can travel a much greater distance as one driver can take a break while the other is driving. The regulation of the Federal Motor Carrier Safety Administration (FMCSA), an agency of the United States Department of Transportation (DOT), restricts the daily driving time for a single-type driver to be at most eleven hours (550 km). In contrast, team-type drivers are allowed to drive more than the aforementioned driving time of eleven hours but less than twenty-two hours (1,430 km). All the drivers are required to get back to DC, or the warehouse, after meeting all the assigned delivery locations.

Furthermore, weight or cubic capacity constraint is enforced per truck to avoid potential financial costs being levied for DOT fines, so each truck can only carry a maximum capacity of 48,000 lb. With several different service times for each store, the vehicle stops to unload the required amount of inventory at the stores visited. Additionally, the average speed for the vehicle driving within the city is assumed to be thirty miles per hour, but when a drive is in a far-away location (surrounding towns), the average speed of fifty miles per hour can replicate interstate speeds. Although the studied multinational pizza restaurant chain has many store locations in the state of Minnesota, our sample data for this work consists of the limited stores located in different cities in Minnesota and neighboring states. Our analysis of the use of simulation-based allocation and optimization methodologies within geographic information system (GIS) maps for vehicle routing of this pizza restaurant aims at finding the best routes that the chain could operate to minimize the total distance among assigned delivery locations. Additionally, we would provide the restaurant chain with a managerial perspective for better servicing costs (if using an additional driver is cost-effective).

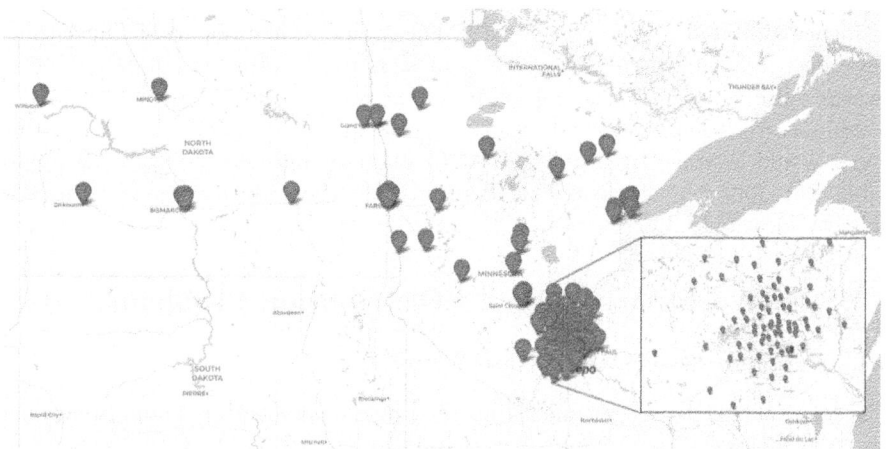

Fig. 1. All store locations map

3.2   Dataset Description

As mentioned, the purpose of the case study is to model and minimize the total distance of delivery locations for each truck in a multinational pizza restaurant chain located in the state of Minnesota. The given real-world dataset includes four consecutive days with specific details of deliveries. However, this paper uses only one-day delivery information as sample data, including a total of 115 unique locations, all of which should be met by nine single-type drivers and two team-type drivers. It is worth noting that the proposed model can be extended to other delivery days to find their optimal delivery paths as well. To illustrate the sparsity of the delivery locations, we provide a map of all the store locations in Fig. 1, where the eleven trucks need to deliver the items. This map clearly shows how the locations are so sparse that even some of them are outside of Minnesota, such as North Dakota. We can infer that this sparsity would increase the distance as well as the delivery time of those locations dramatically; hence, a practical optimal solution can minimize the total distance. Furthermore, the delivery locations for each vehicle can be reassigned in order to optimize the allocated stores and subsequently diminish the total distance.

Table 1. Trucks configuration - pre-determined locations

ID	Type	Number of locations	Total weight (lb)
1000	Single	7	14,596
1001	Team	17	31,440
1002	Team	14	33,171
1003	Single	8	17,289
1004	Single	11	22,014
1005	Single	10	22,981
1006	Single	11	22,393
1007	Single	9	22,380
1008	Single	9	20,346
1009	Single	9	20,196
1010	Single	10	20,685

The features for each delivery consist of the original (initial) sequence, address line, city, state, zip code, weight, service time, and travel time. We use the address line, city, and state details automatically to extract the latitude and longitude of specific locations. In this model, the delivery sequences for each truck would be optimized based on the total distance between the delivery locations. In the given dataset, we have a total of eleven trucks for one day, whose specifications, including their assigned number of locations, total weight, and type, are shown in Table 1. Note that trucks should get back to their starting location, the warehouse (DC), after meeting all the delivery locations.

4 Methodology

4.1 Vehicle Routing Formulation

The vehicle routing problem is essentially an integer linear programming model, which can be formulated as follows:

$$\text{Min} \sum_{i=1}^{n} \sum_{j=1, j \neq i}^{n} c_{ij} x_{ij} \tag{1}$$

subject to:

$$\sum_{i=2}^{n} x_{i,1} = m \tag{2}$$

$$\sum_{j=2}^{n} x_{1,j} = m \tag{3}$$

$$\sum_{i=1,i\neq j}^{n} x_{i,j} = 1 \qquad \forall j = 1, \ldots, n \tag{4}$$

$$\sum_{j=1,i\neq j}^{n} x_{i,j} = 1 \qquad \forall i = 1, \ldots, n \tag{5}$$

$$u_i - u_j + 1 \leq (n-1)(1 - x_{i,j}) \qquad 2 \leq i \neq j \leq n \tag{6}$$

$$u_i \in Z \qquad \forall i = 1, \ldots, n \tag{7}$$

$$x_{ij} \in \{0, 1\} \tag{8}$$

where x_{ij} is a binary decision variable indicating whether the vehicle goes from city i to city j ($x_{ij} = 1$) or not. c_{ij} is the distance from city i to city j. u is added to the model as an auxiliary variable. Additionally, n and m denote the number of cities and the number of vehicles, respectively.

To clarify the formulation, its objective function (1) is to minimize the total distance of each vehicle's route. Constraints 2–3 ensure that exact m vehicles leave and return to the city. Constraints 4–6 guarantee that exactly one vehicle enters and exits each city, and each city is visited only once in the vehicle's route.

4.2   Particle Swarm Optimization (PSO)

Particle Swarm Optimization (PSO) was introduced by Kennedy and Eberhart [20] as a heuristic technique to find applicable near-optimal solutions for diverse optimization problems [29]. The algorithm consists of some particles, swarms of individuals, in the search space, whose actions are inspired by the inherent behavior of a group of animals since they are communicating while looking for food [31]. The technique would find the best solution by moving the entire particles as the swarm toward the best particle location [13]. The particles' location (x) and velocity (v) in the swarm would be updated at each iteration as:

$$v_i(t+1) = v_i(t) + cr_1[x_i^*(t) - x_i(t)] + cr_2[x^g(t) - x_i(t)] \tag{9}$$

$$x_i(t+1) = x_i(t) + v_i(t+1), \qquad i = 1, \ldots, N \tag{10}$$

Algorithm 1: Particle Swarm Optimization (PSO)

$t \leftarrow 1$;

Initialize each particle in the population with random values for position (x_i) and velocity (v_i) for $i = 1, \ldots, N$;

repeat

- Calculate the optimization fitness value of each particle in population.
- Update $x_i^*(t)$ if the fitness value for particle i is greater than its best so-far fitness value.
- Update $x^g(t)$ with the location of the highest fitness value.
- Update v_i according to Eq. 9, $\forall i \in$ population.
- Update x_i according to Eq. 10, $\forall i \in$ population.
- Update the iteration, $t \leftarrow t + 1$.

until $t > t_{max}$;

---

where i, t, t_{max}, and N denote a specific particle, iteration, maximum of total iterations, and population size, respectively. c, r_1, and r_2 are random positive numbers to handle the weight of each particle moving toward its own and the entire group's best position. Additionally, x_i^* is the best solution each particle is obtained by the iteration; and x^g is the best solution of the entire population. The particle swarm optimization procedure is provided in Algorithm 1.

5   Results and Discussion

5.1   Pre-determined Delivery Locations

This section presents the optimal delivery sequences with the corresponding total distance for various vehicle types given their provided pre-determined delivery locations as shown in Tables 2 and 3. They are obtained based on applying the integer programming model to the pre-determined delivery locations. These tables demonstrate that the total distance for each vehicle, except for truck 1010, is improved in comparison with the initial (original) route. Furthermore, trucks 1000, 1003, 1005, and 1006 need to meet more distant locations than other single-type vehicles, resulting in driving and working overtime. Hence, the company should reassign the delivery locations or hire additional drivers for both their team-type and single-type drivers in order to meet the time constraint. Note that "0" denotes the distribution center (DC) in the tables.

To show the sparsity of the locations for each truck, we illustrate the corresponding locations with their optimal routes in the following figures. Regarding the team-type driver shown in Fig. 2, the delivery locations have not been distributed uniformly, and one of the trucks has a farther ride to deliver the items.

Table 2. Team-type driver trucks routes - pre-determined delivery locations

ID	Type	Optimal delivery sequence	Initial route (km)	Optimal route (km)
1001	Team	0, 13, 14, 15, 11, 10, 9, 8, 7, 6, 5, 4, 16, 12, 17, 3, 2, 1, 0	1459.14	1249.71
1002	Team	0, 1, 2, 5, 6, 7, 4, 9, 10, 11, 12, 14, 13, 8, 3, 0	1918.45	1880.95

Figure 3 shows the single-type drivers who have at least one outlier delivery location that they are required to satisfy. Additionally, Fig. 4 illustrates the reasonable pre-determined delivery locations assigned to trucks 1004, 1007, 1008, 1009, and 1010. Their corresponding delivery locations are near each other as well as relatively close to the warehouse location, resulting in low-distance routes. In the next section, we will use the PSO algorithm to re-assign the delivery locations for a more balanced spread of delivery locations.

Table 3. Single-type driver trucks routes - pre-determined delivery locations

ID	Type	Optimal delivery sequence	Initial route (km)	Optimal route (km)
1000	Single	0, 1, 5, 3, 4, 2, 6, 7, 0	1114.73	1098.6
1003	Single	0, 8, 2, 5, 6, 4, 3, 7, 1, 0	334.43	303.59
1004	Single	0, 1, 3, 2, 7, 6, 5, 4, 8, 9, 10, 11, 0	86.33	80.85
1005	Single	0, 1, 2, 3, 10, 9, 4, 8, 6, 7, 5, 0	471.15	440.55
1006	Single	0, 10, 4, 5, 6, 8, 7, 9, 3, 2, 1, 11, 0	461.82	448.73
1007	Single	0, 1, 3, 4, 9, 8, 7, 6, 5, 2, 0	125.62	123.4
1008	Single	0, 9, 5, 6, 7, 8, 4, 2, 3, 1, 0	156.96	122.89
1009	Single	0, 1, 2, 6, 7, 8, 9, 5, 4, 3, 0	146.27	130.22
1010	Single	0, 10, 9, 8, 7, 6, 5, 4, 3, 2, 1, 0	186.8	186.8

5.2   PSO-Based Allocated Delivery Locations

As shown in Table 1, the total weights of various vehicles for both driver types are provided. This given information is used as a constraint to reassign the delivery locations to vehicles while minimizing the total distance. Hence, we use the average of weights for each type as the maximum weight and the minimum value of each type as the minimum weight for the allocation task. In this regard, the weight constraint for team-type drivers is set to be between 30,000 and 33,000. Similarly, single-type drivers have a weight constraint between 15,000 and 21,000. Prior to applying PSO technique individually for both team-type and single-type drivers, we add the distant locations of single-type trucks 1000 and 1003 to the team-type category since they are better suited for team drivers. Then, the PSO algorithm is separately implemented on all remaining single-type locations as well as team-type locations.

Table 4 shows the optimal PSO-allocated distance values for all single-type cases. According to the table, we can conclude that the newly-assigned locations

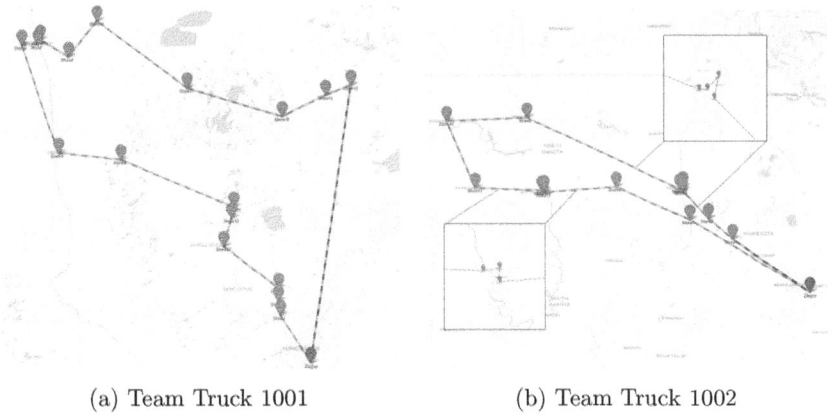

(a) Team Truck 1001   (b) Team Truck 1002

Fig. 2. Optimal routes for team-type drivers and pre-determined delivery locations.

Table 4. Single-type driver trucks routes - PSO-based allocated delivery locations

ID	Type	Initial route (km)	Optimal route (km)	Optimal allocated route (km)
1000	Single	1114.73	1098.6	489.76
1003	Single	334.43	303.59	162.77
1004	Single	86.33	80.85	68.85
1005	Single	471.15	440.55	56.51
1006	Single	461.82	448.73	93.08
1007	Single	125.62	123.4	75.7
1008	Single	156.96	122.89	104.05
1009	Single	146.27	130.22	184.64
1010	Single	186.8	186.8	217.36

by PSO would dramatically improve the routes by decreasing the total distance, specifically for trucks 1000, 1003, 1005, and 1006. Interestingly, the allocation process assigned more delivery locations to trucks 1009 and 1010 and accordingly increased the total distance for these two single-type drivers. Although these trucks locations can be combined with other vehicles that have lower total operating distances, the essential limitation is that all trucks have a maximum weight, which is leveraged as the criteria for the PSO allocation procedure.

Table 5. Team-type driver trucks routes - PSO-based allocated delivery locations

ID	Type	Initial route (km)	Optimal route (km)	Optimal allocated route (km)
1001	Team	1459.14	1249.71	1371.99
1002	Team	1918.45	1880.95	1859.86
1011	**Team**	–	–	**2140.06**

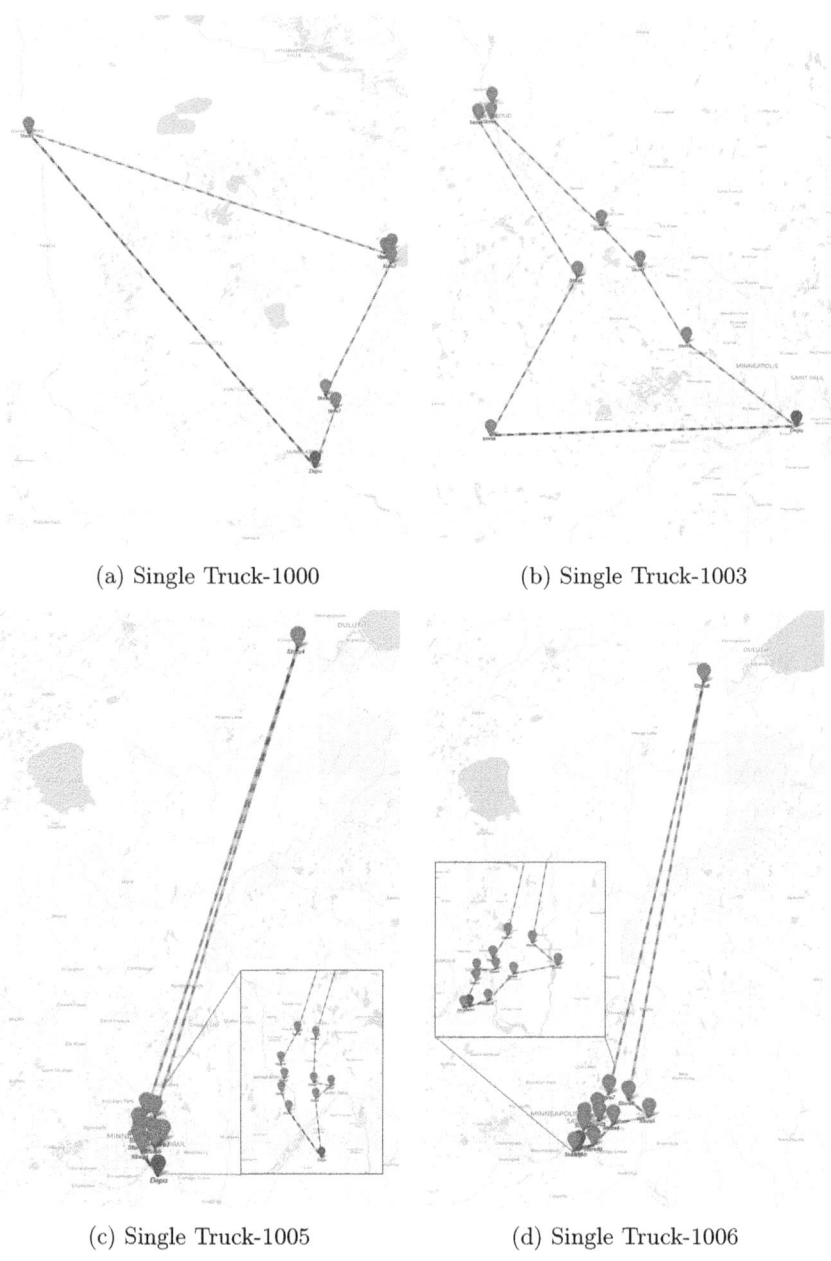

(a) Single Truck-1000   (b) Single Truck-1003

(c) Single Truck-1005   (d) Single Truck-1006

Fig. 3. Optimal routes for single-type drivers and pre-determined delivery locations, which have outlier deliver locations.

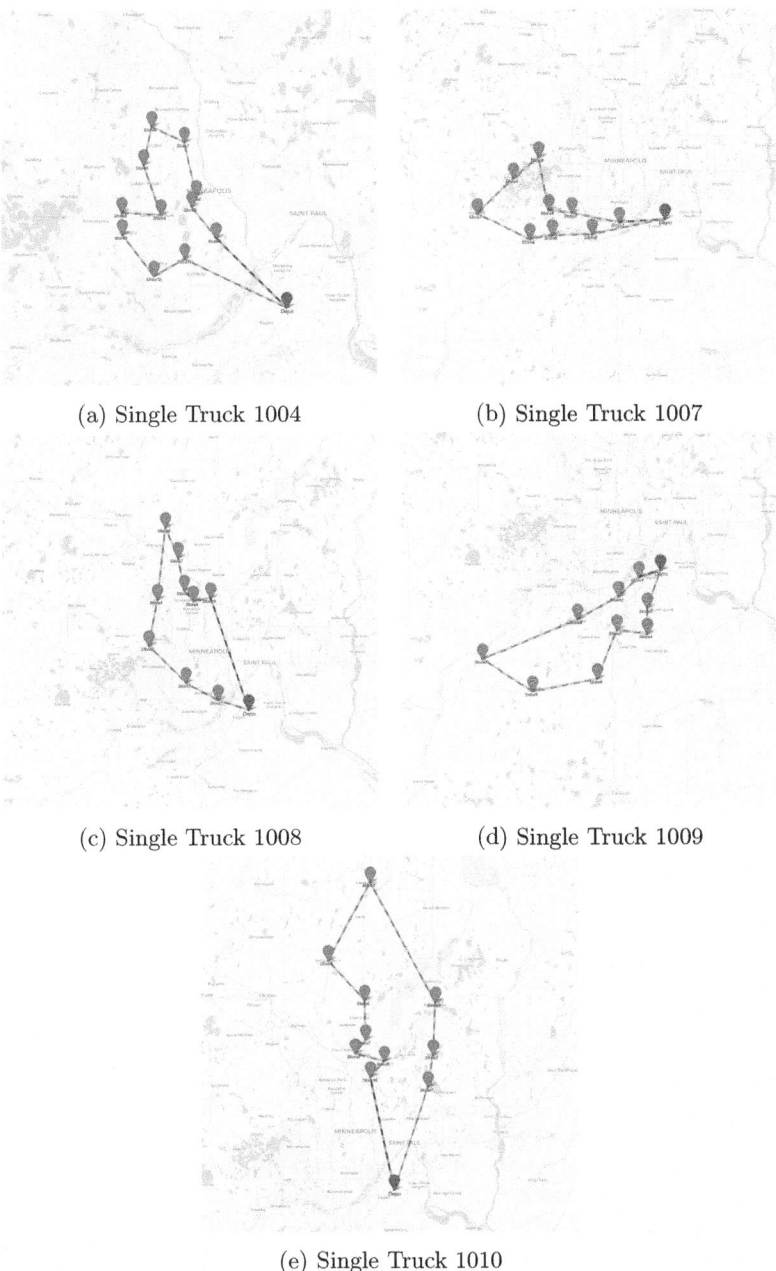

(a) Single Truck 1004

(b) Single Truck 1007

(c) Single Truck 1008

(d) Single Truck 1009

(e) Single Truck 1010

Fig. 4. Optimal routes for single-type drivers and pre-determined delivery locations, which have close delivery locations.

Additionally, the total distances of optimal allocated routes for team-type drivers are summarized in Table 5. There are plenty of delivery locations for only two drivers with a predefined weight restriction, and we also add some distant locations from single-type drivers. Therefore, the optimal allocated route for only two drivers caused a great increase in the total distance values. Hence, we consider adding one more team-type truck (truck 1011) to assign delivery locations to three trucks and reduce the total distance of all team-type drivers as the best decision.

6   Conclusion

This paper studies a real-world delivery application within a multinational pizza restaurant chain to find the optimal routes for its vehicles while the pre-determined delivery locations for both team-type and single-type drivers are given. According to the shown results, the applied optimization procedure improves the route sequences in terms of the total distance. Furthermore, the results show that some of the drivers, especially single-type drivers, are required to drive to distant locations while others have close assigned locations with respect to the warehouse location. Hence, we also simulate and re-allocate the delivery locations to existing vehicles by leveraging the particle swarm optimization (PSO) algorithm and subsequently optimize the newly-assigned delivery locations using the integer programming model. The results show that this optimal allocated scheme can improve the total distance in overall while the locations are assigned in a balanced manner. As a future direction of this study, we will integrate machine learning algorithms with our current scheme to improve the delivery location assignment.

References

1. Abdulkader, M.M.S., Gajpal, Y., ElMekkawy, T.Y.: Vehicle routing problem in omni-channel retailing distribution systems. Int. J. Prod. Econ. **196**, 43–55 (2018)
2. Ahanor, I., Medal, H., Trapp, A.C.: DiversiTree: computing diverse sets of near-optimal solutions to mixed-integer optimization problems. arXiv preprint arXiv:2204.03822 (2022)
3. Alnaggar, A., Gzara, F., Bookbinder, J.H.: Crowdsourced delivery: a review of platforms and academic literature. Omega **98**, 102139 (2021)
4. Amaran, S., Sahinidis, N.V., Sharda, B., Bury, S.J.: Simulation optimization: a review of algorithms and applications. Ann. Oper. Res. **240**(1), 351–380 (2016)
5. Bell, D., Gallino, S., Moreno, A., et al.: Showrooms and information provision in omni-channel retail. Prod. Oper. Manag. **24**(3), 360–362 (2015)
6. Biyouki, S.A., Hwangbo, H.: Blind image deblurring based on kernel mixture. arXiv preprint arXiv:2101.06241 (2021)
7. Boyer, K.K., Prud'homme, A.M., Chung, W.: The last mile challenge: evaluating the effects of customer density and delivery window patterns. J. Bus. Logist. **30**(1), 185–201 (2009)

8. Boysen, N., Emde, S., Schwerdfeger, S.: Crowdshipping by employees of distribution centers: optimization approaches for matching supply and demand. Eur. J. Oper. Res. **296**(2), 539–556 (2022)
9. Brynjolfsson, E., Hu, Y.J., Rahman, M.S.: Competing in the age of omnichannel retailing. MIT, Cambridge (2013)
10. Castillo, V.E., Bell, J.E., Rose, W.J., Rodrigues, A.M.: Crowdsourcing last mile delivery: strategic implications and future research directions. J. Bus. Logist. **39**(1), 7–25 (2018)
11. Chen, H.K., Hsueh, C.F., Chang, M.S.: The real-time time-dependent vehicle routing problem. Transp. Res. Part E: Logist. Transp. Rev. **42**(5), 383–408 (2006)
12. Drexl, M.: Rich vehicle routing in theory and practice. Logist. Res. **5**(1), 47–63 (2012)
13. Du, K.L., Swamy, M., Du, K.L., Swamy, M.: Particle swarm optimization. In: Search and Optimization by Metaheuristics: Techniques and Algorithms Inspired by Nature, pp. 153–173 (2016)
14. Elshaer, R., Awad, H.: A taxonomic review of metaheuristic algorithms for solving the vehicle routing problem and its variants. Comput. Ind. Eng. **140**, 106242 (2020)
15. Figueira, G., Almada-Lobo, B.: Hybrid simulation-optimization methods: a taxonomy and discussion. Simul. Model. Pract. Theory **46**, 118–134 (2014)
16. Goel, A., Kok, L.: Efficient scheduling of team truck drivers in the European Union. Flex. Serv. Manuf. J. **24**(1), 81–96 (2012)
17. Ichoua, S., Gendreau, M., Potvin, J.Y.: Vehicle dispatching with time-dependent travel times. Eur. J. Oper. Res. **144**(2), 379–396 (2003)
18. Janjevic, M., Winkenbach, M.: Characterizing urban last-mile distribution strategies in mature and emerging e-commerce markets. Transp. Res. Part A: Policy Pract. **133**, 164–196 (2020)
19. Juan, A.A., Faulin, J., Pérez-Bernabeu, E., Domínguez, O.: Simulation-optimization methods in vehicle routing problems: a literature review and an example. In: Fernández-Izquierdo, M.Á., Muñoz-Torres, M.J., León, R. (eds.) MS 2013. LNBIP, vol. 145, pp. 115–124. Springer, Heidelberg (2013). https://doi.org/10.1007/978-3-642-38279-6_13
20. Kennedy, J., Eberhart, R.: Particle swarm optimization. In: Proceedings of ICNN 1995-International Conference on Neural Networks, vol. 4, pp. 1942–1948. IEEE (1995)
21. Keskin, M., Çatay, B., Laporte, G.: A simulation-based heuristic for the electric vehicle routing problem with time windows and stochastic waiting times at recharging stations. Comput. Oper. Res. **125**, 105060 (2021)
22. Latifi, K., Ebrahimi, A., Ranjbaran, M., Mirzaei, A., Fakhri, Z.: Efficient customer relationship management systems for online retailing: the investigation of the influential factors. J. Manag. Organ. **29**(4), 763–798 (2022)
23. Lu, S.H., Suzuki, Y., Clottey, T.: The last mile: managing driver helper dispatching for package delivery services. J. Bus. Logist. **41**(3), 206–221 (2020)
24. Macdonald, J.R., Zobel, C.W., Melnyk, S.A., Griffis, S.E.: Supply chain risk and resilience: theory building through structured experiments and simulation. Int. J. Prod. Res. **56**(12), 4337–4355 (2018)
25. Miller, J., Saldanha, J.P., Rungtusanatham, M., Knemeyer, A.M., Goldsby, T.J.: How does electronic monitoring affect hours-of-service compliance? Transp. J. **57**(4), 329–364 (2018)
26. Miller, J.W., Saldanha, J.P., Rungtusanatham, M., Knemeyer, M.: How does driver turnover affect motor carrier safety performance and what can managers do about it? J. Bus. Logist. **38**(3), 197–216 (2017)

27. Miller, J.W., Schwieterman, M.A., Bolumole, Y.A.: Effects of motor carriers' growth or contraction on safety: a multiyear panel analysis. J. Bus. Logist. **39**(2), 138–156 (2018)

28. Nobil, A.H., Sharifnia, S.M.E., Cárdenas-Barrón, L.E.: Mixed integer linear programming problem for personnel multi-day shift scheduling: a case study in an Iran hospital. Alex. Eng. J. **61**(1), 419–426 (2022)

29. Poli, R., Kennedy, J., Blackwell, T.: Particle swarm optimization: an overview. Swarm Intell. **1**, 33–57 (2007)

30. Rhodes, K., Nehring, R., Wilk, B., Patel, N.: Ups helper dispatch analysis. In: 2007 IEEE Systems and Information Engineering Design Symposium, pp. 1–6. IEEE (2007)

31. Sharifnia, S.M.E., Biyouki, S.A., Sawhney, R., Hwangbo, H.: Robust simulation optimization for supply chain problem under uncertainty via neural network meta-modeling. Comput. Ind. Eng. **162**, 107693 (2021)

32. Spliet, R., Gabor, A.F.: The time window assignment vehicle routing problem. Transp. Sci. **49**(4), 721–731 (2015)

33. Ta, H., Esper, T., Hofer, A.R.: Business-to-consumer (B2C) collaboration: rethinking the role of consumers in supply chain management. J. Bus. Logist. **36**(1), 133–134 (2015)

34. Xing, Y., Grant, D.B., McKinnon, A.C., Fernie, J.: Physical distribution service quality in online retailing. Int. J. Phys. Distrib. Logist. Manag. **40**(5), 415–432 (2010)

35. Xu, X., Munson, C.L., Zeng, S.: The impact of e-service offerings on the demand of online customers. Int. J. Prod. Econ. **184**, 231–244 (2017)

36. Zhao, S., You, F.: Resilient supply chain design and operations with decision-dependent uncertainty using a data-driven robust optimization approach. AIChE J. **65**(3), 1006–1021 (2019)

Network Effects from Local Performance Improvements in Europe's Air Transport System

Daniel Lubig[1]([✉]), Hartmut Fricke[1]([✉]), and Bruno Desart[2]([✉])

[1] Technische Universität Dresden, Dresden, Germany
daniel.lubig@tu-dresden.de
[2] EUROCONTROL, Brussels, Belgium

Abstract. The European aviation system is a key enabler in guaranteeing economic welfare and is mandatory to ensure reliable mobility between large cities and metropolises. In 2014, aviation supported 8.8 million jobs in the European Union and contributed over €621 billion to European Union Gross Domestic Product, say 4.7%. A predicted traffic increase of 1.9% per year to reach 16.2 million flights by 2040, in combination with particularly saturated infrastructures, causes a significant challenge to maintain the system's functionality. This paper gives insights into the propagation of local performance improvements in the aviation network. Using a scenario with increased throughput at London Heathrow airport, changes in utilization and delay figures for the EUROCONTROL member state airspace are simulated using a macroscopic airport-airspace simulation model considering the 212 busiest European airports. The actual European airspace layout and equal-sized hexagon airspace portions are considered in this framework. Using a stochastic route generation model, flight trajectories are calculated, and a simulation of the air transportation network based on a flight plan containing over 35,000 flights is performed. The results indicate a significant impact on punctuality and throughput figures on a large part of the airspace in combination with a measurable traffic shift. A decrease in inbound delay can be observed for 2,252 flights (6.4% of the network traffic volume). 39% of them are taking off or landing at London Heathrow airport and benefit directly from the increase in performance. The remaining 61% can be assigned to propagation effects since these affect only flights indirectly impacted by the performance improvement.

Keywords: Air Traffic Network · Airspace Capacity Management · Airport Performance · Propagation Effects

1 Introduction

Prior to the COVID-19 pandemic, the European aviation system was experiencing a steady increase in the number of operated flights. As a result, capacity

© The Author(s), under exclusive license to Springer Nature Switzerland AG 2024
M. Mujica Mota and P. Scala (Eds.): EUROSIM 2023, CCIS 2033, pp. 171–186, 2024.
https://doi.org/10.1007/978-3-031-68438-8_13

was scarce, this worsened by an also increasing volatility of the traffic demand. Airspace design plays a significant role in this context, as each Instrumental Flight Rules (IFR) movement is controlled by Air Traffic Controllers (ATCos) along the complete flight. During busy periods, airspace sectors can become bottlenecks, thus invoking regulatory control measures. One of the main Key Performance Indicators (KPIs) to assess these effects and the resulting network performance is delay, which is highly sensitive when demand approaches capacity. EUROCONTROL evaluations reveal an unsatisfying level of en-route Air Traffic Flow Management (ATFM) delays at individual airspace portions, especially for 2018 (19 million minutes) and 2019 (17.2 million minutes) [1, 2], what represents a significant increase by more than 100% compared to prior years. In more than half of the cases, the delay is caused by a shortage in staffing and so in capacity of Air Traffic Control (ATC). The heterogeneous distribution of the en-route ATFM delay across the European airspace is challenging. Nearly 50% originate and propagate from German and French airspace [2]. ATFM regulations affected nearly 10% of all flights in 2019, pointing to saturation effects of the ATM system.

A traffic increase of 53% by 2040 compared to 2017 is predicted [3], inducing growing congestion in the ATM system. Combined with high service quality standards to be met according to the European Vison "Flightpath2050" or the SESAR Air Traffic Management (ATM) Masterplan, both capacity and performance shall increase to cope with that challenge [4,5]. According to EUROCONTROL, capacity enhancement implemented locally at neuralgic and structural congested elements is considered essential to achieve this ambitious goal [3]. Due to the high degree of connectivity in the aviation network, related propagation effects across linked structures must be considered when pursuing such strategy.

1.1   Status Quo

To ensure safe and efficient air operations, the airspace is truncated into various horizontally and vertically limited areas, regions, and sectors [6]. As per International Civil Aviation Organization (ICAO) Annex 11, the entire Air Traffic Services (ATS) route network shall be covered by a Flight Information Region (FIR). Since airspace sovereignty is largely managed nationwide - though the implementation of Functional Airspace Blocks in Europe aims at overcoming that behavior - the layout of FIR is predominantly determined by current national boundaries across Europe. According to [5,7], a flexible and dynamic organization of airspace is mandatory to further improve system efficiency along the various performance areas such as safety, cost-effectiveness, or ecology. This may be accomplished by shifting from a nation-centric paradigm to an airspace design concept regularly following typical traffic demand patterns. Various approaches are being researched in the literature to achieve this goal. The concept of "dynamic sectorization" aims for a flexible categorization of airspace segments to ensure a balanced ATCo workload across a given region by adjusting airspace borders based on current traffic figures [8–10]. Dynamic sectorization is deemed to improve operational efficiency, resource allocation, and punctuality.

To enhance the airport capacity, a wide range of measures can be applied to improve performance in terms of reduced delay and increased throughput. These measures can be categorized into investment and non-investment opportunities. The first considers activities such as physical extensions [11]. Research validation showed a strong and positive correlation to airport capacity [12–14]. However, regional political constraints and the structural landscape around the airport can complicate or even make physical airport expansions impossible due to the significant consequences on third parties [11]. The second group of measures, non-investment opportunities, tackles improved procedures and processes for air traffic handling. This includes e.g., enhanced usage of slots for runway allocation management, reduced wake vortex separation, or diversion of dedicated traffic to adjacent, less congested airports.

Previous research analyses regarding aviation network effects focus on single aircraft's role and impact on the airline network operation performance. Delay propagation describes the distributed delay caused by an initial delayed flight on other flights operated by the same airline. A well-known metric to measure this effect is the Delay Multiplier (DM) [15–19]. This indicator equals the ratio between the total induced delay in the network and the initial delay of a root flight. The larger the DM, the greater the downstream delay in the network.

In addition to flight-centric views, research papers have addressed network impacts from an airport perspective. Several research papers show different measurement approaches to describe the robustness or connectivity of an airport within a network and methods to identify critical airports [20–22]. Further research activities investigate the influence of the airport status (hub vs. non-hub) on network operations [23], the role of the network structure on delay propagation [24], and the effect of disrupting events on the network performance [14]. The aim is to characterize the network reactions under different conditions and scenarios to get a deeper understanding of the role of airports in a network. Scientific studies differ in the scope of the analyzed network size from a local country view (e.g., Italy [25], or India [26]) up to a holistic global view [27].

1.2   Focus and Structure of the Document

In our contribution, the effect of local performance improvements on the utilization of airspace sectors is analyzed. In the first step, an airport-airspace simulation model is developed based on post-operation data from flights during the busy summer period in 2019. Further, a simulation considering a performance enhancement scenario at London Heathrow Airport (EGLL) is executed to determine the influence on Upper Airspace (UAS) characteristics and related propagation effects on the overall aviation network. The objective airport is one of the major European hubs and a key element within the European air transportation network.

2   Trajectory Model

2.1   Input Data

The input for the model is a data set provided by EUROCONTROL with post-operational data from over 3 million flights performed within the 41 member states between July and September 2019 [28]. In addition to membership, EUROCONTROL concludes Comprehensive Agreements with Morocco and Israel, which are not considered in the analysis. The used information for the model development is summarised in Table 1.

Table 1. Used data set information for developing the airport resp. airspace model.

Information	Description	Example
ADEP	Departure airport	EDDF
ADES	Destination airport	EGLL
SOBT/AOBT	Scheduled/Actual Off-Block Time at the departure airport	11:20:00
Taxi-out duration	–	00:12:36
STA/ATA	Scheduled/Actual Time of Arrival	12:40:00
Taxi-in duration	–	00:08:00
Aircraft type	ICAO designator for used aircraft	A320
Aircraft registration	–	DAIFT
Max. altitude	Maximum recorded Flight Level (FL) for the flight	270

2.2   Airport Model

A comprehensive model representing the majority of European air traffic is implemented to investigate the propagation effects of local performance enhancements. The busiest 212 airports located in the 41 EUROCONTROL member states handle over 90% of the movements and are considered within the model. The airport model includes three major components:

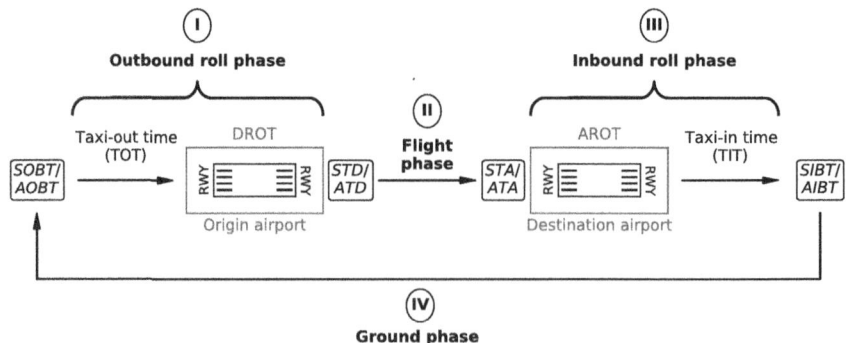

Fig. 1. Exemplary procedure of one flight cycle in the simulation.

(1) Each airport represents a node and is characterized by its number of operated runways as a queuing system. The required separation between aircraft depends on the wake vortex category and the Runway Occupancy Time (ROT), which limits the runway throughput.

(2) The edges correspond to the flights between two airports. A flight contains four elementary phases separated by four timestamps shown in Fig. 1. Waiting times occur if the runway system at an airport is utilized and aircraft have to wait to use the infrastructure. Aircraft can perform multiple flight

Table 2. Information about the processing time for the flight phases used in the airport model.

Information	Description and Determination method
Outbound roll phase	Equals the taxi-out time added by the queuing time at the departure airport runway and the Departure Runway Occupancy Time (DROT). The taxi-out time is recorded for each flight in the EUROCONTROL data set and is converted into a Gaussian distribution for each airport. The queuing time is a result of the implemented runway system. The DROT assumption depends on the aircraft's Wake Turbulence Category (WTC) (Light: 70 s; Medium: 56 s; Heavy: 63 s; Super: 70 s)
Flight phase	Equals the flight time between two airports added by the waiting time at the destination airport runway. The flight time is recorded for each flight within the data set. A Gaussian distribution is determined for each Origin and Destination (OnD) under consideration of the used aircraft type. The waiting time at the destination airport results from the runway queuing system
Inbound roll phase	Equals the sum of an aircraft WTC depending Arrival Runway Occupancy Time (AROT) assumption (Light: 70 s; Medium: 56 s; Heavy: 63 s; Super: 70 s) and the taxi-in time, which is represented by an airport-specific deterministic value in the EUROCONTROL data set
Gate phase	The gate phase includes the turnaround and potential buffer time to mitigate operational uncertainties. Due to missing data, an assumption for the turnaround time determination based on the aircraft WTC is used. For each WTC a reference aircraft (Medium: A320neo; Heavy: A350-900; Super: A380-800) and the indicated full-service turnaround time (Medium: 44 min [31]; Heavy: 61 min [32]; Super: 90 min [33]), published by the manufacturer, is applied as the mean value for a Gaussian distribution. The corresponding standard deviation equals 10% of the mean value. For Light aircraft, a mean of 20 min and a standard deviation of 2 min is assumed

cycles in a day, and delays in prior cycles can cause knock-on delays through-out the network. The determination of the process times for the different phases are shown in detail in Table 2.

(3) Flight demand is obtained from historical flight plan data for each network airport. The designated airport nodes generate and handle the flights using the information described in Table 1.

Due to congestion effects on the day under investigation, operational knock-on delays may occur and are assigned to the flights. The authors give a comprehensive overview of the used airport model in previous studies [29, 30].

2.3   Airspace Layout Grids

The UAS of EUROCONTROL member states comprises 48 FIRs resp. Upper Information Regions (UIRs), as shown in Fig. 2 (left). Segments of equal-size and shape are necessary to ensure more precise performance comparability between airspace. For this purpose, a hexagon airspace grid is used. The advantage of this structure is the equivalent distance between the center of a hexagon to all central points of adjacent hexagons.

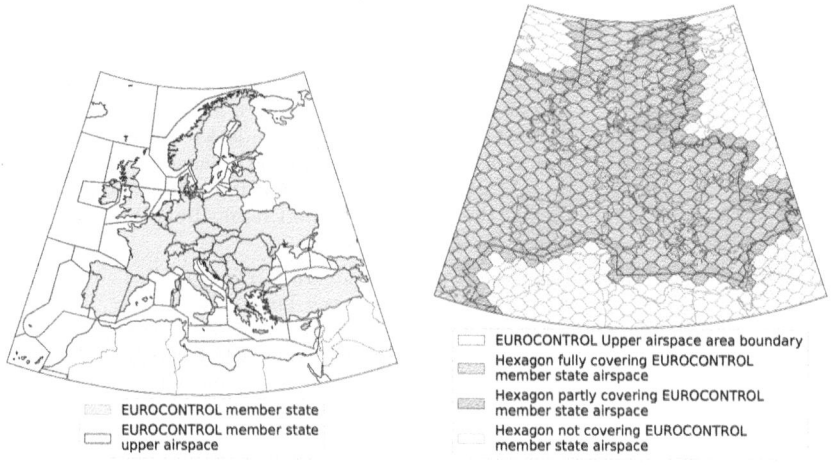

Fig. 2. EUROCONTROL member state airspace design (Left: Upper airspace FIR/UIR; Right: Designed hexagon portions covering the EUROCONTROL member state territory).

Another advantage is the aggregation or splitting of structures into larger or smaller frameworks with the same basic geometric shape. Due to this characteristic, leading companies with business fields in the geodata sector (e.g., Uber [34]) use this geometric shape for indexing or division of spatial zones. In aviation research, hexagonal grids are widely used to classify geographic sectors in different areas (e.g., the definition of weather zones [35], airspace sectorization [8,9,36], and trajectory planning [37,38]).

Figure 2 (right) illustrates the result of the hexagon definition for the European airspace. Each hexagon includes an area of 40,000 km^2. The 303 green-colored hexagons contain EUROCONTROL member states airspace exclusively. Red-colored hexagons only partially represent member state airspace; thus, they are located in the border region to the airspace of other non-EUROCONTROL member states. A total of 105 hexagons are assigned to this category. The 223 hexagons not highlighted are excluded from the analysis since they do not represent EUROCONTROL member state airspace.

2.4   Modeling of Flight Routes

A three-dimensional view consisting of a vertical perspective considering the flight altitude and a horizontal flight path represented through longitudinal and latitudinal coordinates is necessary to describe a generalized procedure for modeling flight routes. The generalized vertical flight profile is shown in Fig. 3. To simplify the profile modeling, three main phases are distinguished. The climbing phase starts at the departure airport at the height of 0 ft until reaching the assigned FL. The cruise level characterizes the second phase at a constant altitude, followed by the descent phase until reaching the height of 0 ft at the destination airport. The profile is divided into two vertical layers. The layer between 0 ft and FL245 represents the vicinity of the origin and destination airport and mainly includes the designed procedures for take-off and landing in the Lower Airspace (LAS). Above FL245, the flight is assigned to the UAS. For instance, this physical threshold limit as the transition between LAS and UAS is applied in Germany, the United Kingdom (UK), and Spain. However, the UAS vertical limit for Italian and France airspace is FL195. The implemented model only considers a fixed transition layer at FL245 for all considered airspace sectors in the EUROCONTROL member state area.

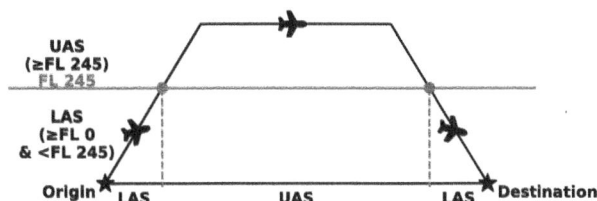

Fig. 3. Simplified assumed vertical flight profile.

The following key assumptions drive the definition of the vertical flight profile:

(1) The paths for the climb and descent phase are assumed by a constant 3° angle resulting in a climb/descent distance of ≈77.5 NM to/from FL245.

(2) The aircraft's optimum flight altitude increases as flight time progress due to weight loss from kerosene combustion. If the current altitude deviates significantly from this value, a step climb is performed for efficiency. However, en-route step climbs are not included in the generalized vertical flight profile as they usually occur at altitudes above FL245, where further airspace classification is unnecessary in this study.

Figure 4 (left) shows an example of the general horizontal flight profile for the route from Amsterdam Airport Schiphol (EHAM) to Copenhagen Airport (EKCH). The red line equals the Great Circle Distance (GCD) route, and the dashed black lines limit an area of 40 NM around the airports known as Airport Sequencing and Metering Area (ASMA), which is important for the calculation of the Horizontal Flight Efficiency (HFE). The GCD between departure airport (λ_O, ϕ_O) and destination airport (λ_D, ϕ_D) is calculated using Eq. (1).

$$\text{GCD}_{O \to D} = \arccos(\sin(\phi_O) \cdot \sin(\phi_D) + \cos(\phi_O) \cdot \cos(\phi_D) \cdot \cos(\lambda_O - \lambda_D)) \cdot 3,444 \text{ NM} \tag{1}$$

Let a Detour Factor (DF) represent the possible fluctuation of the trajectory; two boundaries are defined via a parallel shift of the GCD route, forming a corridor for modeling the flight path. The distance ratio between the constructed shifted trajectory and the GCD equals the DF. Inside the corridor, a position on the perpendicular span line (green illustrated in Fig. 4 (left)) between the corridor boundaries is modeled for each calculated waypoint (WP) using a truncated normal distribution. In contrast to the standard normal distribution, the range of values covers a fixed range, thus avoiding the definition of extreme values with a high deviation from the expected value [39]. The center point of the perpendicular corridor span line refers to the position on the GCD route with the coordinates $\lambda_{r,i}$ and $\phi_{r,i}$ (see Fig. 5 (bottom)). A random number (RN_i) between 0 (corresponding to the position on boundary 1 defined by $\lambda_{1,i}, \phi_{1,i}$) and 1 (corresponding to the position on boundary 2 defined by $\lambda_{2,i}, \phi_{2,i}$) for each WP (WP_n) is calculated, which is afterward transformed into a position (λ_i, ϕ_i) on the perpendicular corridor span line via Eq. (2).

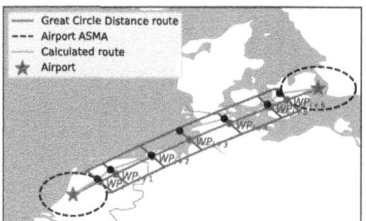 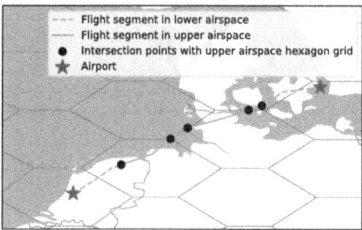

Fig. 4. Left: Example for the route modeling methodology; Right: Exemplary determination of intersection points from the modeled flight route with the hexagonal upper airspace grid.

$$\lambda_i; \phi_i = \begin{cases} \lambda_{r,i} + ((-0.5 + RN_i) * 2) * (\lambda_{2,i} - \lambda_{r,i}); & \text{if } 1 \geq RN_i > 0.5 \\ \phi_{r,i} + ((-0.5 + RN_i) * 2) * (\phi_{2,i} - \phi_{r,i}) & \\ \lambda_{r,i} + ((0.5 - RN_i) * 2) * (\lambda_{1,i} - \lambda_{r,i}); & \text{if } 0.5 > RN_i \geq 0 \\ \phi_{r,i} + ((0.5 - RN_i) * 2) * (\phi_{1,i} - \phi_{r,i}) & \\ \lambda_{r,i}; \phi_{r,i} & \text{if } RN_i = 0.5 \end{cases} \quad (2)$$

This procedure is performed iteratively, from the first WP leaving the origin ASMA to the last WP while entering the destination ASMA. The number of existing WP_n along the route depends on a distance Threshold value (THR) between two following WPs (Eq. (3)). The points are distributed uniformly along the GCD route. The exit and entry points into the ASMA are also defined as WP, resulting in a fixed number of two additional WPs.

$$WP_n = \lfloor \frac{GCD_{O \rightarrow D}}{THR} \rfloor + 2 \quad (3)$$

The expected value of the normal distribution (μ) corresponds to the estimated position on the perpendicular corridor span line. For the first WP along the route, this equals 0.5 since a location near the GCD route is initially assumed to be the most appropriate. In the following iteration process, the expected value ($\mu_{WP_{i+1}}$) for the subsequent segment corresponds to the determined random

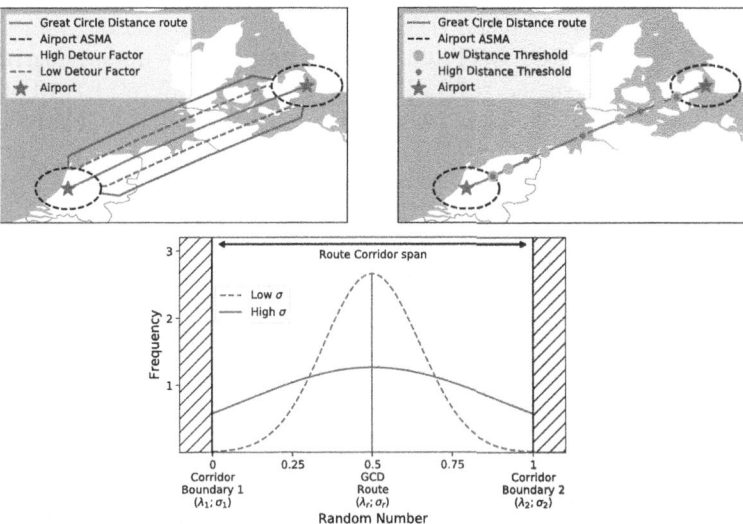

Fig. 5. Parameters influencing the definition of a flight route. (Top left: Detour Factor limiting the corridor width and forms the tube boundary; Top right: Threshold distance defining the number of WPs of the Great Circle Distance route; Bottom: Standard deviation representing the spread around the Great Circle Distance position.)

number from the previous segment (RN_i). This methodology reflects the dependence of the following position based on the current position. For each WP, a location on the respective perpendicular corridor span line is calculated, which is combined with the positions from the origin and destination airport to the modeled route (see Fig. 4 (left)). The modeling of the flight path is thus essentially dependent on three parameters:

(1) Detour Factor (DF) for the definition of the route corridor: The higher the DF, the wider the route corridor.
(2) Threshold distance (THR) for the calculation of the number of WPs: The lower THR, the more WPs are defined for a route.
(3) Standard deviation (σ) for the truncated normal distribution: The higher σ, the wider the spread around μ resp. the GCD route.

The particular relationships are shown in Fig. 5. A sensitivity analysis is conducted to determine the appropriate parameters for the route generation, which refers to the HFE defined by the ratio of the GCD route and the flight path length beyond the 40 NM circles surrounding the departure and destination airports [40]. Both the DF and σ affect the potential divergence from the GCD route along the perpendicular corridor span line within the route corridor. Changing σ while keeping the DF constant results in a stronger effect on the HFE than changing the DF while keeping σ constant. Therefore, the DF is fixed at 1.05, and only σ and THR are varied for the sensitivity analysis. The HFE results for all 34,119 flights with a flight phase outside of the ASMA circles of departure and destination airports with different parameters for σ, and THR are shown in Fig. 6. Target values for the HFE correspond to the average values of the HFE published by EUROCONTROL for the period of the underlying raw data from July (97.02%), August (97.14%), and September (97.16%) 2019 [41]. The results in Fig. 6 reveal a combination of DF = 1.05, σ = 0.07 and THR = 350 km

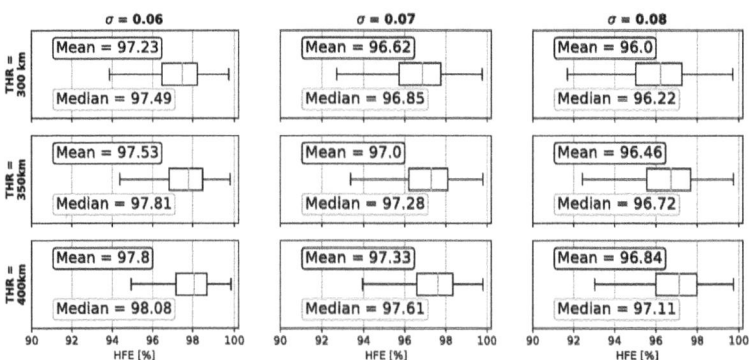

Fig. 6. Results of the sensitivity analysis with varying distance threshold (THR) and standard deviation (σ) under constant Distance Factor (DF) of 1.05 for the route modeling.

to produce the best approximation between the modeled and historical HFEs. Further, the corresponding box plot implies a low spread around the median concerning the 25% and 75% quantiles, which also reflects the actual HFEs properly. A hexagon grid or the actual FIR/UIR airspace layout is laid on the generated flight trajectory, and entry resp. exit points into intersected airspace sectors are calculated based on the flight departure and arrival times (see Fig. 4 (right)). A simplified linear velocity profile is assumed, which is sufficient for the macroscopic flow-oriented analysis.

3   Simulation Setup

Two scenarios are defined to investigate the network performance behavior and the propagation effects within the network:

(1) Baseline scenario - Simulation of the European airport network for a busy weekday during summer 2019 with over 35,000 handled flights.
(2) Scenario I - Similar to the baseline scenario in combination with a moderate local performance improvement at EGLL, which leads to reduced inbound and outbound delay compared to the baseline scenario (-2.4% and -3.6% respectively). Within the simulation, the improvement is achieved by decreasing process times for runway operations (WTC separations and ROTs) at the objective airport. EGLL is chosen as investigated airport due to the high number of operated flights on the scenario day (695 departures and 698 arrivals), the critical role of the airport within the air transportation network, and the comparatively high number of delays in 2019 [2].

Table 3. Used Key Performance Indicators to analyse the simulation results [5,42]

KPI	KPA	Description
Throughput	Capacity	Aircraft volume passing an airspace portion during a defined period
Delay	Operational efficiency	Difference between ATA and STA of a flight. The delay is assigned backwards to the used airspace sectors
Punctuality	Operational efficiency	A flight is punctual when the destination is reached before the STA + 15 min. Every used airspace portion by the flight is also assigned as reached punctually

For all flights, a route is modeled applying the presented methodology, which is used identically for the same run of the baseline scenario and scenario 1 to ensure comparability between the results. Relevant KPIs for evaluating propagation effects are explained in Table 3. The focus of the study is set on the

UAS since most flights (91.7%) are designated at a maximum altitude above FL245 and spend a substantial part of the flight phase in this airspace layer. The network simulation is performed 100 times using an implementation developed with the python programming language. Every run needs around 8 h of processing time with an Intel(R) Core(TM) i5-9500 processor.

4   Simulation Results

Figure 7 displays the sliding hour maximum throughput, calculated for 10-minute intervals, for all runs of the two scenarios at EGLL. When local capacity enhancements are implemented, the throughput is moderately increased and slightly shifted compared to the baseline scenario. During the peak traffic around 8 PM, the throughput exceeds the declared capacity of 88 movements per hour. Since the declared capacity is a planning value for airport coordination tasks, this does not pose a critical operational issue. The maximum observed throughput within the data period of 98 movements significantly exceeds the simulation results for the exemplary day. The local performance increase does not result in implausible throughput and can be considered realistic.

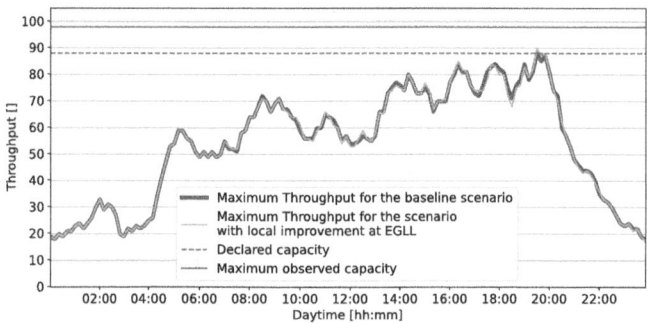

Fig. 7. Max. throughput for EGLL observed in all simulation runs for each scenario.

An indicator for traffic shifts is the deviation of airspace throughput between both scenarios. The critical case corresponds to additional traffic during the same timestamp due to improved performance at EGLL. Figure 8 shows the maximum additional traffic volume for equal periods during the entire operating day between the base case and scenario 1. An increase in throughput can be observed for 392 hexagons (≈96%). A significant increase of more than three additional flights occurs particularly close to EGLL. The highest traffic increase is observed in the hexagon directly above EGLL. Airspace portions representing France, BeNeLux, and Germany are also significantly impacted. These areas belong to actual already congested airspace [2]. Similar results are obtained when considering the UIR layouts in Fig. 8 (left). Secluded airspace from EGLL, which covers essential main traffic routes, are also significantly affected. This refers to

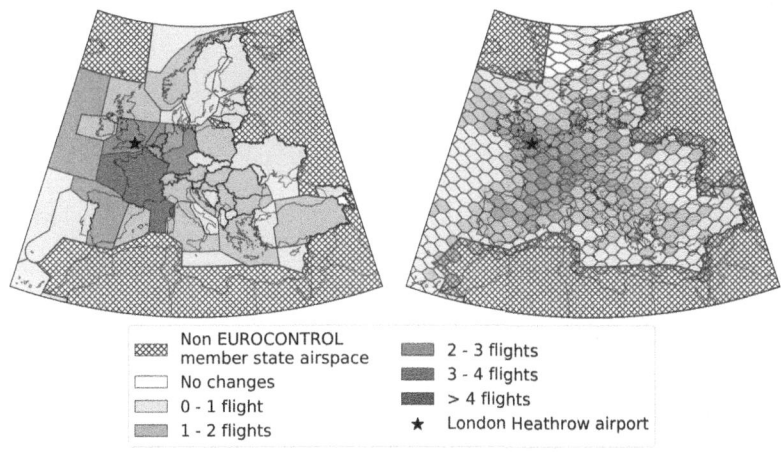

Fig. 8. Increase of maximum airspace throughput for actual UIR (left) and hexagonal airspace portions (right) compared to the baseline scenario.

hexagons over east Turkey for Asia-Pacific routes, eastern from Ireland representing transatlantic routes to North America, and hexagons above the Baltic Sea states used by flights to Asia. It should be noted that the additional traffic does not automatically result in congestion resp. delay by definition. Nevertheless, the responsible airspace operator should be aware of resulting traffic shifts to handle the displaced demand appropriately with the available resources.

Figure 9 indicates the share of flights with decreased inbound delay using a specific airspace portion. Significant improvement with more than 12% improved flights can be derived for 65 ($\approx$16%) hexagons. Similar to Fig. 8 (right) cluster containing hexagon airspace with a high improvement rate increase can be determined around the Baltic Sea, Scandinavia, and the UK. UIRs representing Great Britain and Ireland benefits most from the local performance improvement. Especially flights to North America benefit from the expansion, which equals a performance increase for the airspace hexagons located in these route directions. On average, around 2,252 (6.4%) flights improved their operational efficiency and shortened the inbound delay in scenario 1. 61% of these flights refer to movements without EGLL as origin resp. destination airport. These benefits result from propagation effects on the entire network caused by the local performance improvements. Most inbound delay improvements (69.4%) are below one minute, implying a low impact from a single flight performance view. In scenario 1, only 16 additional flights reach the destination airport punctually. More than 879 min of inbound delay are saved from a macroscopic network perspective due to the moderate local performance improvement at EGLL.

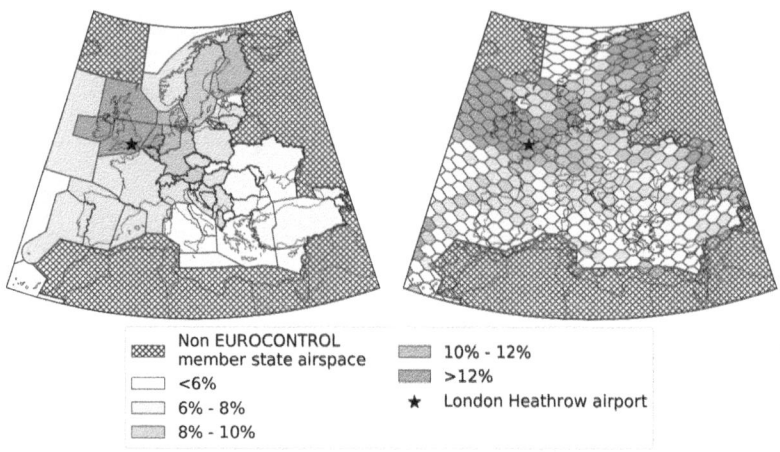

Fig. 9. Rate of improved flights for actual UIR (left) and hexagonal airspace portions (right) compared to the baseline scenario.

5   Discussion and Conclusion

The European aviation network consists of airports and airspace as key infrastructure elements. In this paper, a simulation model comprising an airspace grid and a methodology for modeling flight routes are connected to determine the impacts of local performance improvement at London Heathrow airport on throughput and delay figures for the overall network airspace. A historical traffic sample of an entire day of operation containing over 35,000 flights is simulated, representing a high-traffic demand use case. The results exhibit a measurable positive impact for most of the defined airspace portions in capacity and operational efficiency. This is demonstrated by increased throughput, a shift in traffic peaks, and an improved punctuality rate resp. decreased inbound delay. The highest displacement of movements affects areas above central Europe, which are currently operating close to total capacity, which is a challenge for the Air Navigation Service Providers (ANSPs). Most airspace sections are characterized by improved punctuality of passing flights. For 2,252 flights (6.4%) an inbound delay reduction is obtained. 61% of them are not directly connected to London Heathrow airport as the origin or departure airport, which indicates benefits through propagation effects. A significant improvement can be observed for transatlantic flights, resulting in performance improvements in the airspace sections used by these services. It should be noted that London Heathrow is one of the major hub airports in Europe, with many flights per day and connections to a large number of cities in different European areas. The results are use-case specific and may vary significantly if smaller airports or airports with a diverse network function are enhanced (e.g., leisure-focused airports).

Acknowledgements. EUROCONTROL supported and funded this publication via contract NO.19-220468-C. This paper expresses the authors' views within the scope of this research, and it should not be taken to reflect any affiliated organizations' views and official policy.

References

1. EUROCONTROL. Performance Review Report 2018 (2019)
2. EUROCONTROL. Performance Review Report 2019 (2020)
3. EUROCONTROL. European aviation in 2040: Challenges of growth (2018)
4. EUROPEAN COMMISSION. Flightpath 2050 Europe's vision for aviation: Report of the high level group on aviation research (2011)
5. SESAR JU: European ATM master plan - executive view (2020)
6. ICAO. Annex 11 - Air Traffic Services, 15th edn. (2018)
7. ICAO. 2016 - 2030: Global Air Navigation Plan, 5th edn. (2016)
8. Jägare, P.: Airspace sectorisation using constraint programming (2011)
9. Kulkarni, S., Ganesan, R., Sherry, L.: Dynamic airspace configuration using approximate dynamic programming: intelligence-based paradigm. Transp. Res. Rec. **2266**(1), 31–37 (2012)
10. Gerdes, I., Temme, A., Schultz, M.: Dynamic airspace sectorisation for flight-centric operations. TRC: Emerg. Technol. **95**, 460–480 (2018)
11. Berster, P., Gelhausen, M.C., Wilken, D.: Options to mitigate negative effects of airport capacity constraints - a discussion of potential mitigation effects. In: The 23rd ATRS World Conference, Amsterdam, The Netherlands, pp. 1–21 (2019)
12. Hansen, M.: Post-deployment analysis of capacity and delay impacts of an airport enhancement. ATCQ **12**(4) (2004)
13. Mota, M.M., Scala, P., Boosten, G.: Simulation-based capacity analysis for a future airport. In: APCASE (2014)
14. Baspinar, B., Ure, N.K., Koyuncu, E., Inalhan, G.: Analysis of delay characteristics of European air traffic through a data-driven airport-centric queuing network model. IFAC-PapersOnLine **49**(3), 359–364 (2016)
15. Beatty, R., Hsu, R., Berry, L., Rome, J.: Preliminary evaluation of flight delay propagation through an airline schedule. ATCQ **7**(4), 259–270 (1999)
16. Kondo, A.: Delay propagation and multiplier. In: 51st Annual TRF (2010)
17. Kafle, N., Zou, B.: Modeling flight delay propagation: a new analytical-econometric approach. TRB: Methodol. **93**, 520–542 (2016)
18. Guleria, Y., Cai, Q., Alam, S., Li, L.: A multi-agent approach for reactionary delay prediction of flights. IEEE Access **7**, 181 565–181 579 (2019)
19. Wu, C.-L., Law, K.: Modelling the delay propagation effects of multiple resource connections in an airline network using a Bayesian network model. TRE: Logis. Transp. Rev. **122**, 62–77 (2019)
20. Malighetti, P., Paleari, S., Redondi, R.: Connectivity of the European airport network: "self-help hubbing" and business implications. JATM **14**(2), 53–65 (2008)
21. Lordan, O., Sallan, J.M., Simo, P., Gonzalez-Prieto, D.: Robustness of the air transport network. TRE: Logis. Transp. Rev. **68**, 155–163 (2014)
22. Cong, W., Hu, M., Dong, B., Wang, Y., Feng, C.: Empirical analysis of airport network and critical airports. CJA **29**(2), 512–519 (2016)
23. Pyrgiotis, N., Malone, K.M., Odoni, A.: Modelling delay propagation within an airport network. TRC Emerg. Technol. **27**, 60–75 (2013)

24. Zhang, H., Wu, W., Zhang, S., Witlox, F.: Simulation analysis on flight delay propagation under different network configurations. IEEE Access **8** (2020)
25. Guida, M., Maria, F.: Topology of the Italian airport network: a scale-free small-world network with a fractal structure? Chaos Solitons Fractals **31**(3), 527–536 (2007)
26. Bagler, G.: Analysis of the airport network of India as a complex weighted network. Phys. A **387**(12), 2972–2980 (2008)
27. Guimerà, R., Amaral, L.A.N.: Modeling the world-wide airport network. Eur. Phys. J. B **38**(2), 381–385 (2004)
28. EUROCONTROL. Our member and comprehensive agreement states (2022). https://www.eurocontrol.int/our-member-and-comprehensive-agreement-states
29. Lubig, D., Schultz, M., Fricke, H., Herrema, F., Montes, R.B., Desart, B.: Propagation of airport capacity improvements to the air transport network. In: 2021 IEEE/AIAA 40th DASC, pp. 1–10. IEEE (2021)
30. Lubig, D., et al.: Modeling the European air transportation network considering inter-airport coordination. In: SID 2021 (2021)
31. AIRBUS. A320: Aircraft characteristics, airport and maintenance planning (2017)
32. AIRBUS. A350: Aircraft characteristics, airport and maintenance planning (2018)
33. AIRBUS. A380: Aircraft characteristics, airport and maintenance planning (2016)
34. Brodsky, I.: H3: Hexagonal hierarchical geospatial indexinx system (2022). https://github.com/uber/h3
35. Lindner, M., Rosenow, J., Zeh, T., Fricke, H.: In-flight aircraft trajectory optimization within corridors defined by ensemble weather forecasts. Aerospace **7**(10), 144 (2020)
36. Klein, A.: Airspace partitioning. US Patent App. 11/165,251 (2006)
37. Kramer, K.A., Stubberud, S.C.: Adaptive UAS route planner based upon evidence accrual. In: 2019 IEEE/AIAA 38th DASC, pp. 1–6. IEEE (2019)
38. Zou, B., Buxi, G.S., Hansen, M.: Optimal 4-D aircraft trajectories in a contrail-sensitive environment. Netw. Spat. Econ. **16**, 415–446 (2016)
39. Burkardt, J.: The truncated normal distribution. Department of Scientific Computing Website, Florida State University, vol. 1, p. 35 (2014)
40. EUROCONTROL. Performance Indicator - Horizontal Flight Efficiency Introduction (2022). https://ansperformance.eu/methodology/horizontal-flight-efficiency-pi/
41. EUROCONTROL. Horizontal EN-route flight efficiency (2022). https://ansperformance.eu/efficiency/hfe/
42. SESAR JU. PJ19.04: Performance Framework (2019)

Simulation Applied to a Consulting Team Assignment Problem

Gabriel Armando Rios Esparza[✉] and Esther Segura Segura Pérez

Operations Research Department, National Autonomous University of Mexico, Mexico City, Mexico
gabrielriosez@gmail.com

Abstract. There are intermediary entities that improve the administration of human resources, adapting agents' profiles with companies' requests. This paper presents a university social service program that requires student teams to give consultancies to companies' groups but has a desertion problem derived from an assignment with long distances and incompatibility between schedules and requirements. We use process-oriented simulation to analyze some proposed methodologies and variants of the classic assignment problem as possible solutions while uncovering the need for a flexible method that enables continuous modification and maintains easy user data management. We define our human resources case with multiple attributes (distance, requirements, and schedules) as a multi-criteria assignment model with quadratic (QAP) and qualified semi-assignment characteristics. To solve it, we use a tessellation to facilitate the inclusion and reduction of distances between agents and tasks while evaluating the qualified association of the requirement and schedules as categorical variables. We integrate a modular method based on hexagonal tessellation, reducing the dedicated time from 5 days to 90 min and a total distance of 585 km between students and companies while improving the schedules and requirements compatibility, solving the real problem in its context. We hope this will contribute to other assignment problems with a similar structure or goals.

Keywords: Assignment Problem · Hexagonal Tessellation · Multi-criteria assignment

1 A Summary of the Assignment Problem

The classical assignment problem (AP) establishes the association of a pair x_{ij} of n agents $\{n_1, n_2, ..., n_i\}$ with a set of m tasks $\{m_1, m_2, ..., m_j\}$, stating that each agent must be assigned to a single task and each task must have only one associated agent to perform it, with a cost (c_{ij}) defined by the specific characteristics of each agent-task relationship.

Mathematical representations follow a linear structure to minimize costs or maximize profits. Matrix representation commonly requires a square matrix of costs $n \times n$. . In an unbalanced problem $(n \times m)$, agents or dummy tasks are generated with cost penalties to meet the algorithm requirements. In their graphical representation, bipartite networks

© The Author(s), under exclusive license to Springer Nature Switzerland AG 2024
M. Mujica Mota and P. Scala (Eds.): EUROSIM 2023, CCIS 2033, pp. 187–198, 2024.
https://doi.org/10.1007/978-3-031-68438-8_14

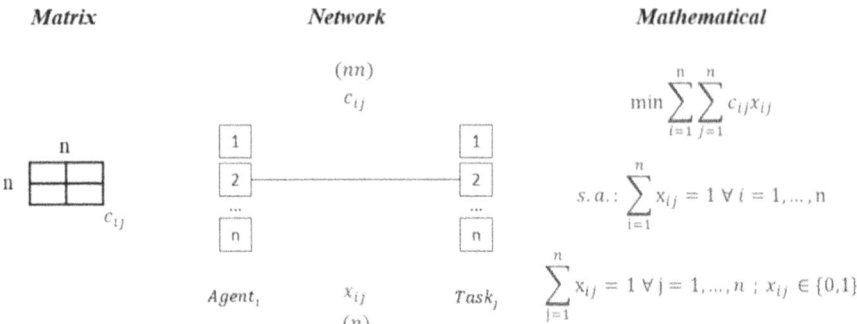

Fig. 1. Three simplified representations of the assignment problem.

are among the most widely used [1], with the n set on one side, the m set on the other, and its associated cost on the union axes (Fig. 1).

In real cases, these representations and their algorithms do not fit naturally. As a result, different areas have developed systems, methodologies, models, or protocols to define a solution for each problem. Moreover, their natural comprehension evolution has generated the need to increase the capacity of requirements and constraints, making algorithms such as the Hungarian method or linear programming less efficient in the shortest time possible [2], necessitating a broader vision to generate integral solutions for more than specific problems in a particular area [3].

1.1   The Assignment Problem in Human Resources

In particular, the human resources assignment was first formalized in 1953 by Votaw and Orden [4], describing the personnel assignment and classification problem as equivalent to the transport problem, later solved with a matrix representation by the Hungarian method of Kuhn in 1955 [5]. Another variant is the qualified assignment problem (Q-AP), which belongs to the one-to-one assignment problems. The Q-AP considers n agents and m tasks. Each agent can perform a task, and each task can be performed by an agent; however, only some agents are equally qualified to complete different tasks [6]. Unlike the AP, the constraints generate binary rating conditions (q_{ij}) with a value of one if the agent is qualified or zero if it is not. Within this problem, there is a possibility that some tasks or agents are left unassigned due to their occupied status, making the number of assignments limited to the agent's or task quantity that requires qualification. Although tasks start simultaneously, and an agent cannot do more than one task, some variations allow task sequencing for a complete allocation.

Other researchers have tested different methodologies for solving variations of the staffing problem. For example, Villagómez [7] studied the problem of assigning work schedules using a hybrid method that combined linear programming, forecasts, and genetic algorithms to assess the necessary number of work schedules and the number of personnel required based on the demand of telephone operators. Naveh, Richter, Altshuler, Gresh, and Connors [8] presented an analysis of the operation of IBM's personnel evaluation system. It uses descriptors, priorities, and constraint scheduling to solve the combinatorial problem, while a shorter-path algorithm reduces the distance between the employee's residence and work location. Widanta, Rizaldi, Setwyohadi, and Riskiawan [9] also compared different weight-based multi-criteria decision methodologies, while Leite, Baptista, and Ribeiro [10] simulated different logical strategies for assigning staff using efficiency and effectiveness metrics.

2 General Characteristics of the Problem

The problem consists of a university social service program with a semester duration that provides consultancies to companies. Its objective is to satisfy companies' requirements without damaging students' academic development, so the student's activities are limited to weekly meetings in the offices or locations of their assigned companies to ensure the creation of diagnoses based on direct observation of their activities.

In the context of a classic problem, students at the university represent agents, who can belong to one of 10 careers, have a defined schedule of attention to companies, and can choose their initial starting point to provide face-to-face service. The tasks represent companies that require support to solve at most five problems they have encountered in their activities, have fixed facilities, and have a schedule of availability to receive attention from the consultants. These problems can incur various issues, so it is required to generate multidisciplinary teams to address them. Therefore, teams of five students are formed with different careers directly related to the five requirements of the companies to improve the follow-up.

To increase the student experience, students must attend at least two companies with problems directly related to their careers. To avoid work overload, they should assist up to three. So, it is necessary to generate groups from two to three companies with similar requirements. The number of accepted students is limited to 50 for budget reasons, allowing the generation of 10 teams because of the multidisciplinary restriction. Making entry limited to 30 companies because of the companies' group restriction.

This investigation originated due to dropout students [11]. The analysis showed that the causes of abandonment were the traveled distances to assist the companies, the incompatibility between the hours of attention, and studied careers with the requirements of the companies, making an assignment problem with three criteria to solve the dropouts.

Other areas present similar multicriteria cases with a dropout problem. Some have concluded that it is because most solutions to the human resource allocation problem do not consider the needs of the personnel because they are treated as a one-time resource acquisition problem, being one of the reasons why the distance between worker and company has not been regarded an important attribute to consider. Some have proposed the creation of clusters to improve this situation for both parties [12].

Although the distance and dropout problem has not been fully addressed in the assignment, selection, or recruitment problem, it has been in the home office analysis versus face-to-face work. In particular, the perception [13] and stress [14] generated by different types of transport on the worker's productivity have been studied, analyzing whether the home office is a viable solution to reduce road congestion and long transport distances [15] while improving flexibility, effectiveness, and environmental problems as possible indirect benefits [16]. Due to the world's status, some of these benefits have been proven despite the context [17].

2.1 Comparison with Variants of the Assignment Problem

The social service problem in terms of the assignment problem consists in associating n students to m companies $(n > m)$, with c_{ij} costs defined by the location, schedule, and career-requirement characteristics. Because of the multidisciplinary restriction of students' teams (divergence) and similarity requirements for groups of companies (convergence), the assignment is a variation of a many-to-many assignment problem. In its general form, it is simplified as a team-to-group assignment problem subject to restrictions of divergence and convergence. In this case, tasks are considered to start and end simultaneously and not to be performed in a particular order.

In particular, the group assignment problem is considered a variant of AP, named the semi-assignment problem [18]. While in the classical problem, agents and tasks are assumed as unique, in the semi-assignment problem, tasks can be repeated. Furthermore, the model considers n agents and m task groups with identical jobs, with the number of groups less than the number of agents $(m < n)$. Another variation of the semi-assignment problem considers the qualification of the agents, creating k teams of agents which share the same category and perform an identical task linked to a group. This consideration simplifies from a many-to-many problem to an AP of one-to-one between teams and groups. This variation could be applied in our context; however, companies grouping will have similar requirements but not identical, and the team generation requires divergence, causing the separation of the qualified assignment from the semi-assignment problem for the relationship between our teams and groups to solve both the career and grouping problems.

The distance must be estimated between agents to form teams, between tasks to form groups, and between teams-groups to solve the grouping problem from the location perspective. This agent-to-agent, task-to-task, and agent-to-task analysis are found in the quadratic assignment problem (QAP), which consists in selecting n facilities through m possible sites $(m \geq n)$, under the consideration of flow f_{ip} between facilities and distance d_{jq} between locations, requiring a flow matrix from facility-to-facility, a distance matrix from site-to-site, and the related analysis between facilities and locations. Three distance matrices are needed to apply the QAP to the consultancies problem, one for the companies, another for students, and the last for the relationship between entities. Groups require closeness to reduce travel distance, teams to increase their efficiency, and both to increase their response time and reduce idle time.

The general assignment problem (GAP) was considered a viable solution perspective. It seeks to minimize the cost of assignment c_{ij} between n tasks and m agents so that each task can be assigned to more than one agent while recognizing the agent's capacity to perform each task and its limitation [19]. With this definition, the student's capacity can be defined under the distance criterion, thereby limiting the maximum trip that can be made for each agent or limiting the number of companies to which it can be assigned.

However, none of the above models can solve the integrated allocation problem by themselves. Most require a balanced structure with univariable conditions, so applying a multicriteria or multidimensional perspective is necessary to generate a viable solution algorithm.

2.2 Multi-criteria Assignment Problems

Unlike the previous univariate problems, the multi-criteria assignments have multiple decision criteria, and finding a valid solution to all of them is more important. There are usually two general structures for decision-making in these cases:

- Combined multicriteria. The different objective functions $(f_1, f_2, \ldots, f_n)$ are joined in a single function (F) capable of representing all of them. Usually, this is simplified by a classification of objectives with relative weights. However, it's common to find cases of this problem with combinations of other variants of the AP [20] or related to the multidimensional problem [21].
- Sequential multicriteria. It consists of prioritizing the objectives to generate a set of the best solutions to obtain an ordered list of viable solutions allowing decision-making based on different scenarios. Volgenant [22] presents the multi-objective lexicographic problem, where he considers the optimization of a first objective, then as long as there is capacity, he will optimize the second, and so on.

Another functional definition to analyze multiple attributes is the multidimensional assignment problem (MAP), which extends the scope of the problem, representing different entities as new dimensions, for example, the assignment of agents, tasks, and machines (3D), or students, teachers, courses, and schedules (4D). In these types of problems, representations differ depending on the context, but their solutions can be represented as combined multicriteria or by a stepwise solution of assignments [23].

3 Some Algorithms to Solve Our Multi-criteria Assignment Problem and the Simulation for Optimization

For our problem, multidimensionality is a viable conceptualization of the attributes, while a sequential multicriteria structure allows its optimization. Different algorithms and methodologies were adapted and simulated with the criteria and data from the original problem, proving functional for the problem: assigning students to companies with specific characteristics.

Although the structure solved the problem, the simulation uncovered a secondary problem in the context affecting its application viability. None of the methods were able to consider the rotation, incompatibility, and complexity associated with the management of data and user procedures, so the methods were redefined with a focus on graphical methods to solve the new problem in search of a flexible structure to reduce time and simplify it for the user [11].

Some methodologies such as linear programming (I), the Hungarian method (II), coloring and figure coding (III), cluster decomposition (IV), municipal segregation (V), database programming (VI), segregated database programming (VIII), and database segregation by coloration and graphs (VIII) were conceptualized in programmable systems, under a simple comparison proposal[1], to verify if the solutions met the characteristics of their context, using four classifications, determined as Adequate (3), achievable/acceptable (2), inadequate (1) and without contribution (0) to qualify nine desired characteristics of the system:

1. Capacity. Amount of processable data.
2. Scheduling. Ease of assessing schedule matching.
3. Career-requirement. Ease of assessing the coincidence between careers-requirements.
4. Distance. The system's capacity to evaluate the distance between two or more points.
5. Accessibility. The system's capacity to assess accessibility between two or more points.
6. Repeatability. The system's capacity to recalculate a new iteration.
7. Data entry. Easy user data management.
8. Interface. Easy to understand by the user.
9. Time. Required time to run the system.

Table 1. Comparative evaluation of methods.

	I	II	III	IV	V	VI	VII	VIII
1	2	2	2	2	3	3	3	3
2	2	1	3	2	1	2	2	2
3	2	1	3	2	1	2	2	2
4	1	1	2	2	2	3	3	3
5	1	1	1	1	1	3	3	3
6	0	0	1	1	3	3	3	3
7	0	0	1	1	1	3	3	3
8	0	1	2	1	1	1	2	2
9	0	1	1	1	1	0	0	1
T	8	8	16	13	14	20	21	22

The evaluations and results allowed a broader view of the original problem, building each new iteration with characteristics from previous methods. Table 1 resumes the improvements in each methodology to meet the institution's criteria.

A hexagonal division was implemented from the graphical analysis and clustering to solve the problem of distance and location. The information scaling is allowed by under-sizing the territory uniformly, simplifying the evaluation of attributes. To solve the repetitive use, database structures were added, which allowed the management of other criteria under a simple coding system. The hexagonal tessellation assignment methodology (HTAM) builds a database of the physical space, containing and unifying the information in symmetric geometric figures capable of supporting repetitive sets that facilitate management.

---

[1] The author and the institution created the algorithms, characteristics, and qualifications to compare the solutions in the context of the problem presented in Sect. 2.

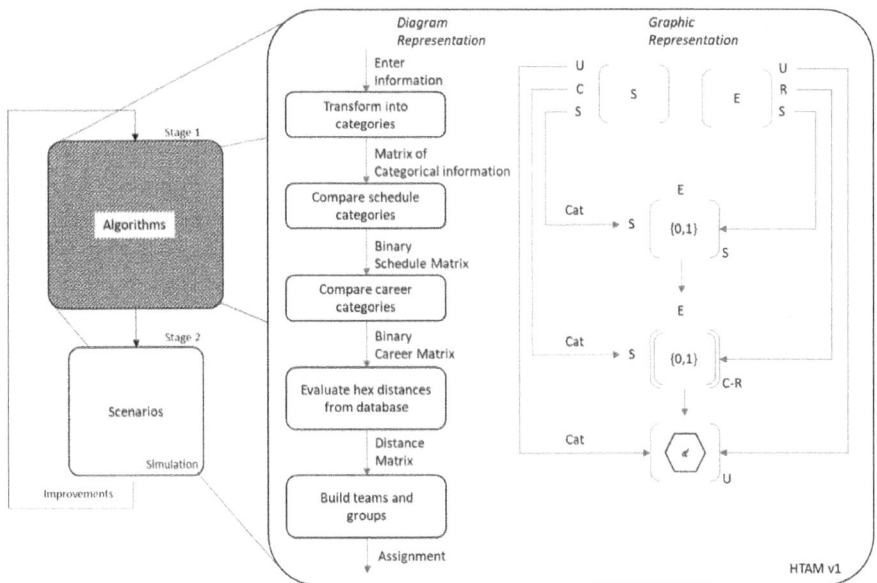

Fig. 2. Simplified diagram for the grouping problem and simulation in the hexagonal tessellation assignment methodology.

We simplify our optimization process into two stages (Fig. 2). The first is the application of the methodology to obtain results, and the second is to simulate scenarios to define the capacity and resilience of the system to improve. The first iteration of the system can be considered a modular sequential multicriteria solution. It transforms the information of students and companies into categorical variables to simplify the process. The schedules are divided into three categories, the careers into ten, and the requirements according to their most relevant career. The system processes the matrix of categorical information to generate a binary matrix of schedules and refine it to a binary matrix of matrices of career-requirements. The universe of possible solutions is reduced with each binary matrix, starting with the lowest level category to reduce the subsequent processing requirements of the system. The locations are contained in hexagonal cells and processed by a database that determines the hexagonal distance between cells, simplifying the assignment problem to one represented by distances.

A simulation model was created to determine the capacity and resilience of the structure, designing different scenarios with modified criteria to define its usual behavior and solution threshold. The first iteration applied to this simulation was developed to mimic the general logic behind the coordination assignment methodology, which later became the test methodology that made possible a comparison of results between a one-attribute assignment method and the hexagonal-based allocation proposal.

The simulation and test methodology used potential populations obtained from data extraction due to the absence of historical information from the program. The student population distribution employed statistical data on the number of admissions to the university. Each career distribution was defined using 2,421 lines of statistical data obtained with web mining related to the last 17 years of admissions. Job boards were sampled on consultancies' job applications to represent the behavior and requirements of the companies, extracting 2,843 lines of data from 395 jobs advertised by 216 different companies.

The test methodology system feeds on distributions to create a population of 1,000 students with ID, free-label, location, schedule, and specific career, from which a non-repetitive random sample is taken as a general list of students, simulating the interview process. Next, teams are created using a counter, which randomly selects the first member of the general list and uses it as a pivot to select the rest of the members, considering the divergence restriction for careers and updating their free label as occupied. When the team is complete, it continues with the next one, taking a new random free element from the general list and repeating the process until all students are marked with an occupied label.

For the companies' group generator, the system creates a population of 400 with free-label, location, schedule, and requirements. First, a non-repetitive selection determines the general list of companies. Next, an affinity matrix evaluates every categorized requirement of every pair of elements and rearranges it in an affinity list for each company. From this point, a procedure like the teams' generator is followed, taking a random element as a pivot and generating groups based on their affinity list.

A new iterative affinity matrix that evaluates the career-requirement attribute is created for groups and teams, qualifying the similarities found in three possible cases:

- Initial relation: Initial state of the matrix.
- Preferential relation: Identifies the assignment with more similarities. It cannot pass to a permanent state until all relations are evaluated.
- Permanent relation: Identifies a perfect affinity between teams-groups or the assignment with more similarities between both sets. The team is divided if there is a low degree of similarity, and the members are reassigned to meet the requirements of the group.

A perfect affinity assumes that all companies in a group have identical categorized requirements. The system compares the group's total requirements list with the team's characteristics to avoid a false match generated by a divergent group.

4   Results

4.1   Original Context

Data from one semester was applied to the methodologies to evaluate the solutions. In addition, the coordination assignment solution was included to find out if there was any improvement.

With a sample of 21 companies and 34 students, the coordination obtained a total unidirectional Euclidean distance of 1,645.9 km, reached a career-requirement match of 48.57%, and achieved 68.57% in schedules considering that the mixed shift matches any shift. The test methodology reduced the distance despite ignoring the location attribute and increased the coincidences between schedules and career-requirements. However, since it takes the next divergent or convergent element to the first, a reshuffle of the same data (R. Test) increased the total distance with losses associated with the matches for the other attributes, making it dependent on the order of arrival. As a result of its programming, HTAM eliminated poor schedule assignments, increased career-requirement matches, and reduced overall distance. Furthermore, taking advantage of the modular multicriteria characteristics of the HTAM, the system was configured to increase the bounded universe and find solutions with reduced distance, maintaining acceptable results in the other attributes (Table 2).

Table 2. Methodologies results.

Methodology	Distance (km)	Career Match (%)	Schedules Match (%)
Coordination	1,645.9	48.57	68.57
Test	1,300.13	94.29	78.1
R. Test	1,383.16	71.43	82.85
HTAM	1,351.62	96.19	100
HTAM (distance)	1,060.98	85.71	94.29

4.2 Scenario Results

Three different scenarios, with ten simulations each, were generated to evaluate the test methodology against HTAM using the same data series belonging to the simulation. The data presents a mismatch error, showing real situations where there are not enough elements with the required characteristics to satisfy a perfect assignment, restricting the maximum matches any method could achieve. For the schedule attribute, the extracted data is adjusted to companies with standard schedules, reducing the comparison between methods.

- The first scenario considers a ratio of 50 students and 30 companies to assess the capabilities of the methods in the original structure.
- The second scenario doubles the values of both populations to a ratio of 100 students and 60 companies, thus increasing the attribute differences between them to determine if they presented errors when processing more significant amounts of data or if their design was limited to the original quantities.
- The third scenario was designed to evaluate the search capacity of the systems in unbalanced spaces, generating 150 students and 30 companies, thus increasing the possibility of finding elements with matching attributes.

In terms of context, the first and second simulations are a closed design that does not consider the possibility of interviews, while the third allows open entry of other students (Fig. 3).

On average, HTAM presented better results for all simulations. However, two unusual results were found in the third simulation of the first scenario and the first simulation of the second. After evaluating these cases, it was verified that the test methodology

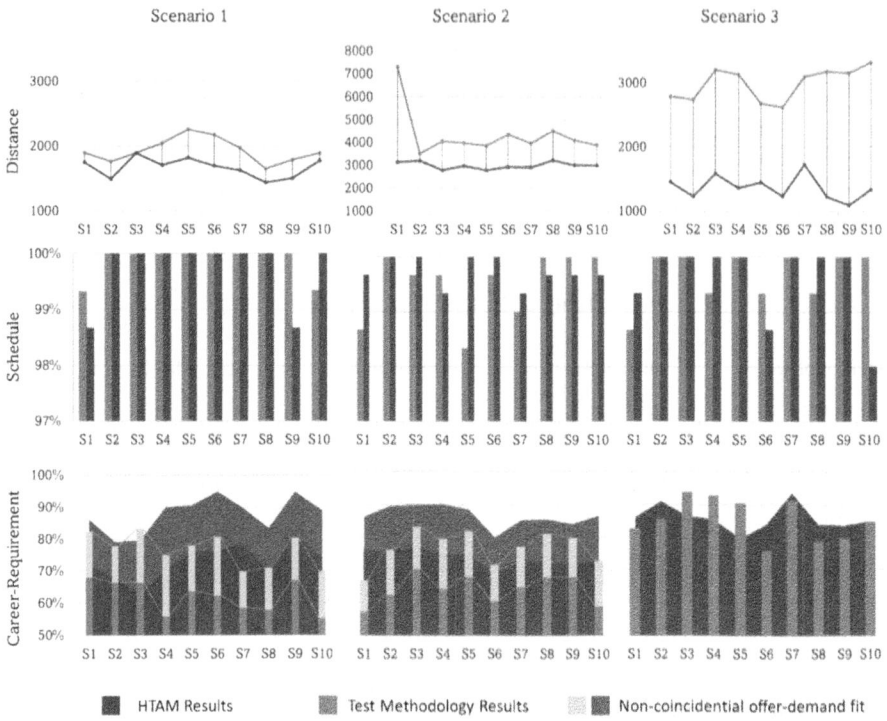

Fig. 3. Scenarios results.

depends entirely on the data order. For the first case, the test methodology obtained a better match in careers and a minimum distance difference compared to HTAM, but in the second, it worsened considerably.

Eliminating non-coincidence through an open offer in the third simulation allowed the methodologies to choose the best elements according to their criteria. Career matching while avoiding the location problem is the test methodology's primary objective, while for the HTAM is the balance of attributes.

5 Conclusions

Problems cannot be completely extracted from their context. Fixed algorithms are inapplicable to different environments without adjusting them on every iteration, requiring more flexibility for their application in different scenarios. We were able to analyze aspects of the problem through various methods, allowing the reduction in execution times and adaptation of a solution in its context.

Our case shows that applying methodologies to generate a result without considering its context can create indirect problems. For example, for the human resources assignment problem, the lack of interest in the characteristics or needs of agents can reflect a lack of interest in the job. If left unattended, it can lead to abandonment and its implications.

While simplifying the problem serves to conceptualize it and obtain an initial result by understanding its fundamentals, its optimization should not only consider achieving better results or adjusting a univariate system. Its context and scenarios must be studied to obtain better results on different attributes and create an integral solution. The simulation allows the preliminary evaluation of algorithms to be improved. Multidimensionality is a way to consider multiple characteristics, separating them initially and understanding their connections to define a better solution.

The system's capabilities continue to be assessed and improved in different aspects. For example, while the initial configuration of the problem considered 21 companies and 34 students, we designed scenarios two and three to evaluate the methodologies responses in larger populations. In addition, a branch of research aimed at larger populations with associated combinatorics led to the application of genetic algorithms, which aim to explore the space of solutions without the need to determine every combination or limit the methodology to rigid solutions.

Although we did not use any optimization algorithm to adapt the MATH, an empirical improvement was obtained from observing the different solutions and their conceptual definition. This observation also led us to categorize variables for each attribute to facilitate their understanding and processing. For example, the careers were originally categorical, the requirements were classified by their affinity to specific careers, the distances using the hexagonal tessellation, and the university defined the schedules to adhere to the standard partition in Mexico. These definitions and modules have allowed the application of new algorithms that alter the sequence and processing of information.

Currently, we divided the assignment and grouping problems to define each solution's scope. This branch aims to define concretely and formally the processing of the different methodologies. Practical research has led to applying genetic algorithms, geographical coordinates, cluster evaluation for auto-generation, and scaling of tessellations, granting the capacity of chaining different attributes to a cell by conceptualizing it as a multidimensionality-based tessellation of planes for modulation and connection of information. The purpose of this system is no longer oriented to solve the original problem. Instead, it continues developing its implementation as a general form in problems, areas, or situations that require a solution with more criteria.

References

1. Edmonds, J.: Maximum Matching and a Polyhedron with 0,1-Vertices. J. Res. Nation. Bureau of Standards, 125–130 (1965)
2. Wang, Y., Zhou, C., Zhou, Z.: The shortest time assignment problem and its improved algorithm. In: from International Conference on Computer Engineering and Networks. Singapore (2020)
3. Hervert, L., López, F., Esquivel, O.: Integrated appoach to assignment, scheduling and routting problems in sales territory bussiness plan. The Int. Conf. Computat. Sci. **80**, 1887–1896 (2016)
4. Votaw, D., Orden, A.: The personnel assignment problem. In: Symposium on Linear Inequalities and Programming, pp. 155–163 (1953)
5. Kuhn, H.: The Hungarian Method for the Assignment Problem. Naval Research Logistics Quarterly, 83–97 (1955)
6. Caron, G., Hansen, P., Jaumard, B.: The assignment problem with seniority and job priority constraints. Oper. Res. **47**, 449–454 (1999)

7. Villagómez, R.O.: Decisions on the allocation of staff to work schedules using linear programming: the case of a company call in Mexico. Investigación Administrativam **44**(115), 1–22 (2015)
8. Naveh, Y., Richter, Y., Altshuler, Y., Gresh, D.L., Connors, D.: Workforce optimization: Identification and assignment of professional workers using constraint programming. IBM J. Res. Dev. **51**, 263–279 (2007)
9. Widianta, M., Rizaldi, T., Setyohadi, D., Riskiawan, H.: Comparison of Multi-Criteria Decision Support Methods (AHP, TOPSIS, SAW & PROMENTHEE) for Employee Placement. Journal of Physics: The 2nd International Joint Conference on Science and Technology, vol. 953, pp. 1–5 (2018)
10. Leite, M., Baptista, A.J., Ribeiro, A.M.: A trap of optimizing skills use when allocating human resources to a multiple projects environment. Team Performance Management **23**(3), 110–123 (2017)
11. Rios Esparza, G.A., Alonso Ventura, P.: Optimization of personnel allocation between support groups and companies. DGOSE UNAM case, Mexico City (2016)
12. Maw-Shin, H., Yung-Lung, L., Lin, F.-J.: The impact of industrial clusters on human resource and firms' performance. J. Model. Manag. **9**(2), 141–159 (2014)
13. Brutus, S., Javadian, R., Panaccio, A.J.: Cycling, car, or public transit: a study of stress and mood upon arrival at work. Int. J. Workp. Health Manage. **10**(1), 13–24 (2017)
14. McLennan, P., Bennets, M.: The journey to work: A descritptive UK case study. Facilities **21**(7), 180–187 (2003)
15. Tunyaplin, S., Lunce, S., Maniam, B.: The new generation office environment: the home office. Ind. Manag. Data Syst. **94**(4), 178–183 (1998)
16. Pérez-Pérez, M., Martínez-Sánchez, A., de Luis-Carnicer, M.P., Vela-Jimenez, M.J.: The environmental impacts of teleworking. Manage. Environ. Q. **15**(6), 656–671 (2004)
17. Yang, E., Kim, Y., Hong, S.: Does working from home work? Experience of working from home and the value of hybrid workplace post-COVID-19. J. Corp. Real Estate, vol. Ahead of print (2021)
18. Kennington, J., Wang, Z.: A shortest augmenting path algorithm for the semi-assignment problem. Oper. Res. **40**(1), 178–187 (1992)
19. Ross, G., Soland, R.M.: A branch and bound algorithm for the generalized assignment problem. Math. Program. **8**, 91–103 (1975)
20. Ross, T.G., Zoltners, A.A.: Weighted assignment models and their application. Manage. Sci. **25**(7), 686–696 (1979)
21. Lee, S.M., Schniederjans, M.J., Cole, J.P.: A multicriteria assignment problem: a goal programming approach. Interfaces **13**(4), 75–81 (1993)
22. Volgenant, A.: Solving some lexicographic multi-objective combinatorial problems, European J. Operat. Res. **139**, 578–584 (2002)
23. Pierskalla, W.P.: The Multidimensional Assignment Problem. Oper. Res. **16**(2), 422–431 (1968)

Applying Simulation System to Compare the Use of Electric and Diesel Truck Fleets

Irina Yatskiv (Jackiva)[(✉)], Jurijs Tolujevs, Vladimirs Petrovs, and Aleksejs Vesjolijs

Transport and Telecommunication Institute, Lomonosova 1, Riga 1019, LV, Latvia
{Jackiva.I,Tolujevs.J,Petrovs.V,Vesjolijs.A}@tsi.lv

Abstract. Computer simulation is one of the methods for investigation of processes in transport systems. Due to the emergence of electric vehicles, new tasks related to the charging stations placement can be solved with its help. This task became especially relevant when electric goods vehicles appeared on the transportation market. The paper devoted on the development and application of a simulation model to analyse the process of regional transportation, which is carried out using electric and diesel trucks of similar carrying capacity. The model was created using the TraPodSim simulation system, recently developed by the authors of this paper. The core of the TraPodSim system is a universal multi-agent model implemented using the AnyLogic package. The model presented in the paper uses the real geography of the transport network and the real technical parameters of electric and diesel trucks.

Keywords: Cargo Transportation · Electric Trucks · Multi-agent Simulation · Indicators

1 Introduction: Motivation and Research Actuality

Computer simulation is one of the methods for investigation transportation processes. Due to the emergence of electric vehicles, new tasks related to their using can be solved with simulation modelling help. For instance, the task of optimal placement of charging stations became especially relevant when electric goods vehicles (eTrucks) appeared on the transportation market. eTrucks can consume up to 1000 kWh of energy per day even for intracity and regional transportation.

Research [19] describes theoretical foundations for building the TraPodSim simulation system and the main properties of this system. The core of the system is a universal multi-agent model implemented using the AnyLogic package. The main purpose of the work is to test the possibility of using the TraPodSim simulation system to solve practical problems of analysing the use cases of electric or diesel trucks. The idea was not to demonstrate simulation based on numerical data related to a specific business process but to discuss the possibilities of the developed tool, which allows considering the most various features of a particular case of electric trucks, and as the model results provide the user with the maximum possible set of indicators of the transportation process. Legend was developed based on the live geography of the transport network and the fleet of real electric and diesel trucks to demonstrate this possibility.

© The Author(s), under exclusive license to Springer Nature Switzerland AG 2024
M. Mujica Mota and P. Scala (Eds.): EUROSIM 2023, CCIS 2033, pp. 199–214, 2024.
https://doi.org/10.1007/978-3-031-68438-8_15

The review of simulation examples for transport research in [19] highlights that most of the examples are of the use of trucks up to 26 tons for intracity freight transportations [4, 13]. Very often, the electric truck does the entire trip with a battery that was only charged once before leaving the depot. Charging stations' sharing options at the depots of partner carriers are discussed in [15] including placing individual fast charging stations in the service area. Use of medium capacity electric and diesel trucks (from 7.5 to 28 tons) to transport goods between regional cities pair 30–50 km apart is one of the features of the simulation example described in this paper. The distance of the selected day route is 491.6 km. The average energy consumption of the considered electric vehicles class is 1.1 kWh/km, and the consumption for the entire daily route is 540.8 kWh. Battery capacity for the simulated electric trucks is 336 kWh, so each of them needs at least 200 kWh of additional energy to complete a daily transportation plan. The example below considers two ways to perform battery charging for an electric truck: a) at loading/unloading points that are part of the route, and b) at separate charging stations that are owned by the carrier and can be used without wasting time to wait a free spot.

The second feature of the proposed simulation scenario is the use of fuel consumption per unit of cargo as one of the technical parameters of the simulated trucks. For electric trucks, this parameter is kWh/kg/km, and for diesel trucks it is l/kg/km. Its precise numerical values are almost never found in open sources. The usual parameter, which is calculated by dividing the spent energy by the total mass of truck and the distance travelled, has the same dimension, but it does not reflect the effect of the load on fuel consumption. A 10 ton truck would use significantly more energy if it were to carry 16 tons of cargo instead of empty. This is especially true for routes with frequent stops at intersections, when a lot of energy is wasted due to accelerating the truck from zero to cruise speed. This parameter has never been used in models known to the authors. It caused due to obtaining numerical values of this parameter, and it is necessary to measure it under various driving conditions of an electric vehicle, which in practice is too rarely done. However, in the case of regular transportation on a permanent route, it is advisable to make such measurements. In the frame of the ePIcenter project [12] in which the authors participate, the company that produces electric trucks provided a list of technical parameters of a vehicle, among which was this parameter, but not its numerical value. The article shows the fundamental possibility of considering this parameter when modelling the transportation process, which makes this model innovative and useful for real business.

Next section considers new publications on the topic of modelling the road transportation processes, which are not presented in [19]. In connection with the task of comparing the use of electric and diesel trucks of medium capacity, this section provides references to sources about the prospects of using this type of vehicles. The section pays special attention to the above problem of taking into account the effect of the load on fuel consumption. Section 3 provides a brief description of the developed TraPodSim system. Sections 4 and 5 describe an example of using this system to compare the performance of electric and diesel trucks when performing the same tasks for transportation and the simulation results, respectively. The last part concludes on the results and possible further research.

2   Current Facts in Relevant Areas

Progress in the Field of Simulation of Road Transportation Processes. Work is in progress on the creation of not only models, but also new software systems for the simulation of the processes of using of BEVs (battery electric vehicles). For instance, [5] describes the development of a tool for simulating the daily journey of a car that considers real route parameters and the driving style. The route is divided into segments corresponding to the acceleration, deceleration, and constant speed and moreover the movement is also taken into account on the ascent and descent sections. Shape and size of BEV data is also used as model parameters. Route is set using the Google Maps application. The model is used to predict the driving range of BEV given all the parameters listed above. With its help, it's possible to predict the remaining energy in the battery at the end of the day and schedule its recharging. The model is analytical, allowing to estimate energy consumption indicators based on the inputs of the model, but the model is not dynamic and does not show the movement of BEV OVER TIME.

Review [7] provides an overview of specific tools for urban simulation, and research papers on corresponding issues. The authors studied 37 urban simulation platforms and noted that 41% of them are focused on traffic simulation. [6] is noted as the most significant work in this area. The tool orientation for modelling only some specific processes in the city (for example, traffic), the authors consider it a disadvantage and justify the need to create integrated platforms for urban simulation. An example of such platform is described in [9]. The models created with its help are dynamic simulation models that are implemented programmatically using the open source package SUMO. Road network is modelled by calling the OpenStreetMap service. One of the layers of the model is called "emissions", and with its help all emissions created by traffic are determined. To model the uncertainty the authors describe the random parameters of the model using empirical distributions or a theoretical Gaussian distribution.

Electric Trucks Application Prospects and Their Comparison with Diesels. A national laboratory of the U.S. Department of Energy (NREL) in their report [14] paid a lot of attention to the prospects for the eTrucks (electric trucks) emergence in the transportation market. A significant increase in sales through 2050 forecasted for BEVs that are Light-Medium and Medium in terms of payload, i.e. with a Gross Vehicle Weight < 26,000 lbs (< 11.8 tons). They are classified as short-haul according to the range of transportation, i.e., with a range of < 499 miles (< 803 km). Forecasts related to the Heavy and long-haul classes indicate that in 2050 another 20% of the entire fleet will be accounted for by ICEV (internal combustion engine vehicle), and this group will consume more than half of the total energy attributable to road transport. The economics of using BEVs by 2030 will roughly match those of ICEVs for almost all Light-Medium and Medium truck classes. For trucks of the Heavy class, by 2035 this situation will only be achieved when they are used for short-haul. An extensive analysis of all economic aspects of introducing Medium and Heavy-Duty BEVs is also given in the Roush Industries report [16]. Unlike NREL [14], the authors of this report believe that the economics of using BEVs will equal those of ICEVs as early as 2027.

Scania company [17] conducted an in-depth comparative analysis of the emissions size observed at the stages of production, use, maintenance, and recovery of ICEV and

BEV vehicles. Two trucks with a Gross Vehicle Weight of 28 tons were selected to improve the accuracy of the comparison. It is assumed that such trucks can be success-fully used in urban and regional delivery cycles. The following statement was made as a result of the analysis: "Even though the impact from production phase is almost doubled for the BEV compared to the ICEV, it is the use phase that is the clearly dominant phase for both ICEV and BEV. In the base line scenario (EU baseline), the BEV can reduce the total life cycle GHG emissions with 38% in comparison to ICEV" [17].

Evaluation of Fuel Consumption Per Unit of Cargo. Recently, Finnish experts have published in [3] the results of a study on fuel consumption and CO_2 emissions for timber trucking. direct fuel consumption measurements were taken for 13 diesel trucks under various driving conditions (distance, weather conditions and gross weight of truck). One of the indicators of fuel consumption has the dimension $l(t \cdot km)^{-1}$, i.e. when calculating it, the influence of both the distance travelled and the total mass of the truck is taken into account. A large number of measured values for this indicator were processed statistically. For example, for trucks with a Gross Vehicle Weight (GVW) of 68 tons driving when loaded, an average value of $0.012\ l(t \cdot km)^{-1}$ was obtained. It should be noted that this indicator considers only the total weight of a truck, and it cannot be used to calculate the impact on fuel consumption of the transported cargo weight.

The mentioned above task results are reflected in [8]. Figure 1 shows the regression lines obtained for diesel trucks with engine power of 340 ps and 380 ps. These lines show the decrease in fuel efficiency, measured in mpg (miles per gallon), as the weight of payload increases, and this indicator can be converted to l/100 km. For example, for a truck type 340, when driving without a load, the value is 21.9 l/100 km, and when transporting 10 tons of cargo – 24.5 l/100 km. Of course, numerical data on Fig. 1 cannot be directly used for practical calculations, since many other factors affect fuel consumption, and not just the payload value. But it is possible to obtain empirical data for each specific transport process and construct a group of diagrams like shown in Fig. 1.

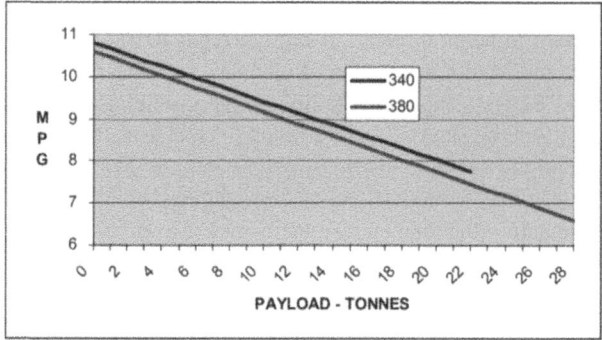

Fig. 1. The dependence of the productivity of a gallon of fuel on the value of payload [8].

3 TraPodSim Simulation System Characteristics

The TraPodSim system is designed to create simulation models of the transportation process using electric or diesel trucks along the transport network routes specified for a specific region [19]. This system includes two models created using the AnyLogic package (Fig. 2):

- an auxiliary model based on a GIS map (AnyLogic GIS Model) for placing geographic points and selecting sections of real roads that will be included in the simulated transport network;
- the main executable model (AnyLogic Main Model) based on the graphical copy of the GIS-map obtained using the auxiliary model.

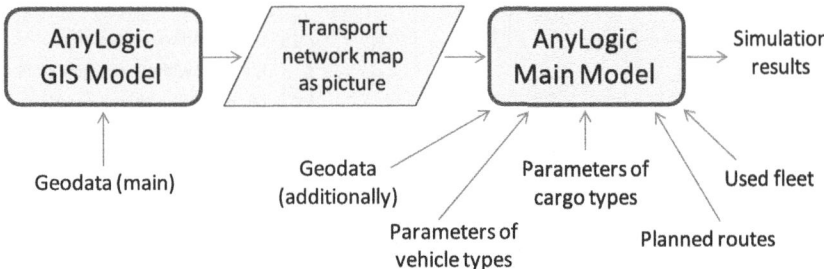

Fig. 2. Structure of the TraPodSim simulation system.

The system is built in a way that there is no need to change the program codes in the original GIS and Main models during specific model creation. Users must perform certain actions in the AnyLogic environment only when working with the GIS Model, when they define the transport network configuration by accessing the OpenStreetMap service via the Internet. In the Main Model, instead of a GIS map, its graphical copy is used in the form of a picture stored as a PNG file. This approach ensures the repeatability of simulation experiments and reduces the model processing time. To store the input and output data of the model, the database of the AnyLogic program is used in the form of a set of MS Excel tables. Most of these tables are presented below when describing an example of using the TraPodSim system to model a specific transportation system. It is possible to use animation to observe the movement of vehicles in the transport network when processing the model. Planned routes for all simulated vehicles can be created using any external program such as Route Planning. A simplified version of a program with corresponding functionality is created using the VBA programming language, and it is also part of the TraPodSim system.

The challenge often arises of choosing the optimal location for charging stations when modelling an eTrucks fleet. eTrucks batteries can be charged either while performing operations at the loading/unloading point (L/U point) or by visiting separate Charging stations. By generating Geodata (additionally), the user of the TraPodSim system can implement any option for placing charging stations with a focus on the methods shown in Fig. 3. The meaning of each method is shown in the figure.

The user of the TraPodSim system could prepare and conduct experiments with the created model, which can be divided into two groups of experiments:

- without changing the location of charging or filling stations;
- to determine the optimal number and location of separate Charging stations.

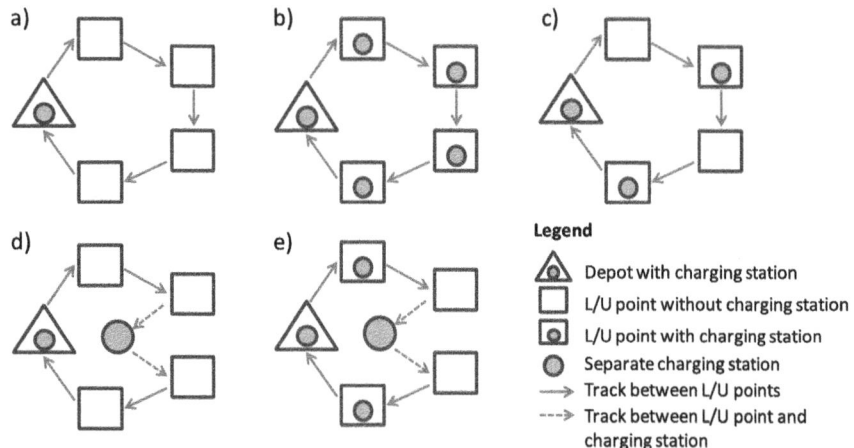

Fig. 3. Five Ways to Place Charging Stations: a) the charging station is only available at the depot; b) there is a charging station at each L/U point; c) charging stations are placed only in some L/U points; d) the carrier itself equips one or more high-capacity charging stations in the region in which it will serve customers; e) case is a combination of cases c and d.

Experiments are prepared only with the help of the Route Planning program in the first case. Result is data of the Planed routes and Used fleet types. In the second case, the user must change some graphic elements of the AnyLogic Main Model, which are labelled as Geodata (additionally) on Fig. 2.

4   Developed Model Description

The purpose of the simulation is to compare medium-duty electric and diesel trucks (7.5 to 28 tons) for transporting goods between cities, the distance between which is between 30 and 50 km. The structure of the transport network, the modelled route, and the characteristics of the cargo are fixed initial data, allowing for the reduction of the spread of the simulation results for various scenarios and making the differences between them more contrasting. These data do not reflect a live business process but belong to a class of synthetic data that adequately describes the transportation type selected for analysis using simulation. Experiments' feature carried out with the model is the use of a deterministic approach to simulate the transportation conditions. Same routes of movement between the main geographical points of the transport network are used in all experiments, as well as the same volumes and characteristics of the transported goods. This approach facilitates the comparison of both primary and secondary indicators obtained by modelling different scenarios.

4.1 Transportation Process Model

North Rhine-Westphalia region in Germany was selected for modelling. It includes 8 cities, marked on the map with numbers from 1 to 8 (Fig. 4). There is a depot in Düsseldorf where all trips within one working day start and end. It is assumed that the cargo can be transported between any cities, excluding Düsseldorf. This means that cities numbered 2 to 8 have L/U points. Cities (1–8) are the main points of the transport network. Numbers 9 to 12 refer to additional locations that may have high-capacity charging stations or filling stations for diesel vehicles. The paths shown in red in Fig. 4 are obtained using AnyLogic GIS Model, which accesses the OpenStreetMaps service. Map was transferred as a PNG file to AnyLogic Main Model in accordance with the TraPodSim modelling system application described above.

The model uses usual square matrix of distances between the main points (1–8), and a special table with the distances between the main and additional points (Table 1). This decision has a reason behind it. The square matrix of distances is created in AnyLogic GIS Model and remains unchanged while working with the model. A table with distances between main and additional points is created in the AnyLogic Main Model, and users can change it when creating\eliminating additional points during experiments.

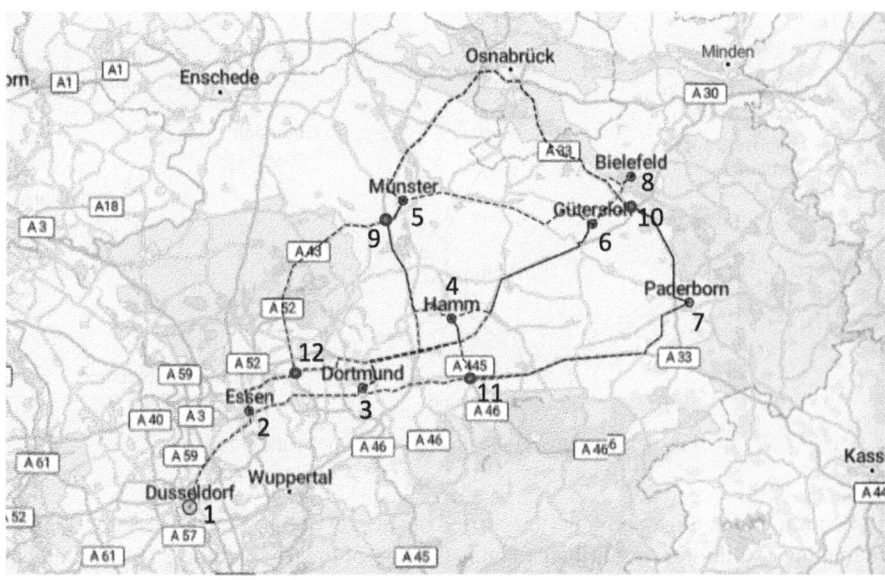

Fig. 4. Transport network created with OpenStreetMaps.

The structure of the transport network is shown as a graph on Fig. 5 where thick lines correspond to direct links between cities. This means, for example, one can drive from Dortmund to Münster without visiting Hamm. The arrows indicate the daily route followed by each vehicle leaving Düsseldorf. The code of the main route, which is implemented in all experiments with the model, has the form: 1-2-3-4-5-6-7-8-6-4-3-2-1. Deviations from the route may occur when the vehicle enters one of the additional points (9–12) to charge the battery or refuel. The length of the main route is 491.6 km.

Table 1. Distances between the main and additional points of the transport network.

to from	9	10	11	12
Düsseldorf	111.27	175.38	98.32	52.27
Essen	77.00	141.10	64.04	17.99
Dortmund	57.03	106.52	29.46	29.89
Hamm	37.47	64.40	15.89	56.15
Münster	5.26	69.07	63.19	65.51
Gütersloh	63.48	10.84	69.46	109.71
Paderborn	110.56	31.29	68.99	133.71
Bielefeld	80.43	9.94	86.40	126.66

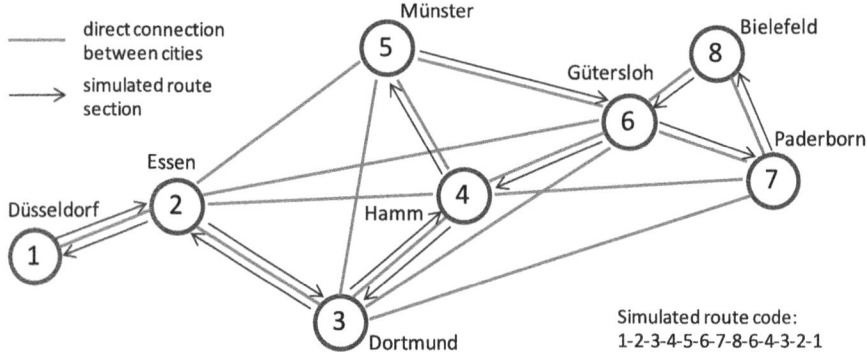

Fig. 5. Simulated route on a transport network graph.

Loading/unloading operations are carried out at every point except Düsseldorf. Only the loading operation is performed at the beginning of the trip in point 2 (Essen), and at the end of the trip, the unloading operation is performed. Both operations are performed in other points. Each vehicle on every section of the route carries 8 or 16 tons of cargo depending on the carrying capacity. Vehicles average speed was 72 km/h in the model experiments described below.

4.2 Vehicle and Cargo Parameters

The eTrucks eActros and eEconic (both Mercedes-Benz) (see Table 2) were selected for modelling. While the eEconic has so far been planned as a dedicated garbage truck, it may have other uses in the future. Diesel trucks Actros 1835 and Actros 2543 (both Mercedes-Benz) (Table 3) were chosen as analogues against eTrucks. The main parameter, which is approximately the same for both pairs, is Loading Capacity (tons). The number of axles is also similar: two axles (4 × 2) for eActros and Actros 1835, three axles (6 × 2) for eEconic and Actros 2543.

Table 2. eTrucks' parameters.

Vehicle type name	Loading capacity (pallets), number	Loading capacity (weight) tons	Battery capacity, kWh	Energy consumption (empty), kWh/km	Energy consumption (per weight unit), kWh/kg/km	Energy consumption (idling), kW	Charging speed, kWh/h
eActros	18	10.6	336	0.95	0.00001048	1	161
eEconic	20	17.8	336	1	0.00001048	1	161

Table 3. Diesels' parameters.

Vehicle type **Actros**	Loading capacity pallets, number	Loading capacity (weight), tons	Fuel tank volume, litres	Fuel consumption (empty), l/km	Fuel consumption (per weight unit), l/kg/km	Fuel consumption (idling), l/min	Refueling time, min
1835	18	11.4	380	0.22	0.00000262	0	15
2543	20	16.5	380	0.24	0.00000262	0	15

The first three parameters for both groups of trucks can be found in Mercedes-Benz reference materials [1, 2, 10, 11]. The Energy or Fuel Consumption (empty) parameter, i.e. consumption when driving without a load, was determined by approximate calculation using reference data, which usually refers to driving with an average load.

Fuel Consumption (per weight unit) parameter value for diesel trucks value can be derived from the diagram shown in Fig. 1. Fuel efficiency decreased from 10.75 mpg to 9.6 mpg when the load increased from 0 to 10 tons. This data corresponds to a change in fuel consumption from 21.88 l/100 km to 24.5 l/100 km. This means that with an increase in load by 1 kg, an increase in fuel consumption per 1 km of track occurs, which is determined by the formula:

$$(24.5 - 21.88)/10/1000/100 = 2.62 * 10^{-6} = 0.00000262 [l/kg/km]. \quad (1)$$

Since the value of the Energy Consumption (per weight unit) parameter for eTrucks could not be found in sources, it was considered that the numerical value of electricity consumption in kWh/km is approximately 4 times greater than l/km (refer to the data shown in Tables 2 and 3: 1 kWh/km and 0.24 l/km). On this basis, the value:

$$0.00000262 * 4 = 0.00001048 [kWh/kg/km]. \quad (2)$$

The Charging Speed value for eTrucks refers to the linear portion of the battery charge curve (from 20 to 80% capacity) and was derived from the manufacturer's data (battery capacity 336 kWh and charging time - (20–80%) 1 h 15 min) using next formula:

$$(0.8 * 336 - -0.2 * 336)/1.25 = 161.28\,[kWh/h]. \tag{3}$$

Charge level was brought to 50% in the experiments described below with the model during battery charge in L/U points, and when charging at an external station, up to 80%. The Refuelling Time parameter for diesel trucks has a symbolic meaning, since with a tank capacity of 380 L, trucks can travel at least 1200 km. This means that with a route length of 491.6 km there is no need for additional refuelling.

Any number of cargo types can be described in the model (Table 4). In the experiments only TCU1 cargo was transported with a fixed weight of cargo unit - 500 kg.

Table 4. Cargo parameters.

Cargo type name	Cargo type	Minimum weight, kg/unit	Maximum weight, kg/unit	Average weight, kg/unit	Average loading time, sec/unit	Average unloading time, sec/unit
TCU1	general	500	500	500	30	30
TCU2	pallet	100	500	300	60	60

In all experiments, a daily transportation plan was set, according to which 32 cargo units, i.e. 16 tons of cargo, had to be taken out of each L/U point. This cargo had to be left at L/U point, which is the next one along the route. It was assumed that trucks with a lower carrying capacity would take on board exactly 16 cargo units, i.e. 8 tons of cargo, and trucks with a larger carrying capacity - all 32 cargo units, i.e. 16 tons of cargo. It follows from this that the daily transportation plan can be carried out using two trucks with a lower carrying capacity or one truck with a larger carrying capacity.

4.3 Description of Simulated Scenarios

Figure 6 shows a diagram of the formation of six scenarios for the transportation of goods, the simulation results of which should be compared with each other.

Two options for the location of charging stations were used for modelling transportation using eTrucks, which in Fig. 3 are labelled Case b and Case d. These options correspond to two ways of implementing charging, shown in Fig. 6.

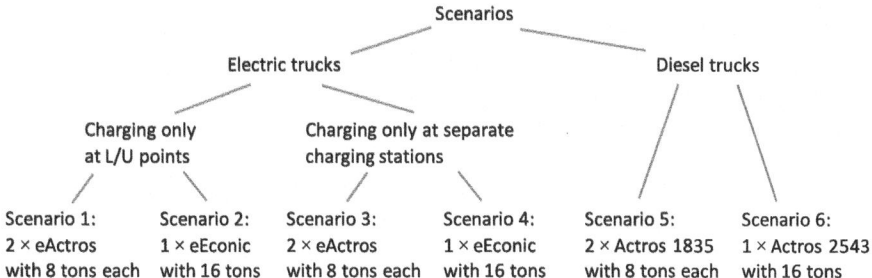

Fig. 6. Six scenarios for cargo transportation.

5 Results of Scenario Modelling

The structure of the transport network, the modelled route and the characteristics of the cargo are fixed initial data, allowing to reduce the spread of the simulation results for various scenarios and make the differences between them more contrasting. These data do not reflect a real business process but belong to a class of synthetic data that adequately describe the transportation type selected for analysis using simulation.

Figure 7 shows the eTrucks driving diagrams for scenarios 1 to 4. Although two eActros trucks are used in scenarios 1 and 3, they both follow the same routes, so the diagram only shows this trip once. The start time of each trip is chosen so that the route is clearly visible on the diagram. Routes for scenarios 1 and 2 can also be seen in Fig. 8. On Fig. 7 scenarios 3 and 4 assume the use of external charging stations, which are indicated by numbers 9–12 on the route map (Fig. 3). In scenario 3 simulation, each eActros used charging stations numbered 10 and 12. In scenario 4 simulation, one eEconic used charging stations numbered 10 and 11. In Fig. 7 and 8, it's possible to visually estimate the execution time of route each operation and its duration as a whole, but the exact times can be seen in the tables below with numerical simulation results.

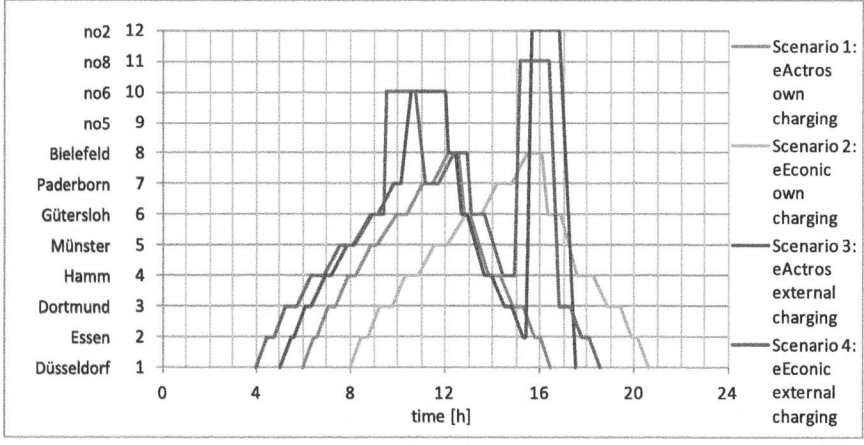

Fig. 7. eTrucks traffic patterns for scenarios 1, 2, 3, and 4.

Figure 8 shows diagrams for both eTrucks and Diesel trucks movement. Scenarios 1 and 2 are chosen without the use of external charging stations for eTrucks, so that the routes of all trips are the same. The charts clearly show that the Diesel truck Actros 1835 travels almost 1 h faster than the competitor, while the Diesel truck Actros 2543 is only 0.5 h faster. This is because a large load increases the loading/unloading time, during which the eTruck manages to recharge the battery up to 50% of its capacity, and therefore it needs only a little extra time to finalise re-charge.

Numerical simulation results in the form of the standard output of the TraPodSim simulation system are shown in Table 5 [19]. For scenarios 1, 3, and 5, numbers are shown that are the sum of the trip results of the same trucks. The value of the "Volume of freight traffic" indicator will always be 6768.89 tons*km if the truck has not made a trip with cargo to the charging station. In Scenario 3, both eActros made these trips, but they used alternative routes to the charging stations, resulting in a total "Drive with cargo" trip by some kilometres shorter. Exactly the same effect is observed in scenario 4, where eEconic made two trips to charging stations with cargo on board, so the value of the "Volume of freight traffic" indicator decreased slightly.

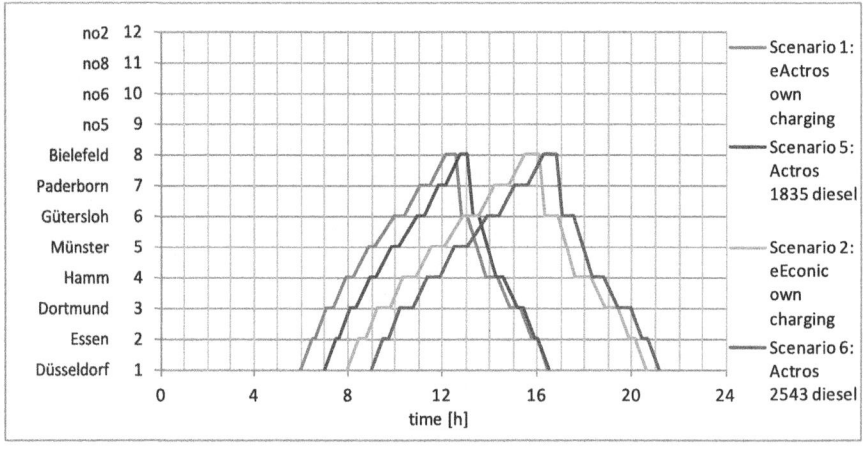

Fig. 8. Comparison of driving time eTrucks and Diesel trucks.

If "Total trip duration" and "Energy or Fuel expended" are considered the main physical indicators of the transportation process, then scenarios 2 and 4 are the best when using eTrucks, and scenario 6 when using Diesel trucks. When comparing scenario 1 (charging only at L/U points) with scenario 3 (charging only at separate charging stations), we observe that the "Energy or Fuel expended" values are almost the same, but the value of the "Total trip duration" indicator increases by 4 h when using scenario 3. The general conclusion is also obvious: one truck with a larger load capacity in all studied scenarios performs better than two trucks with a load capacity of half that.

There are many regulations and calculators on the Internet for calculating CO_2 emissions. For example, the calculator [17] offers the following data:

- production of 1 kWh of electricity in the EU creates 347 g of CO_2 (average);

- combustion of 1 L of diesel fuel creates an average of 2.63 kg of CO_2.

The price of one kWh at a high-capacity charging station in Germany was 79 Ct/kWh at the beginning of December 2022, while the price of one liter of diesel was 1.85 EUR/l. The Table 6 was created based on these data, which shows that the smallest amount of CO_2 emissions is expected in scenarios 2 and 4 (Charging only at L/U points, 1 × eEconic), while scenario 6 is the least costly (1 × Actros 2543).

Table 5. Standard Output of Six Scenario Simulation Results.

Scenario	Route lenght, km	Volume of transported cargo, tons	Volume of freight traffic, tons*km	Total trip duration, h	Empty drive, h	Empty drive, kWh	Drive with cargo, h	Drive with cargo, kWh
eTrucks								
1	983.21	160	6768.89	20.92	1.90	130.25	11.75	874.74
2	491.61	160	6768.89	12.62	0.95	68.55	5.88	493.99
3	1044.55	160	6683.74	24.99	2.90	198.63	11.60	863.74
4	489.89	160	6741.50	14.52	0.95	68.56	5.85	491.99
Diesel								
5	983.21	160	6768.89	18.29	1.90	30.16	11.75	203.88
6	491.61	160	6768.89	12.16	0.95	16.45	5.88	119.27

Table 5. (continued).

Scenario	L/U time, h	L/U time, kWh	Charging duration, h	Number of charging	Initial energy, kWh	Energy added, kWh	Energy expended, kWh	Residual energy, kWh
eTrucks								
1	5.33	5.33	3.74	14	672	603.87	1004.99	270.88
2	5.33	5.33	2.24	8	336	360.27	562.54	133.72
3	5.33	5.33	5.15	4	672	828.65	1062.36	438.29
4	5.33	5.33	2.39	2	336	384.29	560.54	159.75
Diesel								
Scenario	L/U time, h	L/U time, litres	Refueling duration, h	Number of refueling	Initial fuel, litres	Fuel added, litres	Fuel expended, litres	Residual fuel, litres
5	5.33	0	0	0	760	0	234.04	525.96
6	5.33	0	0	0	380	0	135.72	244.28

Table 6. Calculation of CO2 emissions end costs for simulated scenarios.

Scenario	Energy expended, kWh	Fuel expended, litres	CO2 emissions, kg/kWh	CO2 emissions, kg/l	CO2 emissions per Scenario, kg	Energy price, EUR/kWh	Diesel price, EUR/l	Energy/Diesel costs per Scenario, EUR
1	1004.99		0.347		348.73	0.79		793.94
2	562.54		0.347		195.20	0.79		444.41
3	1062.36		0.347		368.64	0.79		839.27
4	560.54		0.347		194.51	0.79		442.83
5		234.04		2.63	615.53		1.85	432.98
6		135.72		2.63	356.94		1.85	251.08

6 Conclusion

As a result of deterministic data application on the conditions of daily transportation application (route, volume of cargo, average speed) it was possible to identify significant differences between the main indicators of the processes specified by the corresponding scenarios. Only some of the conditions could be made random variables in the case of applying the stochastic approach. But the route should be saved, otherwise the common base for comparing scenarios would be completely lost. The result of comparing the indicators of several completely independent random processes is unlikely to be of practical importance. If as a result of statistical modelling in the considered example, a confidence interval is determined for the indicators, for example, ±10% of the average, then the qualitative result of the comparison of scenarios will remain the same. It should be noted that the factor of subjectivity in the case of statistical modelling would be more significant, since all the deterministic parameters used in the model are absolute, and the nature of the distribution of random parameters in the absence of reliable statistical data, the model developer will simply have to assume.

Use of real data as input parameters of the model is a feature of the described modelling example, which includes: (1) configuration of the transport network with exact distances between the main geographical points; (2) technical parameters of all types of trucks used in the planned scenarios.

Fixed load size for each truck was chosen to improve the accuracy of scenario comparison equal to 8 or 16 tons, depending on it carrying capacity. Randomness automatically appears in scenarios in which the eTruck must select an external charging station, since it is not known in advance at what time and in what place this event can occur.

The main result of the work is confirmed possibility to apply the TraPodSim simulation system to solve practical problems of analysing options for using electric or diesel trucks. Such tasks can be solved for any geographic region, any type of transportation demand and any fleet of electric or diesel trucks. The users can compare any options for the placement of charging stations both in L/U points and in other geographical locations. The CO_2 emission and economic indicators can be expanded and converted to annual figures if user of the model has sufficient information about the business process within which the transportation is carried out.

The TraPodSim system is a versatile but non-trivial non-commercial product [19]. This means only the developers are ready to provide such a service to any interested party and implement new task as quickly and accurately as possible. Possible limitations of the system usage may arise only due to the lack of initial data necessary for the full functioning of the model.

Acknowledgements. This work has been supported by the "Enhanced Physical Internet-Compatible Earth-frieNdly freight Transportation answER (ePIcenter)" project (https://epicenter project.eu/) and has been funded from the European Union's Horizon 2020 research and innovation programme under grant agreement No 861584.

References

1. Actros-1835: https://www.mercedes-benz-trucks.com/de_DE/buy/mercedes-benz-truck-experience/verteilerverkehr/actros-5-1835-l-4x2-pritschenaufbau.html. Last accessed 28 December 2022
2. Actros-2543: https://www.mercedes-benz-trucks.com/de_DE/buy/mercedes-benz-truck-exp erience/verteilerverkehr/actros-5-2543-l-6x2-schwenkwandaufbau.html. Last accessed 28 December 2022
3. Anttila, P., et al.: Effect of vehicle properties and driving environment on fuel consumption and CO2 emissions of timber trucking based on data from fleet management system. Transport. Res. Interdiscip. Perspect. **15**, 15 (2022)
4. AnyLogic Electric Vehicle Route Optimization, https://www.anylogic.com/resources/case-studies/electric-vehicle-route-optimization-delivery-with-simulation-software. Last accessed 28 December 2022
5. Armenta-Déu, C., Cattin, E.: Real driving range in electric vehicles: influence on fuel consumption and carbon emissions. World Electr. Veh. J. **12**, 166 (2021)
6. Bragard, Q., Ventresque, A., Murphy, L.: Self-balancing decentralized distributed platform for urban traffic simulation. IEEE Trans. Intell. Transp. Syst. **18**, 1190–1197 (2017)
7. Chang, Y.T., Pal, A., Hackl, J., Hsieh, S.H.: The needs and trends of urban simulation platforms – a review. In: 38th International Symposium on Automation and Robotics in Construction (ISARC 2021), pp. 122–128 (2021)
8. Coyle, M.: Effects of Payload on the Fuel Consumption of Trucks. UK (2007)
9. Dong, S., Ma, M., Feng, L.: A smart city simulation platform with uncertainty. In: ACM/IEEE 12th International Conference on Cyber-Physical Systems (ICCPS '21), May 19–21, 2021, pp. 229–230. Nashville, TN, USA. ACM, New York, NY, USA (2021)
10. eActros, https://www.mercedes-benz-trucks.com/de_DE/emobility/world/our-offer/eactros-and-services.html#root/content/headline_489846305. Last accessed 28 December 2022
11. eEconic: https://special.mercedes-benz-trucks.com/fileadmin/user_upload/Documents/eEc onic/E_Produktblatt_eEconic_ENG_screen.pdf. Last accessed 28 December 2022
12. ePIcenter: ePIcenter consortium, 'Enhanced Physical Internet-Compatible Earth-frieNdly freight Transportation answER (ePIcenter). EU Horizon 2020 RAI programme, grant agreement No 861584' (2021). https://epicenterproject.eu/. Last accessed 28 December 2022
13. Lebeau, P., Macharis, C., Van Mierlo, J., Maes, G.: Implementing electric vehicles in urban distribution: A discrete event simulation. World Electric Vehicle Journal **6**(1), 38–47 (2013)

14. Ledna, C., Muratori, M., Yip, A., Jadun, P., Hoehne, C.: Decarbonizing Medium- & Heavy-Duty On-Road Vehicles: Zero-Emission Vehicles Cost Analysis. NREL, US (2022)
15. Martins-Turner, K., Grahle, A., Nagel, K., Göhlich, D.: Electrification of urban freight transport - a case study of the food retailing industry. Procedia Comp. Sc. **170**, 757–763 (2020)
16. Nair, V., Stone, S., Rogers, G., Pillai, S.: Medium and Heavy-Duty Electrification Costs Evaluation. Final Report. Roush Industries Inc, USA (2022)
17. Scania: https://www.scania.com/content/dam/group/press-and-media/press-releases/docume nts/Scania-Life-cycle-assessment-of-distribution-vehicles.pdf. Last accessed 28 December 2022
18. SunEarthTools: https://www.sunearthtools.com/tools/CO2-emissions-calculator.php. Last accessed 28 December 2022
19. Yatskiv (Jackiva), I., Tolujevs, J., Petrovs, V., Vesjolijs, A.: A modelling system for evaluating options for building and using a fleet of battery electric trucks. Transport and Telecomm. J. **23**(4), 334–343 (2022)

A Causal STAM Model to Increase Airspace Network Capacity

Gonzalo Martin and Miquel Angel Piera[✉]

Universitat Autònoma de Barcelona, Barcelona, Catalonia, Spain
{Gonzalo.Martin.Lopez,MiquelAngel.Piera}@uab.cat

Abstract. ATM digitalization paves the way for new efficient solutions that overcomes present inefficiencies caused by spatial fragmented airspace. Lack of airspace capacity is an important factor that impacts on a sustainable and efficient air transport system with aggravated indicators in future growing demand scenarios. Despite research on new ATM digitalized services to improve airspace capacity, the mitigation of latent capacity by enhancing synergies among adjacent sectors has not been addressed yet. In this paper, spatio-temporal sector interdependencies are analyzed to quantify the topological interdependencies and evaluate the increment of capacity that can be achieved in those sectors that cannot fit the dynamic demand requirements. A sector network model has been implemented formalizing the ATC sectors as network nodes, and traffic flows at different levels of granularity as time-stamp perishable edges. The dynamic evolution of the occupancy in adjacent sectors together with the inverse correlation between saturated sectors, paves de way for a Short Term ATM Mechanism to improve the capacity invulnerability at sector level while at the same time provides a mechanism to improve the airspace capacity at network level.

Keywords: Demand-Capacity Balance · sector occupancy · early handover · STAM

1 Introduction

Air Traffic Management (ATM) is formally defined as the dynamic, integrated management of air traffic and airspace safely, economically and efficiently, through the provision of facilities and seamless services in collaboration with all parties and involving airborne and ground-based functions [3]. Among the different factors with a negative impact on the ATM KPI's, lack of airspace capacity is considered an important penalty to airspace users that are forced sometimes to important delays due to ground regulations, or fly less efficient alternative routes. The expected increment on air traffic demand would worse considerably present capacity problem.

In order to coordinate all the resources necessary to make different aircraft operations, shared airspace and airports compatible, in a safe, efficient and orderly manner, the ATM services are grouped in 3 main functionalities: Airspace Management (ASM), Air Traffic Flow and Capacity Management (ATFCM) and Air Traffic Services (ATS). It should be

© The Author(s), under exclusive license to Springer Nature Switzerland AG 2024
M. Mujica Mota and P. Scala (Eds.): EUROSIM 2023, CCIS 2033, pp. 215–226, 2024.
https://doi.org/10.1007/978-3-031-68438-8_16

noted that these 3 groups of services are structured considering the time gap between the decisions and actions to be provided with respect to the operational context when the flights will take place. Thus, ASM services plan at a strategic level the organization and structure of sectors with a certain level of flexibility to be responsive to short-term changes. The ATFCM services provides functionalities to balance the traffic demand with the capacity of system resources (ie. ATCo's). Finally, ATS implement the services to ensure all tactical activities supporting safety and efficiency of air traffic flights, and it consists of Air Traffic Control (ATC) service, Flight Information Service (FIS) and Alerting Services (ALSs).

The demand-capacity balance (DCB) of ATFCM [1] services is performed considering the planned traffic to restructure, group or split the sectors and managing available personnel to adapt to the sector's capacity for the required traffic load by opening additional sectors.

Worthwhile to note that the evolution of traffic demand is quite dynamic and the fragmentation between strategic and tactical decision support systems cannot absorb efficiently the differences between the planned and the flown traffic, causing inefficiencies such as latent capacity or ground regulations.

To mitigate the barriers between strategic and tactical fragmented decision supporting tools, ATFCM services has been extended with a set of services grouped in STAM (Advanced short-term ATFCM measures) trying to offer a more agile mechanism between the Local Traffic Manager (LTM) and the Nework Manager (NM).

1.1 STAM Measures

DCB service implemented by ATFCM rely mainly on planned traffic. In Europe, the Central Flow Management Unit (CFMU) operates the Enhanced Tactical Flow Management System (ETFMS) to monitor current situation and the traffic demand. Computer Assisted Slot Allocation (CASA) is used to detect control sector overloads and to propose and assess regulations with two possible degrees of freedom: postponing takeoff and re-routing. The model proposed in this paper could be used to extend the CASA software by extended the control actions to a third degree of freedom by balancing adjacent sectors with an Early Hand Over mechanism.

Dynamic DCB (dDCB) service is triggered on the day of operation to manage at real-time imbalance situations that use to arise due to discrepancies between planned traffic and the traffic at the day of operations [14]. Below are summarized the main STAM measures to preserve network stability by slight refinements on both flights in execution phase and on ground to address residual problems of limited magnitude:

- Capacity Management: The main STAM action to increase the capacity rely on the service "Airspace Volume Configuration" which can trigger a dynamic capacity tuning by a short-notice configuration. This service is mainly oriented to act on en-route sectors and it can be used to dynamically adjust (both increase or decrease) the sector capacity. Thus, a capacity adjustment could be achieved considering the available resources (ATCo's), weather predictions and incoming traffic flow.
- Tactical Rerouting: At tactical level, ATCo's can provide instructions to divert traffic from a saturated sector, or at preflight level by the Network Manager suggesting alternative routes to those flights subject to network delays.

- Level Capping: Overloaded sectors can be tackled by triggering a flight level limitation to shift excessive flights to less loaded sectors. This dDCB service can be applied at certain flights or at flow level, but AU's should be informed before taking-off.
- Application of Minimum Departure Intervals: This dDCB service is applied to on-ground aircraft by assigning departure time intervals which are implemented by the ATC TWR. This service is one of the first actions considered in STAM measures to decrease ATCo's workload at TMA sectors.
- Miles in Trail: En-route radar air traffic controllers can trigger this service by issuing to individual aircraft flying at the same flight level a clearance to maintain the same speed/Mach number to avoid separation minima infringements among them. This service is considered a dDCB since it reduces the ATC workload.
- SID STAM: A slight change in the planned Standard Instrument Departure Route (SID) triggered by the ATC TWR can contribute to offload an overloaded departure sector, avoiding in this way to reduce capacity by ground regulation.
- Slot Swapping: This STAM service provides support to ATCFM with the identification, assessment and request of eligible flights for slot swapping to reduce the impact of delays on airspace user operations. Thus, when a ground aircraft predicts a problem to finalize its turnaround on time, this service provides the opportunity to swap the slot with another AU through a cooperative process with the Network Manager (NM).

Despite the benefits of these STAM services, there are several aspects such as the unbalanced capacity among adjacent sectors that is still generating latent capacity and requires a much more agile mechanism to tackle the gap between the sector planned entry time and the sector predicted entry time once the flight is en-route.

In this paper a new mechanism is presented using a causal model to offload an overloaded sector by an "early handover" procedure, avoiding the mentioned inefficiencies caused by re-routing or ground delays. The mechanism implemented opens the door for an opportunistic synergy among adjacent sectors increasing the overall capacity of the sector network.

In Sect. 2, research on STAM measures is presented, while in Sect. 3 a new STAM mechanism is described to benefit from sector latent capacity. Section 4 describes the methodology and results achieved. Section 5 summarizes the conclusions.

2 State of the Art

A critical aspect to trigger a STAM service is to detect those airspace regions where an imbalance between demand and capacity can be predicted, or where the ATC could be overloaded due to the complexity of traffic. Usually these airspace regions with complex traffic are known as hotspots.

Despite the importance to predict air traffic complex areas, it is recognized that air traffic complexity or hotspots are still nowadays an open research question with no consensus to what constitutes a complex traffic scenario in the context of ATM.

Some authors [12] associate air traffic complexity with the workload of controllers, however, this approach rise several issues about how to measure ATCo's workload with respect to traffic. A recent paper [7], illustrates a new approach based on the analysis of spatio-temporal interdependencies among aircraft in a given area which overcomes shortages of traditional complexity indicators relying on density, or dynamic density [15] considering the complexity to manage traffic in evolution. Some other works try to define a more subjective measure of ATC workload by monitoring the cognitive activity of ATCo's by analyzing psychophysiological data, such as Electro-Encephalogram (EEG) and/or Electro-Cardiogram signals [6]. A more recent alternative to evaluate the workload of human operators, is the use of socio-technological simulation models [13] that rely on Hierarchical Task Analysis [2] approach to formalize the different actions and tasks an ATC must perform according to traffic evolution.

A recent work [11] introduces the concept of Dynamic Airspace Configurations (DAC) which is a STAM alternative to restructuring airspace [4]. In [5, 10] the Dynamic Airspace Sectorization (DAS) is well described as a STAM measure, considering Flight Centric, providing the required flexibility for a sector to be re-sized, reconfigured and reassigned as traffic conditions change in real time. Finally, in [9] a Sectorless STAM measure is also evaluated considering the tactical and strategic management of aircraft trajectories.

Despite the benefits that these STAM measures provides to compensate DCB shortages caused by discrepancies between planned and flown trajectories, some of these methods requires the coordination of different actors through the integration of multiple systems and corrective actions should be anticipated to the overcapacity by means of predictive models subject also to different sources of uncertainty.

3 Network Resilience

A local, agile, efficient, effective and tactical STAM measure should take benefit of the real time information of present digitalized airspace, such as the location of the different aircraft in the sector. This paper introduces a mechanism to compute the dynamic evolution of the sector occupancy, and avoids overcapacity coordinating the the entry of a new aircraft with the "virtual" exit of another aircraft.

The implementation of such a control mechanism requires a trajectory predictor to compute the entry time of the aircraft in adjacent sectors heading to the sector under study, but also to compute the exit time of the aircraft already in the sector. Speed regulation delaying the entry time of one of the candidates aircraft coordinated with speed regulation for an early exit could be easily implemented for a particular pair of entry/exit aircraft, however this mechanism rises some drawbacks:

- Performance: Airspace users sets the cruise speed according to the cost index of the company. Any change in the speed with respect to cost index will have a negative impact on the fuel consumption and in consequence in the cost.
- Delays: The delay applied to the entry time of an aircraft will be propagated to the downstream sectors and to the TMA. This delay could be mitigated by speeding in the sector under study, but again with a negative impact on the performance.
- Scale: In case the sector overload is not caused just by 1 aircraft, rather is the consequence of a traffic increment, slowing down several aircraft while speeding up other aircraft can rise several issues on safety and performance. Furthermore an increment on the ATCo workload would be expected to coordinate all these speed change instructions.

As it can be observed, the 4 functional requirements mentioned would fail with this mechanism:

- Local: Delays applied to an entry aircraft are downstream propagated to subsequent sectors.
- Agile: It requires de coordination of several pilots, each one with a different reaction delay to the ATC instruction.
- Efficient: Any speed change with respect to airline cost-index will introduce a cost penalty.
- Effective: Speeding up or slowing down an aircraft can spark conflicts that would require ATC instructions such a heading, with an impact on the sector occupancy

3.1 Early Hand-Over Mechanism

To overcame those drawbacks, it has been implemented a simulator with an early handover mechanism that fulfill the 4 requirements. An ATC handoff occurs when an executive controller transfers an aircraft to next executive controller along the aircraft's route. Under nominal conditions (Fig. 1 left hand side), the"Transfer of control" occurs once the aircraft crosses the boundary (blue area) between sectors, and the receiving controller is in complete control of the flight.

Figure 1 (right hand side) illustrates two adjacent sectors, but with Sector_i at the limit of its capacity, while the adjacent Sector_k has a low occupancy. Aircraft with red colour is used to identify the next entry aircraft at sector_i which will increase the sector occupancy above the declared capacity. The Early Handover procedure provides a STAM mechanism to virtually transfer the control of an aircraft in Sector_i to the ATCo at the adjacent sector (Sector_k) if 2 preconditions are satisfied:

- The control of an Aircraft can be transferred to the control of the adjacent Sector if it is conflict free.
- The Adjacent sector occupancy is bellow the capacity and the controllers can accept the extra workload for the earliness time-window duration.

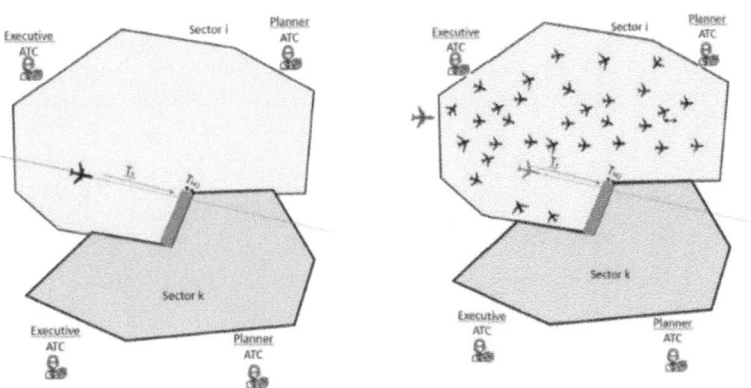

Fig. 1. HandOver and Early Handover mechanism

As it can be observed, the application of an early handover preserves the 4 STAM requirements: local, agile, effective and efficient when applied to individual aircraft, however this mechanism rises several open questions if scale-up: how to compute the maximum increment of traffic that could be absorbed by adjacent sectors.

3.2 Causal Model for Spatio-Temporal Interdependencies Among Adjacent Sectors.

Adjacent sectors could support DCB by implementing cherry picking techniques over a small number of flights on an overloaded sector to meet capacity constraints while avoiding to apply ground regulation measures to the entire traffic flow. However, it is essential to estimate the latent capacity of adjacent sectors to avoid a deadlock when flows moves among both sectors in opposite directions.

To illustrate the STAM mechanism and its sensitivity to the spatio-temporal sector interdependencies, Fig. 2 shows a simple simulation exercise representing the evolution of the occupancy indicators of 2 adjacent sectors in which the full flow of aircraft entering one sector (Sector_i) feeds the adjacent sector (Sector_k).

Figure 2 a) describe a peak increment on the occupancy in sector_i in less than 1 h, increasing from an occupancy of 20 aircraft to 30 aircraft. The average dwell time in sector_i is 20 min (t_{dw})which is used as a deterministic value in this illustrative exercise. In consequence, the observed increment on the occupancy in the adjacent sector_k is similar to the occupancy evolution in sector_i with a delay of 20 min. The same dynamic is replicated in Fig. 2 b) but it been highlighted in green the latent capacity in sector_k that could be used by the proposed early handover mechanism once the occupancy in sector_i reaches the maximum capacity (30 aircraft). Figure 2 c) and d) illustrates the same dynamics but considering a peak demand that last for 30 min. The computation of the latent capacity is sensible to several factors such as the dwell time of the own sectors with respect to adjacent sectors, the complexity of the traffic and the demand dynamics.

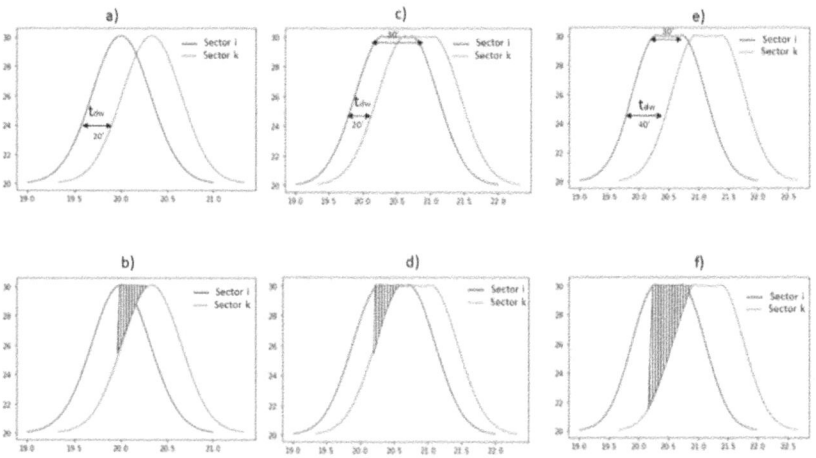

Fig. 2. Dynamic Occupancy in adjacent sectors considering different sector dwell times

Figure 2 e) and f) shows the latent capacity when the dwell time of sector_i is 40 min rather than 20 min. As it can be observed, the latent capacity of the adjacent sector increases considerably.

A simplified version of the causal model for Early Handover has been represented in Fig. 3 using the Coloured Petri Net formalism, in which the clock mechanism has not ben described for a better understanding of the dynamics. Transition "HandOver" formalizes the flow of aircraft between adjacent sectors if the entry sector occupancy is bellow its capacity. In case the occupancy is above the sector capacity the transition cannot be fired and a deadlock is forced. The alternative Transition to avoid a deadlock is to fire transition "Early Hand Over" in which the control of an aircraft is transfered to the adjacent sector if the current sector occupancy is close to the accepted capacity and the aircraft is located less than 7 min to the exit border of the sector. The place node Fl_seq provides the information for the planned flow of an aircraft through the different sectors together with the expected dwell time in each sector.

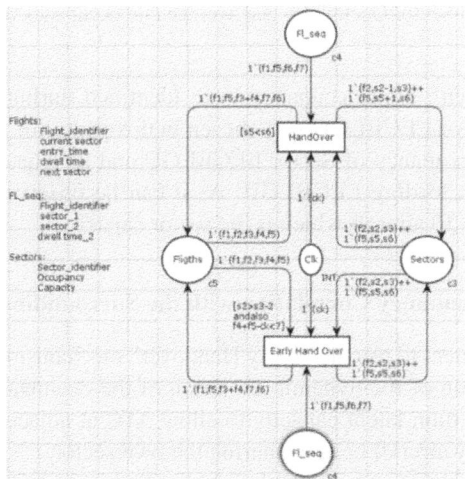

Fig. 3. Early Handover Colored Petri Net model

The model used in this paper is an extension of the model formalized in Fig. 3, in which the clock and the undertaities that affect the dwell time are specified together with a policy to avoid deadlocks.

4   Methodology and Results

The air traffic in Spain airspace during 27/6/2022 has been used to simulate the early handover mechanism in those sectors with a demand close to the declared capacity.

4.1   Identify the Sectors with Higher Occupancy

Traffic provided from "The OpenSky Network" (http://www.opensky-network.org) [16] has been used to identify the occupancy of the different sectors in Spain during 27/6/2022.

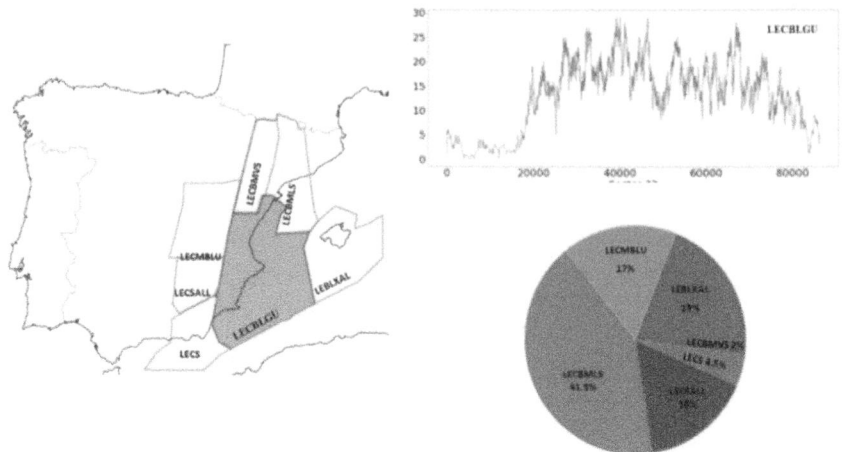

Fig. 4. LECBLGU Sector Occupancy and average rate occupancy in adjacent sectors

The few sectors with high occupancy were identified during the traffic analysis, and among them, Sector LECBLGU was chosen in this paper for illustrative purposes. Figure 4 shows the occupancy of Sector LECBLGU and its adjacent sectors, together with the rate of flows feeding LECBLGU. As it can be observed, sector occupancy reaches values around 30 aircraft which is the sector capacity.

4.2 Analyze the Occupancy Correlation with the Surrounding Sectors

An important requirement to use the "Early Handover" mechanism as a STAM measure is an inverse correlation of the dynamic evolution of the occupancy with the adjacent sectors or at least a certain latent capacity to allow ATC at adjacent sectors to assume the control of the entry aircraft before entering the own sector.

Figure 5 illustrates the occupancy of the surrounding sectors. All of them shows low occupancy, except for sector LECBMLS which has a peak value of 25 aircraft but an average value bellow 15 aircraft.

4.3 Evaluate the Dwell Time of Trajectories in the Sector Under Study

The dwell time of trajectories in LECBLGU has been analyzed considering the feeding sectors.

In Fig. 6 the dwell time in seconds for trajectories from sectors LEBLXAL, LECM-BLU, LECBMLS is represented. It is easy to observe from the histogram the different flows from the feeder sectors to the exit adjacent sectors through LECBLGU sector. Thus, for example, flows from sector LEBLVAL can be grouped in flows to LECBMLS (less than 4 min), to LECMBLU (average 13 min) and flows to LECS (average 20 min).

A Discrete Event Simulator has been implemented in python to test the early handover mechanism. Thus, rather than simulate the full traffic, the dwell time is used for each entry aircraft to predict the exit time stamp and compute the occupancy. Thus, to improve

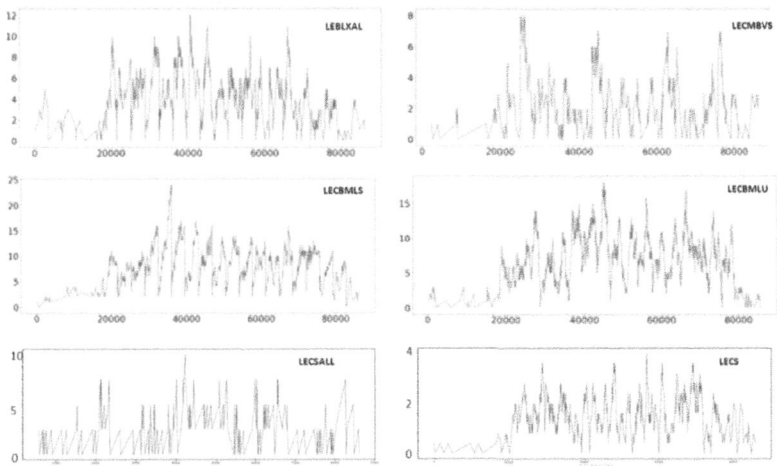

Fig. 5. Occupancy dynamic evolution in LECBLGU adjacent sectors in a nominal traffic

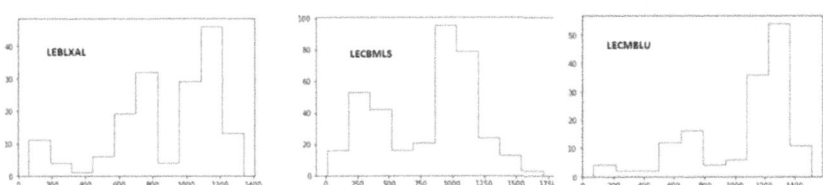

Fig. 6. Histogram of the dwell time in LECBLGU adjacent sectors: LEBLXAL, LECBMLS, LECMBLU.

the accuracy of the dwell time considering the entry and exit adjacent sectors, the data in the histograms shown in Fig. 6 has been split to obtain homogeneous samples. Figure 7 shows the Probability Density Functions used to simulate the dwell time of trajectories entering from sector LECBMLS.

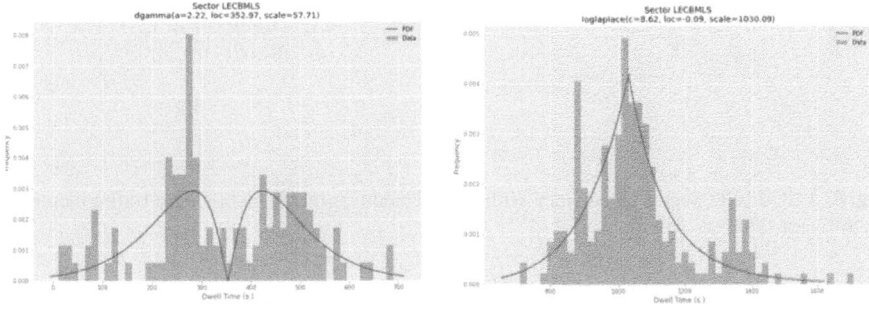

Fig. 7. Histogram and the Probability Density Functions to Simulate the Sector Dwell Time

The variability around the mean value is caused by the performance of the different aircraft, while the extreme values is caused by the entry/exit points of the sector LECBLGU.

The simulator implemented assigns an exit time-stamp to each entering aircraft by using the identified pdf's. This aircraft attribute is used by the early handover mechanism to identify those aircraft that are closer in time to the exit point. Thus, if all the chosen aircraft to maintain the occupancy below a certain threshold can be located less than 5 min to the sector boundary, the probability that these aircraft are conflict free is higher than if chosen aircraft for the early handover are located 10 min to the sector boundary.

4.4 Synthetic Traffic

To evaluate the increment of traffic that could be tackled by the Early Handover mechanism, synthetic traffic must be generated considering different requirements:

- Preserve the same flows: A change in the trajectory dwell time in sector LECBLGU would have an impact on the occupancy. Thus, preserving the same flows, ensure similar dwell time conditions.
- Preserve the same occupancy dynamics: It would not be realistic to accept an increment of traffic that affects only on the valley periods of loaded sectors.

Figure 8 a) and b) illustrates the occupancy evolution of Sector LECBLGU with a 20% increment of traffic without Early Handover mechanism active (figure a)) and with the early handover mechanism active (Figure b). As it can be observed, the peak traffic above the capacity is filtered to adjacent sectors. Similar results have been obtained with a traffic increment of 35% (FIGURE 8 c) and d)). The red line shows the sector capacity.

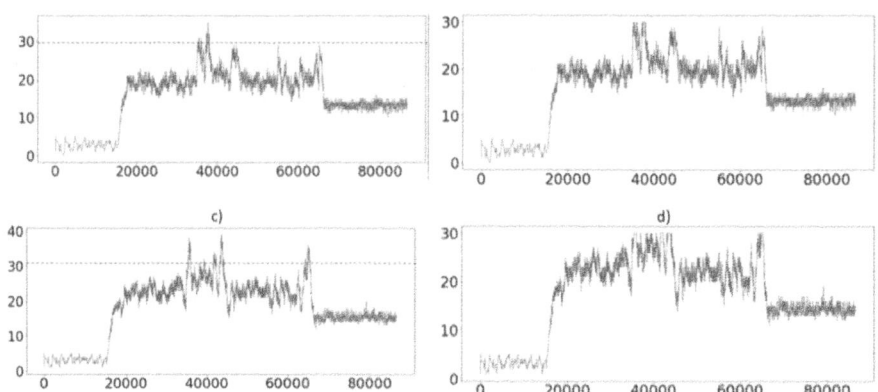

Fig. 8. LECBLGU Sector Occupancy with Early Handover mechanism with a traffic increment of 20% and 35%

Since the fast time simulation implemented in python do not consider the performance of the aircraft, the traffic scenario that required early handover has been simulated in an air traffic simulator (bluesky) to validate that chosen aircraft are conflict free. In the scenario with a 20% traffic increment, all the chosen aircraft were conflict free. Furthermore, the chosen aircraft were located less than 5 min to the sector boundary which suggest that no important impact or issues would arise on the workload of the receiving ATCo's.

In the simulation with a 35% traffic increment, 3% of the flights transferred to the adjacent sector had a separation minima infringement applying the early handover to those aircraft closest to the exit point.

5 Conclusions

Balancing sector capacities is crucial for an efficient air traffic management, but several barriers involving different actors arise in the design of an agile, local, effective and efficient tactical mechanism. In this paper a method to improve sector capacity by enhancing the synergies among adjacent sectors at network level has been described.

The occupancy correlation analysis among adjacent sectors provides a valuable information about the potential to balance the demand on saturated sectors by taking advantage of latent capacity from adjacent sectors.

The Early Handover mechanism has been applied to a particular sector with a traffic increment of 20% in the demand, however further research on traffic complexity is required to predict the maximum demand increment that can be balanced using the proposed mechanism.

The policy implemented in the simulation exercises, always chosen the aircraft that was closer to the exit point. Future work will consist to implement an algorithm to identify the maximum time threshold to the sector boundary in which all aircraft are conflict free. The time threshold will determine the pool of aircraft that are candidates for an early handover.

An alternative that is also under development is the integration of the STAM mechanism in the air traffic simulator, in order to aircraft candidates considering not only the time to the sector boundary but also the complexity of its trajectory, and prioritizing the less complex aircraft [8].

Acknowledgements. This research is partially supported by the national Spanish project: "A Multi-Agent negotiation framework for planning conflict-free U-space scenarios" (ref. PID2020-116377RB-C22). Opinions expressed in this article reflect the authors' views only.

References

1. Bertsimas, D., Patterson, S.S.: The air traffic flow management problem with enroute capacities. Oper. Res. **46**(3), 406–422 (1998)
2. Diaper, D., Stanton, N.: The Handbook of Task Analysis for Human-Computer Interaction, Boca Raton, FL. CRC Press, USA (2003)
3. Ralvi 3 ICAO. Procedures for Air Navigation and Air Traffic Management Pans-atm Doc 4444. ICAO (2016)
4. Florencia Lema-Esposto, M., et al.: Optimal Dynamic Airspace Configuration (DAC) based on State-Task Networks (STN). SESAR Innovation Days (2021)

5. Gerdes, I., et al.: Dynamic airspace sectorisation for flight-centric operations. Transportation Research Part C: Emerging Technologies **95** (2018)
6. Hernández-Sabaté, A., Yauri, J., Folch, P., Piera, M.À., Gil, D.: Recognition of the Mental Workloads of Pilots in the Cockpit Using EEG Signals. Appl. Sci. **12** (2022)
7. Isufaj, R., et al.: Spatiotemporal Graph Indicators for Air Traffic Complexity Analysis. Aerospace **8** (2021)
8. Isufaj, R., et al.: From single aircraft to communities: a neutral interpretation of air traffic complexity dynamics. Aerospace **9** (2022)
9. Korn, B., et al.: Sectorless Atm – Analysis And Simulation Results. ICAS Congress (2010)
10. Melgosa, M., et al.: Capacity Management based on the Integration of Dynamic Airspace Configuration and Flight Centric ATC solutions using Complexity. ICRAT Conference (2020)
11. Kopardekar, P., et al.: Initial Concepts for Dynamic Airspace Configuration. 7th AIAA conference (2012)
12. Pawlak, W., et al.: A framework for the evaluation of air traffic control complexity. In: Guidance, Navigation, and Control Conference (1996)
13. Piera, M.A., Muñoz, J.L., Gil, D., Martin, G., Manzano, J.: A socio-technical simulation model for the design of the future single pilot cockpit: an opportunity to improve pilot performance. In: IEEE Access **10** (2022)
14. Sesar, J.U.: Advanced short-term ATFCM measures (STAMs) (2016)
15. Laudeman, L.V., et al.: Dynamic density: An air traffic management metric. NASA (1998)
16. Schäfer, M.: Martin Strohmeier, Vincent Lenders, Ivan Martinovic, Matthias Wilhelm. Bringing up OpenSky: A large-scale ADS-B sensor network for research. ACM/IEEE International Conference on Information Processing in Sensor Networks (2014)

Supervised Machine Learning for Input Modelling of an Agent-Based Simulation Model for Autonomous On-Demand Shuttle Services

Maylin Wartenberg, Marvin Auf der Landwehr$^{(\boxtimes)}$ ⓘ, Laura H. M. Nguyen, and Christoph von Viebahn ⓘ

Hochschule Hannover, Ricklinger Stadtweg 120, 30459 Hannover, Germany
{maylin.wartenberg,marvin.auf-der-landwehr}@hs-hannover.de

Abstract. The quality of simulation-based experimentation is directly related to the estimation of its key input parameters. Yet, especially when it comes to innovative transportation concepts that are characterized by a multiplicity of influencing factors, such as autonomous on-demand transport systems, reliable and realistic input parameters are difficult to obtain. In order to tackle this challenge, we propose an automated machine learning integration to estimate contextualized parameters for an agent-based simulation model. To demonstrate the effectiveness of the proposed approach, a test scenario is conducted on the estimation of mobility patterns for an agent-based simulation model of autonomous on-demand shuttle operations. Here, the results of several simulation experiments prove the viability of the proposed input modelling approach, showing that an automated machine learning integration can generate more accurate estimations of input parameters in innovative, highly uncertain systems in an efficient manner.

Keywords: Automated Machine Learning · Computer Simulation · Agent-based Modelling · Autonomous On-demand Transport · Supervised Learning

1 Introduction

In our modern society, innovative, flexible forms of mobility and transportation are highly relevant for both users and operators. On the one hand, individuals require transportation services that meet their specific requirements in terms of convenience, affordability and accessibility [1]. On the other hand, traditional transportation services are characterized by low profit-margins (e.g., in the case of public transport) or a high level of environmental pollution (e.g., in the case of individual motor car traffic) (e.g., [3]). In consequence, there is a growing demand for innovative concepts that are capable of ensuring inclusive, environmentally friendly, expedient, profitable and convenient transportation routines [34]. To assess the effects and implications of new transportation strategies in a cost-efficient and low-risk manner, simulation can be a viable, impactful approach that has been demonstrated to be effective in a variety of application settings such as autonomous [26] and shared mobility [4]. Due to their immanent complexity and the high degree of interconnectedness across the involved entities, the operations of mobility

© The Author(s), under exclusive license to Springer Nature Switzerland AG 2024
M. Mujica Mota and P. Scala (Eds.): EUROSIM 2023, CCIS 2033, pp. 227–241, 2024.
https://doi.org/10.1007/978-3-031-68438-8_17

systems and the associated demand patterns are usually nontrivial [22, 28]. In light of this intricacy, simulation can be considered as powerful tool to evaluate the impacts of such systems on different performance metrics such as mileages or emissions [3]. Yet, innovative transportation systems, which are often derived on a conceptual rather than an operational level with actual real-world piloting or instantiation (e.g., [26]), typically suffer from uncertainty in terms of input parameters and modelling assumptions because their implications cannot be directly observed or measured. Hence, it is utterly important to analyze historic or estimated data in an effective and holistic manner to derive valid parameter assumptions for simulation modelling [40].

With input data models being a major determinant for the credibility of simulation models as well as the accuracy of simulation results, input parameter modeling has become a popular domain in simulation research [29]. A central concern when it comes to the construction of data input models is the approximation of realistic parameters from the finite empirical or theoretical samples, which "induces undertenancies and errors of estimation [that are] subsequently propagated to the simulation output [22, p. 609]. To solve this issue, simulation scholars have proposed several sampling and approximation procedures, which predominantly opt to detect the statistical distributions of the input variables (e.g., [19, 27]). Yet, these approaches often fall short on considering all underlying influencing factors, which is particularly relevant for the case of transportation, where behavioral system characteristics such as trip counts are highly dependent on contextual factors such as weekdays or holiday seasons [1, 33]. Machine learning (ML) techniques can aid in investigating relevant patterns and factors that impact the distribution of input data and consequently derive more accurate and dependable parameters for simulation modelling endeavors [36]. Consider the case of a public transport operator that seeks to estimate the expected performance of an on-demand mobility service based on prospective simulation experiments. While the operator may be able to sample data from historical records, it will not be sufficient to merely derive statistical distributions from these records to derive relevant input parameter such as anticipated penetration/usage rates, as this would neglect the complex context that might be responsible for fluctuations in the historical data. Instead, it is still likely to assume that the historic mobility patterns are directly related to the interplay of contextual influencing factors like the respective weekday, the daytime, and the distance that is to be traveled to reach a given target location. Thus, to account for the heterogeneous nature of data sources and ensure credible as well as accurate simulation results, input parameters would need to be aligned with these factors instead of being determined deterministically or based on restrictive empirical or theoretical distributions [39].

Based on structural or contextual influencing factors as well as their inherent interdependencies, ML techniques such as supervised learning are able to predict a specific target variable for unknown data sets, which makes them highly suitable for input data modelling in simulation development [39]. Consequently, this paper seeks to demonstrate how ML models can be incorporated in a transportation simulation to improve parameter modeling by means of more contextual, realistic, and diverse inputs. The model is integrated into an agent-based simulation model that has been designed to investigate the implications of an autonomous on-demand transportation system for in rural areas. Moreover, the proposed approach is validated by comparing the experimental results of several simulation runs, where demand values have either been set in accordance with a statistical distribution or a supervised ML procedure. In doing so, this study contributes to the extant knowledge on the interplay of simulation and big data and provides practical guidance for the integration of (supervised) ML models into agent-based simulation artifacts.

In the remainder of this paper, we first present related work on the interplay of simulation and ML for input modelling before we elaborate on the simulation environment that serves as reference architecture for the parametrization. Subsequently, we present a ML model that has been trained based on mobile communication data for mobility streams and integrate this model in our simulation environment to derive exemplary results on the autonomous on-demand transport system. Finally, we conclude with a brief discussion.

2 Theoretical Background

Traditionally, the construction of input models for simulation is done by means of three different approaches that opt to generate a representation that accurately replicates the behaviors of the processes within a given system [20, 39]: (1) A trace-driven approach, where the observed data is directly used in the simulation, (2) an empirical distribution method, where collected data values are employed to define an empirical distribution function, and (3) a theoretical distribution method, which seeks to fit a theoretical distribution to the data and perform hypothesis tests to determine the goodness of fit. Yet, these approaches feature various disadvantages such as limited prediction ability (1), exhibition of irregularities such as extreme kurtosis (2), and underestimation of the uncertainty in stochastic systems (3) [22, 39].

To mitigate these disadvantages and improve the quality of input modelling, simulation scholars have started to adopt various ML-based approaches to input modelling (e.g., [22, 36, 39]). Von Rueden et al. [36] and Giabbanelli [12] summarize general approaches for combining simulation with ML. By the very nature of its methodology, ML is rather useful for interpolations than extrapolation. Accordingly, it is particularly useful for settings and problems that require prediction of new cases on the basis of historical or fictional information [25]. Overall, ML can be used for both, simulation input modelling as well as calibration. While "input modelling uses real-world input data to fit families and parameters of input distributions that are then used in the simulation [...], calibration, on the other hand, adjusts simulation parameters [...], so that the simulation output closely matches the real-world system output" [25, p. 8]. In this context, Negahban [24] employed deep neural networks for simulation model calibration to reduce the uncertainty in agent behaviors (i.e., calibration), while Kavak et al. [17] elaborated on the integration of ML techniques for the instantiation of agent attributes based on individual level data (i.e., input modelling). Similarly, Hayashi et al. [15] modelled an integration between agent-based simulation and ML to predict behavioral patterns in a more comprehensive way. In Elbattah et al. [10], different ML-based clustering method are used to analyze patient characteristics and facilitate the design of care pathways, whereas Li and Ji [21] investigated the role of Bayesian deep neural networks for estimating input data distributions in simulation models that are concerned with road pavement operations. Using the example of a manufacturing process and the associated production, transportation, and merchandise handling tasks, Marmolejo-Saucedo et al. [23] investigate the potential of recursive artificial intelligence for improving the parametrization of discrete-event simulation models, finding the approach to be highly viable in generating high-quality parameters and accurate results. Finally, Kho et al. [18] apply k-means clustering and gradient descent optimization ML techniques on real-time production data that has been captured via RFID technology, finding that valid predictions about the expected total manufacturing time for a given number of batch inputs can be obtained. Yet, feasible application examples that outline the use of ML for simulation input modelling and compare the effects of ML-based approaches with rather traditional techniques are still scarce, with only few studies providing demonstrable use-cases of ML applications in the transportation sector [5, 33].

Thus, in this contribution, we seek to combine ML-based demand prediction with an agent-based simulation model for autonomous shuttle services to highlight the potential of ML in generating more accurate and reliably simulation results.

3 Simulation Environment

To demonstrate how ML can be used for input modelling, we developed an agent-based simulation model in AnyLogic (v. 8.7.12). The model studies the operations of shared autonomous vehicles (SAV) that are part of a rural autonomous on-demand transportation service [11]. SAVs can carry up to five passengers at a time and can operate on an electric battery load for a maximum of 4 h. Since such a system touches upon multiple interrelated decision variables such as the locations of pickup and drop-off points, vehicle capacities, and service areas, it is not conducive to analytical closed form solutions and can reasonably be analyzed by means of simulation instead. Still, several important complications need to be considered for simulation modelling. For example, transportation requests from passengers fluctuate during a day and there are peak and off-peak periods. Moreover, service levels, which are mainly characterized by waiting times and delays, are directly contingent on the demand patterns of the passengers. Different service requests result in a variety of target destinations and trip characteristics. Therefore, it is likely to assume that feeding an aggregate input model to the simulation may not be sufficient to realistically mimic the operational behaviors and implications of the on-demand transportation system.

Figure 1 shows the user interface of the agent-based simulation model, which has been designed for the prototypical case of a rural area (i.e., Sarstedt, Pattensen, Nordstemmen) in the region of Hanover, Germany. The green buildings indicate pickup and drop-off points, while the black 'person' icons depict the number of passengers that have been dropped at a given destination over the course of a working day. In turn, the car icons represent the SAVs in service, whereby the color is aligned with the current status of the vehicle (yellow = occupied; black = unoccupied). Our simulation approach combines agent-based modelling properties with discrete-event simulation characteristics, whereby the synchronous time advancing mechanism is triggered by sequential behavioral state changes of agents and the resulting interactions in the specified agent networks. Agent behaviors were modelled through state charts that define the logical system flows and interactions based on the modeled state. This approach allows a representation of the autonomous and heterogenous behaviors of individual system entities (e.g., passengers, SAVs), while considering collective interdependencies and emerging reciprocations [13]. Routing procedures are conducted in line with a cluster- and time-window-based k-Nearest-Neighbor algorithm [9], whereby individual routes are chosen by a distance-based cost function. The agent types of the simulation include a Main agent, which contains the user interface of the simulation, a Depot agent, which serves as home location and is responsible for recharging SAVs at simulation runtime, a Location agent, which instantiates 50 pickup and drop-off points across all clusters, an Order agent, which represents service requests by (potential) customers, as well as a SAV agent, which contains the population of autonomous vehicles that executes the autonomous transportation service. Moreover, mobility requests (i.e., orders) for the SAV service are generated by a Cluster agent, whereby the individual probability of an order at a given location is either set based on a fixed distribution (base case) or a supervised ML model.

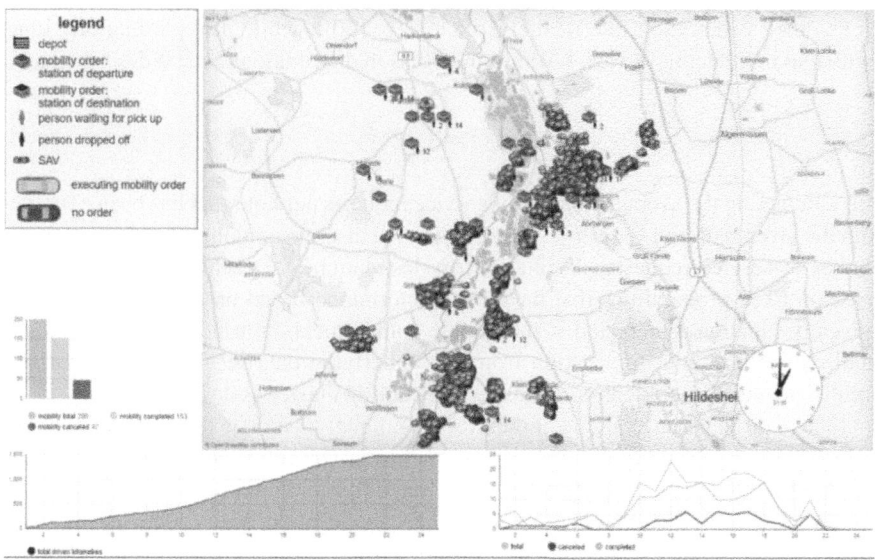

Fig. 1. User-interface of the agent-based simulation model for SAV services.

Each service request is associated with a maximum waiting time. If a SAV agent is unable to collect a passenger within this time interval, the request is automatically cancelled within the Order agent. The aim of this simulation model is to help decision-makers in identifying a minimum number of SAVs that can cope with the demand patterns in the area of investigation without resulting in an unwanted level of order cancellations. Furthermore, we opt to quantify the mileages that result from the service as a proxy for its environmental value (i.e., less mileages = less emissions).

4   Data and Machine Learning Model

To derive mobility patterns that can serve as demand blueprint for the operations of an autonomous on-demand shuttle service, we have collected mobile communication data on mobility streams for the given area of investigation. The final data set contains 1,639,546 individual trips for the period January 2019 until March 2020 and features information on the starting point, distance, purpose, and number of trips as well as the corresponding weekday, daytime, and holiday season. Based on the given mobile communication data, which depicts mobility streams based on signals across cell towers [37, 38], the area of investigation has been sub-divided into 50 clusters, each covering an area between 800 × 800 to 1,500 × 1,500 m. Moreover, every cluster contains 0 to 3 pick-up and drop-off point locations for the SAV service that have been determined in line with existing infrastructures in the area of investigation (e.g., existing bus stops).

In the base case, the number of service requests per hour are scheduled based on fixed Poisson distributions for each cluster. The parameters for these distributions have been derived from the mobile communication data set and represent a cluster's hourly frequency of outgoing trips multiplied with an adjustment factor that accounts for the assumed penetration rate of the service (2%, [30]), while disregarding any other contextual influencing factors. Overall, the Poisson distributions can be described as:

$$P_\lambda(k) = \frac{\lambda^k}{k!} e^{-\lambda}, \tag{1}$$

with λ depicting the average number of service requests per hour that has been extracted from the given data and $k = 0,1,...$ representing the discrete random variable (i.e., actual number of service requests). Similarly, trip destinations for each cluster are selected based on fixed probabilities that have been determined based on the share of trips to a given destination as depicted in the mobile communication data. Thereby, due to the small cluster size and information from the underlying data set, it is assumed that a mobility request cannot feature its respective starting cluster as destination.

Table 1 provides an example of the average number of service requests per hour and the respective trip destinations for a given start cluster. The number of service requests for an individual cluster is then transferred to the individual pickup/drop-off locations within this cluster. For clusters that contain more than one location, service requests are split equally across all locations, while for clusters that do not contain any pickup/drop-off points, the nearest location is enriched with the respective number of service requests.

Table 1. Probabilities for mobility requests.

Start cluster	Service requests (λ)	Destination cluster 01	Destination cluster 02	...
01	0.225	0.0%	1.6%	...
02	0.21	2.1%	0.0%	...
03	0.225	3.2%	19.5%	...
...	...	...	...	...

The second case opts to predict service requests via a supervised ML approach. In doing so, the ML model is trained with the mobile communication data under consideration of the respective cluster, weekday, daytime, trip purpose, and holiday season. Moreover, following the recommendations of Dieterich [8], the given data set is split into a train (70%) and a test set (30%), and K-fold cross validation is used. To autonomously translate top-down knowledge (i.e., classification of the algorithms based on a priory knowledge of the problem) into bottom-up reasoning (i.e., inferring a value, function, or algorithm via a method of learning) [35], we adopted an automated machine learning (AutoML) approach, where the proposed algorithm is required to find a suitable combination of operations for each segment of the ML pipeline to minimize the errors. In mathematical terms, given a data set $\mathcal{D}$, our AutoML can be expressed as follows [16]:

$$\theta^* = argmin_{\theta \in \Theta} \mathbb{E}_{(D_{train}, D_{valid}) \sim \mathcal{D}} V(\mathcal{L}, \mathcal{A}_\theta, D_{train}, D_{valid}), \tag{2}$$

where $V(\mathcal{L}, \mathcal{A}_\theta, D_{train}, D_{valid})$ measures the loss of a model generated by algorithm $\mathcal{A}$ with hyperparameters θ on training data D_{train} and evaluated on validation D_{valid}. Since we only have access to finite data $D \sim \mathcal{D}$ in practice, we need to approximate the expectation in Eq. 2. On a workflow-level, AutoML covers (1) data preparation, (2) feature engineering, (3) model selection, (4) model training, (5) (hyper-) parameter optimization, and (6) visualization [16].

The AutoML approach has been conducted via H2O Driverless AI (DAI) (v. 1.10.2), which is equipped with evolutionary algorithms to perform feature engineering and selection [6]. Features from the dataset that have been employed to predict the number of trips within the area of investigation (which serve as basis for the determination of service requests) are weekday (i.e., Monday to Sunday), daytime (i.e., 12 am to 11 pm), infrastructure (i.e., number of residential houses, workplaces, shopping facilities, schools, and recreational facilities) and holiday season (i.e., yes or no). Model training settings consist of standard settings like the experiment type, the target column, as well as the model type, which is set to 'regression' in this case, and DAI specific expert setting that need to be configured by the user, including accuracy, time tolerance, interpretability, and the selection of a scorer. Here, accuracy controls the search efforts of the AutoML process to produce pipelines that are as accurate as possible. After initial exploratory experiments, we observed the best validation and test scores for our dataset with an accuracy score of 6 (out of 10). In turn, time tolerance, which has been set to 4 (out of 10) in this study, relates to the duration of the search process, allowing the use of heuristics for stopping the search process early when it is set to low values. Interpretability in DAI is understood as measure that can be improved if the features used by the model are understandable to the domain expert and if the relative number of features is kept at a minimum. Accordingly, this setting controls several factors such as the use of filtering features selection on the raw features and the number of feature engineering methods used. By assigning it with a value of 6 (out of 10) in this study, co-linear and uninformative features are partially filtered out, while a reasonable number of new features is constructed via different feature engineering methods. Finally, scoring describes the functions are to be optimized by the underlying search of the AutoML process. These functions are selected in correspondence with the given problem type (i.e., regression or classification) and include Classification Accuracy or Log-Loss for classification problems, and gini coefficient (GINI), mean absolute error (MAE), and Root Mean Square Error (RMSE) for regression problems. In this work, in line with the underlying regression problem, we chose the RMSE scorer, where optimal performance is achieved with a value of 0.

In terms of ML algorithm, DAI supports the following options: XGBoostGBM, LightGBM, Generalized Linear Models, Follow Regularized Leader, and RuleFit Models. To improve the predictive performance of the model, a stacked ensemble method has been employed to find the optimal combination of these algorithms, ultimately yielding a combination of the XGBoostGBM (41%) and the LightGBM (59%) algorithm. Both algorithms are implementations of gradient boosted decision trees that are considered among the most recent and efficient ML algorithms in a variety of domains (e.g., [6, 31]) and provide more regularized model formalization and better over-fitting control [7, 31]. Further information on the mathematical approach related to these algorithms can be found in Al Daoud [1]. Finally, robustness, accuracy, and validity of the prediction models were tested by means of 4-fold cross-validation.

Table 2. Summary of AutoML results (testing performance shown as averages and standard deviation in parentheses)

Testing performance			Model size		
RMSE	*MAE*	*GINI*	*Features*	*Ensemble*	*Time*
9.23 (0.118)	6.06 (0.055)	0.94 (0.001)	28	XGBoost: 41% LightGBM: 59%	11.5

Training settings: Accuracy = 6; Time tolerance = 4; Interpretability = 6; Scorer = RMSE

Table 2 provides a synopsis on the results of the AutoML Pipeline, focusing on RMSE, MAE, and GINI, as major performance metrics, which are commonly regarded as standard measures for regression problems in the ML literature (e.g., [32]). These metrics are presented as averages over all test folds of in the cross-validation process, while the respective standard deviations are shown in parentheses. Given the total number of trips for the examined case and area of investigation (~74,200), a RMSE of 9.23 and a MAE of 6.06 indicate reasonably strong predictive ability. Similarly, the GINI coefficient of 0.94 signifies that the elements are randomly distributed across various classes, which in turn indicates that the model possesses a high level of discriminatory power. Model size characteristics encompass the number of features used, the number and share of algorithmic components within the final model ensemble, as well as the total computation time required for data preparation, shift and leakage detection, model and feature tuning, feature evolution, final pipeline training and scorer building (in minutes).

Ultimately, we adapted a two-phase procedure to integrate the proposed ML model into the simulation environment, featuring (1) the integration of relevant DAI components into the simulation, and (2) the integration of the ML model into the model logic. While weekdays and holidays can be specified manually within the user interface of the simulation model, other input values for the ML model (e.g., daytime, infrastructure of a given cluster) are determined and integrated automatically at simulation runtime. The predicted number of trips for a given combination of input values is then transformed to a parameter for each cluster that changes on an hourly basis to reflect daytime-dependent mobility requirements. Since these requirements depict actual mobility patterns and do not directly reflect (potential) service requests for the on-demand SAV service, we have multiplied the number of mobility requirements for each cluster with an (assumed) penetration rate of 2% (Poisson distributed), which has been defined based on existing studies on the user acceptance of demand-responsive transport services in rural areas (e.g., [30]). Eventually, the resulting value determines the rate of hourly service requests and is integrated into the behavioral logic (i.e., state chart) of the cluster agent to model the SAV demand logic for each cluster and its pickup/drop-off locations. Finally, destination clusters as well as individual pick-up and drop-off locations for a given cluster were determined in accordance with the procedure that has already been described for the base case.

5   Exemplary Results

To determine the total number of service requests, we conducted 140 simulation runs for each case (10 runs per weekday and holiday/non-holiday season).

Table 3. Average daily service requests per case.

Case	Replications	Mean	Standard deviation
Base case	70	259	20.63
ML case	70	302	80.58

Table 3 provides an overview of the total number of service requests per case across all clusters that have resulted from the simulation experiments. Interestingly, the ML case yields both a higher average of service requests across all clusters as well as significantly higher standard deviation. This can be explained by the fact that the ML case accounts for additional contextualities in the data such as holidays. For example, in the base case, cluster 1 features an average of 0.225 service requests per hour and 5.4 service requests per day. Even though the data is slightly varied by means of a Poisson distribution for each simulation run, apart from daytimes, (potentially relevant) influencing factors are not considered. In contrast, in the ML case, cluster 1 features different numbers of service requests based on daytime, weekday, trip purpose, cluster infrastructure and holiday season (e.g., 0.388 requests on a Tuesday at 8 am during non-holiday season, equaling 350 requests during the whole day, and a spread of 83 requests between the clusters with the highest and the lowest number of requests). This phenomenon can also be illustrated by looking at the hourly differences in service requests (Fig. 2), where the base case seems to slightly overestimate demand during holiday seasons, while it significantly underestimates potential mobility requests during non-holiday seasons.

To compare the individual influence of the service request input modelling approaches of these two cases on order cancellations (Fig. 3) and mileages (Fig. 4), we

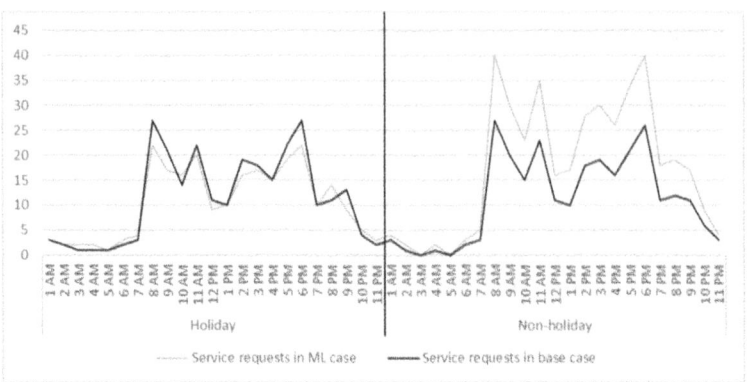

Fig. 2. Mean number of hourly service requests on Fridays (holiday vs. non-season) per case.

conducted simulation experiments for different parameter configurations of the number of SAVs (15 and 20) as well as the maximum waiting times (15 and 20) for two exemplary days (Friday and Sunday during holiday and non-holiday season).

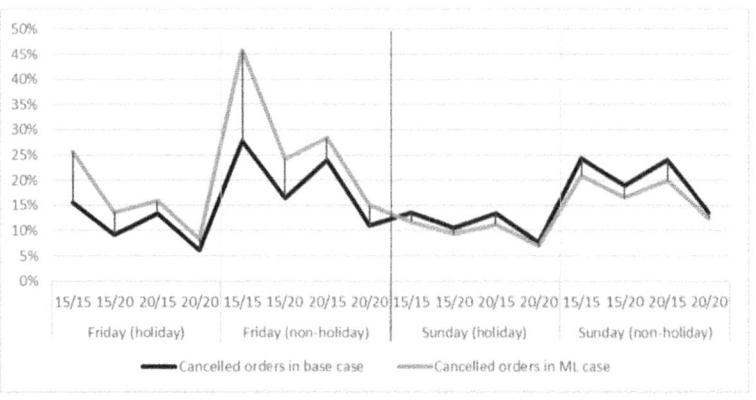

Fig. 3. Mean share of order cancellations (in %) per case, number of SAVs, waiting times, weekday, and holiday season.

Both, in terms of order cancellations as well as mileages, we can observe that the simulation results of the two cases significantly vary. While on high-traffic days such as Fridays the base case predicts less order cancellations and mileages than the ML case, it seems to overestimate these key performance indicators (KPI) on low-traffic days such as Sundays. Moreover, it seems noteworthy that the spread between the prediction made for the base case and the ML case is significantly large for non-holiday seasons. This can mainly be attributed to the fact that the fixed probability distributions in the base case do not account for weekday-, trip purposes (e.g., shopping) and season-based demand differences. Local minima for order cancellation shares are reached with 20 SAVs and a maximum waiting time of 20 min on Fridays and Sundays in the base case, whereas mileages are lowest for 20 SAVs and 15 min waiting time. Similarly, the ML case shows local minima for order cancellations with 20 SAVs and 20 min waiting time on both days. Yet, in terms of mileages, a minimum is reached with 20 SAVs and 20 min waiting time on Fridays, whereas mileages are lowest for 20 SAVs and 15 min waiting time on Sundays.

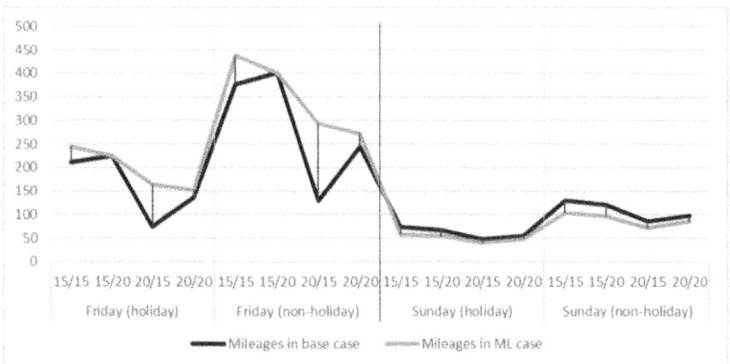

Fig. 4. Mean mileages (in km) per case, number of SAVs, waiting times, weekday, and holiday season.

Finally, to determine the individual impact of fleet size and waiting time on cancellations and mileages, we evaluated the simulation experiments based on a 2k-factor-design [14]:

$$e_{l(m)} = \frac{\sum_{i=1}^{2^k} \pm E_{i_m}}{2^{k-1}} \tag{3}$$

Here, e is the impact of a parameter l (i.e., number of SAVs, waiting time), which results from the mean result value of a KPI m (i.e., cancelled orders, mileages) for a parameter value i (i.e., 15, 20), which is subtracted for low (i.e., 15) and summated for high parameter values (i.e., 20). Ultimately, E represents the result values for parameter value and the associated KPI, while k depicts the total number of factor combinations, equaling 4 in our example. Table 4 synopsizes the experimental results (please note that these results must be interpreted proportionally to the result values). In the ML case, the outcomes suggest that both, an increasing fleet size as well as longer waiting times result in a decrease of order cancellations and mileages. In contrast, the results from the base case indicate that an increase in waiting times is likely to result in additional mileages. Even though longer waiting times allow for more efficient vehicle allocation and routing procedures, this is phenomenon is likely to occur due to the case that a vehicle may have to fulfill several service requests which have a large geographical distance to each other, so that a minimum number of SAVs results in additional mileages that would not occur if several vehicles are assigned to different trips with individually closer destinations.

Table 4. Impact of fleet size and delays on cancellations and mileages.

	Base case		ML case	
	Fleet	*Delay*	*Fleet*	*Delay*
Cancellations	−0.037	−0.101	−0.079	−0.117
Mileages	−117.789	35.149	−78.684	−12.309

6 Discussion and Conclusion

Simulation has become equally valuable and important in order to study complex systems. Thereby, input modelling can be regarded as major influencing factor for the accuracy and reliability of simulation results [29]. Traditional methods that rely on fixed statistical distributions typically underestimate the uncertainty in the input data and fail to implement contextual influencing factors in a holistic manner. In this article, we have presented an AutoML-based approach on how to predict customer demand for SAVs and integrate it into an agent-based simulation model. To show the differences between traditional and ML-based input modelling, we conducted various simulation experiments for both cases. Looking at the results of this study, we can observe system parameters and KPIs can highly vary based on the selected input modelling approach. Therefore, our exemplary results highlight the importance of holistic and contextualized input modelling in simulation development to enable scholars and transportation planners to take more informed and efficient decisions on relevant system parameters such as fleet size (i.e., number of SAVs) and service proposition (i.e., maximum waiting time). As shown by our results, simply fitting a distribution to available data is likely to underestimate the uncertainty in the input data. While the ML-integration allows for more a fine-grained analysis, it is difficult to mimic contextually realistic system behaviors by the mere use of (fixed) statistical distributions. Yet, this is particularly important as "the quality of input data modeling significantly impacts the accuracy of simulation results" [22]. Hence, ML-generated input data provide more realistic representations of input processes by taking into consideration underlying system characteristics such as weekdays, daytimes, and holidays, ultimately mitigating the limitation induced by a lack of real-world performance data and providing a more balanced approach between fidelity and tractability. In consequence, the results of the ML case can be used by decision makers to determine influential system parameters in a more reliable fashion (e.g., by adapting fleet size and service proposition based on prospected peak times and days).

Input modelling via supervised ML can be adopted for any complex socio-technical system or phenomenon that is driven by complex input processes, such as the demand for autonomous on-demand services as proposed in this publication. Yet, it also needs to be mentioned that ML-based input modelling features high requirements in terms of data and computational costs [36]. Hence, when opting to adopt this approach, scholars need to assess its potential benefits based on their individual research context.

References

1. Al Daoud, E.: Comparison between XGBoost, LightGBM and CatBoost using a home credit dataset. Int. J. Comp. Info. Eng. **13**(1), 6–10 (2019)
2. Atasoy, B., Ikeda, T., Song, X., Ben-Akiva, M.E.: The concept and impact analysis of a flexible mobility on demand system. Transport. Res. Emerg. Technol. **56**, 373–392 (2015)
3. Auf der Landwehr, M., Trott, M., von Viebahn, C.: Environmental Sustainability as Food for Thought! Simulation-Based Assessment of Fulfillment Strategies in the E-Grocery Sector. In: Kim, S., Feng, B., Smith, K., Masoud, S., Zheng, Z., Szabo, C., Loper, M. (eds.) Proceedings of the 2021 Winter Simulation Conference (WSC), 1-12. IEEE, Piscataway (2021)
4. Becker, H., Balac, M., Ciari, F., Axhausen, K.W.: Assessing the welfare impacts of Shared Mobility and Mobility as a Service (MaaS). Trans. Res. Part A: Policy and Practice **131**, 228–243 (2020)
5. Cavalcante, I.M., Frazzon, E.M., Forcellini, F.A., Ivanov, D.: A supervised machine learning approach to data-driven simulation of resilient supplier selection in digital manufacturing. Int. J. Inf. Manage. **49**, 86–97 (2019)
6. Cerrada, M., et al.: AutoML for Feature Selection and Model Tuning Applied to Fault Severity Diagnosis in Spur Gearboxes. Math. Computat. Applicat. **27**(1), 6 (2022)
7. Chi, S., Suk, S.J., Kang, Y., Mulva, S.P.: Development of a data mining-based analysis framework for multi-attribute construction project information. Adv. Eng. Inform. **26**(3), 574–581 (2012)
8. Dieterich, T.G.: Approximate statistical tests for comparing supervised classification learning algorithms. Neural Comput. **10**(7), 1895–1923 (1998)
9. Dudani, S.A.: The distance-weighted k-nearest-neighbor rule. IEEE Trans. Syst. Man Cybern. **4**, 325–327 (1976)
10. Elbattah, M., Molloy, O., Zeigler, B.P.: Designing care pathways using simulation modeling and machine learning. In: M. Rabe, A.A.J., Mustafee, N., Skoogh, A., Jain, S., Johansson, B. (eds.) Proceedings of the 2018 Winter Simulation Conference (WSC), pp. 1452–1463. IEEE, Piscataway (2018)
11. Fagnant, D.J., Kockelman, K.M.: The travel and environmental implications of shared autonomous vehicles, using agent-based model scenarios. Transport. Res. Part C: Emerg. Technol. **40**, 1–13 (2014)
12. Giabbanelli, P.J.: Solving challenges at the interface of simulation and big data using machine learning. In: Mustafee, N., Bae, K.-H.G., Lazarova-Molnar, S., Rabe, M., Szabo, C., Haas, P., Son, Y.-J. (eds.) Proceedings of the 2019 Winter Simulation Conference (WSC), pp. 572–583. IEEE, Piscataway (2019)
13. Gómez-Cruz, N.A., Saa, I.L., Hurtado, F.F.O.: Agent-based simulation in management and organizational studies: a survey. Eur. J. Manag. Bus. Econ. **26**(3), 313–328 (2017)
14. Gutenschwager, K., Rabe, M., Spieckermann, S., Wenzel, S.: Simulation in Produktion und Logistik. Springer (2017)
15. Hayashi, S., Prasasti, N., Kanamori, K., Ohwada, H.: Improving behavior prediction accuracy by using machine learning for agent-based simulation. In: Nguyen, N.T., Trawiński, B., Fujita, H., Hong, T. (eds.) Asian Conference on Intelligent Information and Database Systems, pp. 280–289. Springer, Heidelberg (2016)
16. Hutter, F., Kotthoff, L., Vanschoren, J. (eds.): Automated Machine Learning: Methods, Systems. Springer, Challenges (2019)
17. Kavak, H., Padilla, J.J., Lynch, C.J., Diallo, S.Y.: Big data, agents, and machine learning: Towards a data-driven agent-based modeling approach. In: Frydenlund, E., Jafer, S. (eds.) Proceedings of the Annual Simulation Symposium, pp. 1–12. SCS, San Diego (2018)

18. Kho, D.D., Lee, S., Zhong, R.Y.: Big data analytics for processing time analysis in an IoT-enabled manufacturing shop floor. Procedia Manufacturing **26**, 1411–1420 (2018)

19. Law, A.M.: A tutorial on how to select simulation input probability distributions. In: Pasupathy, R., Kim, S.-H., Tolk, A., Hill, R., Kuhl, M.E. (eds.). Proceedings of the 2013 Winter Simulations Conference (WSC), pp. 306–320. IEEE, Piscataway (2013)

20. Law, A.M., Kelton, W.D.: Simulation modeling and analysis, Vol. 3. McGraw-Hill, New York (2007)

21. Li, Y., Ji, W.: Enhanced input modeling for construction simulation using bayesian deep neural networks. In: Mustafee, N., Bae, K.-H.G., Lazarova-Molnar, S., Rabe, M., Szabo, C., Haas, P., Son, Y.-J. (eds.) Proceedings of the 2019 Winter Simulation Conference (WSC), pp. 2978–2985. IEEE, Piscataway (2019)

22. Liu, Y., et al.: Enhancing input parameter estimation by machine learning for the simulation of large-scale logistics networks. In: Bae, K.-H., Feng, B., Kim, S., Lazarova-Molnar, S., Zheng, Z., Roeder, T., Thiesing, R. (eds.) Proceedings of the 2020 Winter Simulation Conference (WSC), pp. 608-619. IEEE, Piscataway (2020)

23. Marmolejo-Saucedo, J.A., et al.: Improving a Manufacturing Process using Recursive Artificial Intelligence. In: Dolgui, A., Bernard, A., Lemoine, D., von Cieminski, G., Romero, D. (eds.) IFIP International Conference on Advances in Production Management Systems, pp. 266–275. Springer, Cham (2021)

24. Negahban, A.: Neural networks and agent-based diffusion models. In: Chan, W.K.V., D'Ambrogio, A., Zacharewicz, G., Mustafee, N., Wainer, G., Page, E. (eds.) Proceedings of the 2011 Winter Simulation Conference (WSC), pp. 1407-1418. IEEE, Piscataway (2017)

25. Nelson, B.L.: 'Some tactical problems in digital simulation' for the next 10 years. Journal of Simulation **10**(1), 2–11 (2016)

26. Oh, S., Seshadri, R., Azevedo, C.L., Kumar, N., Basak, K., Ben-Akiva, M.: Assessing the impacts of automated mobility-on-demand through agent-based simulation: A study of Singapore. Transport. Res. Part A: Policy and Practice **138**, 367–388 (2020)

27. Poropudas, J., Pousi, J., Virtanen, K.: Multiple input and multiple output simulation meta-modeling using Bayesian networks. In: Jain, S., Creasey, R.R., Himmelspach, J., White, K.P., Fu, M. (eds.) Proceedings of the 2011 Winter Simulation Conference (WSC), pp. 569–580. IEEE, Piscataway (2011)

28. Rabe, M., Ammouriova, M., Schmitt, D.: Improving the performance of a logistics assistance system for materials trading networks by grouping similar actions. In: Rabe, M., Juan, A.A., Mustafee, N., Skoogh, A., Jain, S., Johansson, B. (eds.) Proceedings of the 2018 Winter Simulation Conference (WSC), pp. 2861–2872. IEEE, Piscataway (2018)

29. Rabe, M., Scheidler, A.A.: An approach for increasing the level of accuracy in supply chain simulation by using patterns on input data. In: Tolk, A., Diallo, S.Y., Ryzhov, I.O., Yilmaz, L., Buckley, S., Miller, J.A. (eds.) Proceedings of the 2014 Winter Simulation Conference (WSC), pp. 1897–1906. IEEE, Piscataway (2014)

30. Schasché, S.E., Sposato, R.G., Hampl, N.: The dilemma of demand-responsive transport services in rural areas: Conflicting expectations and weak user acceptance. Transp. Policy **126**, 43–54 (2022)

31. Shehadeh, A., Alshboul, O., Al Mamlook, R.E., Hamedat, O.: Machine learning models for predicting the residual value of heavy construction equipment: An evaluation of modified decision tree, LightGBM, and XGBoost regression. Automation in Construction 129 (2021)

32. Singh, P.: Fundamentals and Methods of Machine and Deep Learning: Algorithms, Tools, and Applications. John Wiley & Sons (2022)

33. Torre-Bastida, A.I., Del Ser, J., Laña, I., Ilardia, M., Bilbao, M.N., Campos-Cordobés, S.: Big Data for transportation and mobility: recent advances, trends and challenges. IET Intel. Transport Syst. **12**(8), 742–755 (2018)

34. Trott, M., Baur, N.F., Auf der Landwehr, M., Rieck, J., and von Viebahn, C.: Evaluating the role of commercial parking bays for urban stakeholders on last-mile deliveries–A consideration of various sustainability aspects. Journal of Cleaner Production, 312 (2021)
35. Vaccaro, L., Sansonetti, G., Micarelli, A.: An empirical review of automated machine learning. Computers **10**(1), 1–27 (2021)
36. von Rueden, L., Mayer, S., Sifa, R., Bauckhage, C., Garcke, J.: Combining machine learning and simulation to a hybrid modelling approach: Current and future directions. In: Berthold, M.R., Feelders, A., Krempl, G. (eds.) International Symposium on Intelligent Data Analysis, pp. 548–560. Springer, Cham (2020)
37. Wang, X., et al.: Spatio-temporal analysis and prediction of cellular traffic in metropolis. IEEE Trans. Mob. Comput. **18**(9), 2190–2202 (2018)
38. Xu, F., Li, Y., Wang, H., Zhang, P., Jin, D.: Understanding mobile traffic patterns of large scale cellular towers in urban environment. IEEE/ACM Trans. Network. **25**(2), 1147–1161 (2016)
39. Zhang, J., et al.: Dynamic time warp-based clustering: Application of machine learning algorithms to simulation input modelling. Expert Systems with Applications 186 (2021)
40. Zouaoui, F., Wilson, J.R.: Accounting for parameter uncertainty in simulation input modeling. IIE Trans. **35**(9), 781–792 (2003)

Monitor, Control, and Theoretical Systems

The Impact of Adding Interaction-Driven Evolutionary Behavior to the Schelling's Model

Yakup Turgut[1] and Sanja Lazarova-Molnar[2,3][(✉)]

[1] Faculty of Engineering, Industrial Engineering Department, Kırklareli University,
39100 Kırklareli, Turkey
yakupturgut@klu.edu.tr
[2] Institute of Applied Informatics and Formal Description Methods,
Karlsruhe Institute of Technology, 76131 Karlsruhe, Germany
lazarova-molnar@kit.edu
[3] Mærsk Mc-Kinney Møller Institute, University of Southern Denmark,
5230 Odense, Denmark
slmo@mmmi.sdu.dk

Abstract. Schelling's model is thoroughly researched because it highlights a crucial human phenomenon known as segregation. The term "segregation" is used to describe the practice of dividing up the human population into several groups. The policies of governments, environmental concerns, economic concerns, familial dynamics, and so on, all have a role in shaping the ways in which people are separated from one another. All of these factors contribute to segregation by influencing people's preferences in different ways. The studies have evaluated each factor separately and demonstrated how each variable influences segregation dynamics. In this study, we investigate the effect of agents' evolutionary (dynamic) behavior on segregation dynamics. Behavioral patterns of agents are made interaction-dependent, and the results of this interaction-driven evolutionary model are compared to those of the original model. The results demonstrate that evolutionary behavior leads to a higher number of subgroups when the model converges than the number of subgroups in the original model. Furthermore, because the evolutionary model is more complex than the original model, it converges later and produces results that are more variable than those of the original model.

Keywords: Segregation · Evolutionary · Agent Based Simulation

1 Introduction

People tend to live with those who are similar to them on a variety of aspects, including cultural, social, religious, and others. As a result of these tendencies, people begin to surround themselves with those who share their interests and values, ultimately leading to segregation. Schelling [15] proposed a simulation

© The Author(s), under exclusive license to Springer Nature Switzerland AG 2024
M. Mujica Mota and P. Scala (Eds.): EUROSIM 2023, CCIS 2033, pp. 245–258, 2024.
https://doi.org/10.1007/978-3-031-68438-8_18

model to demonstrate how segregation happens when people only associate with those who are similar to them. As a metric for how similar different agents are to one another, Schelling only considered one dimension (color). The specifications of the model are presented in Sect. 2.

From a societal point of view, segregation could be a major problem for countries because it could disrupt their ethnic structures and cause social inequality in their communities. For example, some natural resources might not be reachable for some groups, and education or job opportunities might be different for different groups because of the segregation [11,14]. All of these disruptions and social inequalities threaten the long-term social sustainability of the countries [12].

As the first to demonstrate that individuals' preferences lead to segregation, Schelling's study spurred a flurry of subsequent research. Researchers have begun to investigate how segregation might occur if people's true preferences are considered [10]. In [17], to make Schelling's model more plausible, the author incorporated regulatory constraints. Specifically, Schelling's approach allows agents to relocate to any vacant area if they are unhappy with where they are. In reality, however, people may be unable to relocate any empty space at all due to rules, restrictions, and other circumstances. In [4], the authors expand the scope of the model by treating agents' movement decisions as strategic behavior rather than just movements. This makes a lot of sense, considering that in the actual world, people may be unable to readily relocate owing to financial constraints, family responsibilities, and the like, even if they are unhappy in their current living situation. In [5], the authors modify the original model by incorporating considerations of fairness and altruism to demonstrate how these factors influence segregation dynamics. In [3], the author studies the effects of networks on the dynamics of Schelling's model. The conclusion is that segregation dynamics are profoundly influenced by network patterns and that a more exact analysis is required if the model is to be applied to a big area, such as a metropolis.

In [8], the author focuses on the original model's two key assumptions: finite resources and individual tolerance. The author investigates how the dynamics of segregation change when resources are not scarce, as this reduces agents' tolerance level. The conclusion is that extreme intolerance of others causes an artificial society's degree of segregation to rise at first, but eventually become unsustainable, resulting in the disintegration of the highly separated clusters. In [2], the author analyzes how policies may affect residential segregation and the intricate urban system as a whole. Using empirical data, he creates a model that accurately represents the spatial organization of a city's neighborhoods and the resulting segregation. The results of the simulation are encouraging, showing that the model may be used to explore problems of concrete policy interest and gain useful insights into the future of the urban landscape and the factors behind the dynamics of residential segregation. In [16], the authors consider whether and under what conditions Schelling's "self-organized" unintended segregation may apply to school segregation. They draw the conclusion that parental preferences for neighboring schools may play a significant role in reducing the impact of school choice on segregation dynamics. As a result, the Schelling's model has prompted a great deal of research into the dynamics of segregation, and future studies on the Schelling's model will continue to apply this model to real-life scenarios.

The Schelling's model has also been criticized for its static structure, which prevents agents' behavior in simulations from evolving or changing over time. This is because all of the model's parameters remain constant and do not vary as time progresses, which in turn leads to the agents' behavior remaining constant. In [13], the authors focused on the threshold value parameter of the model and made it dynamic to determine how it affects segregation results. They make it dependent on time and investigate how the dynamics of segregation vary over time. In [1], the authors included a time-dependent aging effect and made the agent's movement decisions and behavior dynamic over time. In this study, we also consider the ever-changing behaviors of agents. For this, we also used the threshold value to make agents' behavior dynamic in simulation. Our approach, however, relies more on the agents' interactions with one another than on the time that has passed. The reasoning behind this is that interactions are often the factors that strongly impact behavior patterns and, thus, make the agents evolve [6,7]. The term "evolutionary behavior" refers to the way in which people alter their actions in response to both internal and external stimuli in order to adapt to their surroundings [18]. We would like to explore the effects of this type of evolution. We compared our model to the original one to highlight the differences and shed light on the impact of interaction-driven evolution on segregation dynamics.

2   Model Description

Schelling [15] developed the first model demonstrating that spatial segregation occurs as a result of individuals' desire to live with others who share their preferences. Schelling's Segregation model is one of the earliest agent-based models. The models' specifics and underlying assumptions are as follows:

- Assume that there exists an artificial society made up of N agents. Each agent is a member of either group A or group B. This hypothetical society is a simplified representation of a real one, in which only the pertinent behaviors are recorded and the rest are disregarded.
- Agents are placed on a lattice or grid, with the possibility that M other agents will surround them at any given time. Each agent in a square lattice of N × 2 cells would have 8 in-group or out-group neighbors (known as the Moore neighborhood).
- One of the model's assumptions is that individuals have a preference, denoted by P, to be located near other individuals who are just like them, i.e., those who are part of the in-group. This preference is stated as the ratio of in-group agents to the total number of agents surrounding an individual. This is termed a threshold value. If the threshold value is met for the agent, the agent is considered "happy" and is content to remain in the same place. If this condition is not met, however, and an agent is surrounded by members

of the other group beyond his or her tolerance level, the agent will leave the area in search of more solitude. One agent will keep moving until his or her preference is met.

For the sake of simplicity, we assume that all agents have the same tolerance threshold, even if they may have varying preferences. Schelling's model demonstrated that segregation can arise even if individuals are initially distributed at random between the two groups and agents have just a weak preference to be positioned near members of the in-group.

In the Schelling's model studies, the number of happy agents is employed as a measure of segregation. In this study, we also evaluate the results of our evolutionary model based on the number of happy agents. In addition, we employ a second metric, the Segregation Index (SI), to more precisely evaluate the simulation's segregation status at each simulated iteration.

$$SI = \sum_{i=1}^{N} SR_i \tag{1}$$

where SR_i indicates the similarity ratio of agent i, N is the number of agents in the system. SR_i is calculated as follows:

$$SR_i = \frac{N_S}{N_S + N_{DS}} \tag{2}$$

where N_S indicates the number of similar agents in agent i's neighborhood, N_{DS} is the number of dissimilar agents.

One of Schelling's model's assumptions that has not been sufficiently challenged in the literature is the constant threshold value of agents during simulation runs. We focused on challenging this assumption to make the model more realistic, and we describe our model in the subsequent section.

3   Evolutionary Schelling's Model

Our proposed model illustrates the incorporation of the dynamic character of human preferences throughout the simulation. As such, it can be seen as an extension of the original model, with the goal of making it more realistic by representing the natural evolution of human preferences over time. While formulating behaviors of agents that evolve over time, we wanted to remain loyal to the initial model specifications as much as possible. Thus, we considered a dynamic threshold value rather than a static one to enable evolving agents' movement decisions. For alteration of the threshold value, we adopted an *interaction-dependent* strategy that we describe in the following.

Interaction-Dependent Threshold: Individual movement decisions vary over time as a result of interactions with other individuals. For example, if a person lives with individuals who are similar to her/him, she/he would be more willing to live with people who are similar to her/him, which could increase her/his threshold value. On the other hand, if they live with individuals who are unlike her/him, they may become accustomed to living with others, which could reduce the threshold value. As a result, a person's interactions with others can alter their aversion to living with dissimilar groups and groups similar to themselves. In this manner, we iteratively update the threshold value for each individual as follows:

$$threshold_{t+1} = threshold_t + (2 \times N_S^t - N_{DS}^t) * r_t \tag{3}$$

where N_S^t indicates the number of similar agents in agent i's neighborhood at time t, N_{DS}^t is the number of dissimilar agents at time t. r_t is the rate of threshold change at time step t. We use two strategies for the value of r_t:

- Fixed threshold change rate: For simplicity reasons, we arbitrarily set r_t at 0.001 for all agents in our experiments.
- Randomized threshold change rate: To incorporate heterogeneity in the rates of change of tolerance among agents, we make r_t randomized and uniformly distributed between 0.0001 and 0.01 for each agent in our experiments.

4   Experimental Results and Discussion

We programmed the evolutionary model in Python 3 using the Mesa library [9]. As described in Sect. 2, we created a 20-by-20 grid with varying densities and

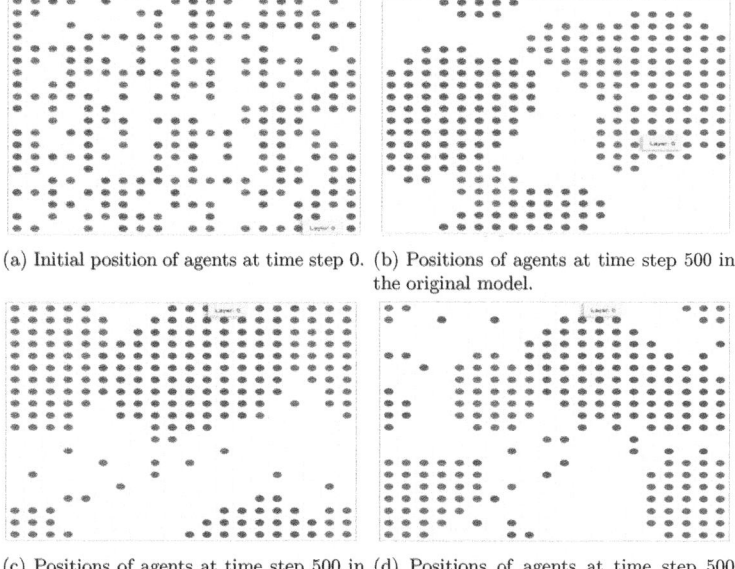

(a) Initial position of agents at time step 0.   (b) Positions of agents at time step 500 in the original model.

(c) Positions of agents at time step 500 in the evolutionary model (for fixed-threshold change rate).   (d) Positions of agents at time step 500 in the evolutionary model (for randomized threshold change rate)

Fig. 1. Occurrence of segregation in both models with density of 0.6 and threshold of 0.5.

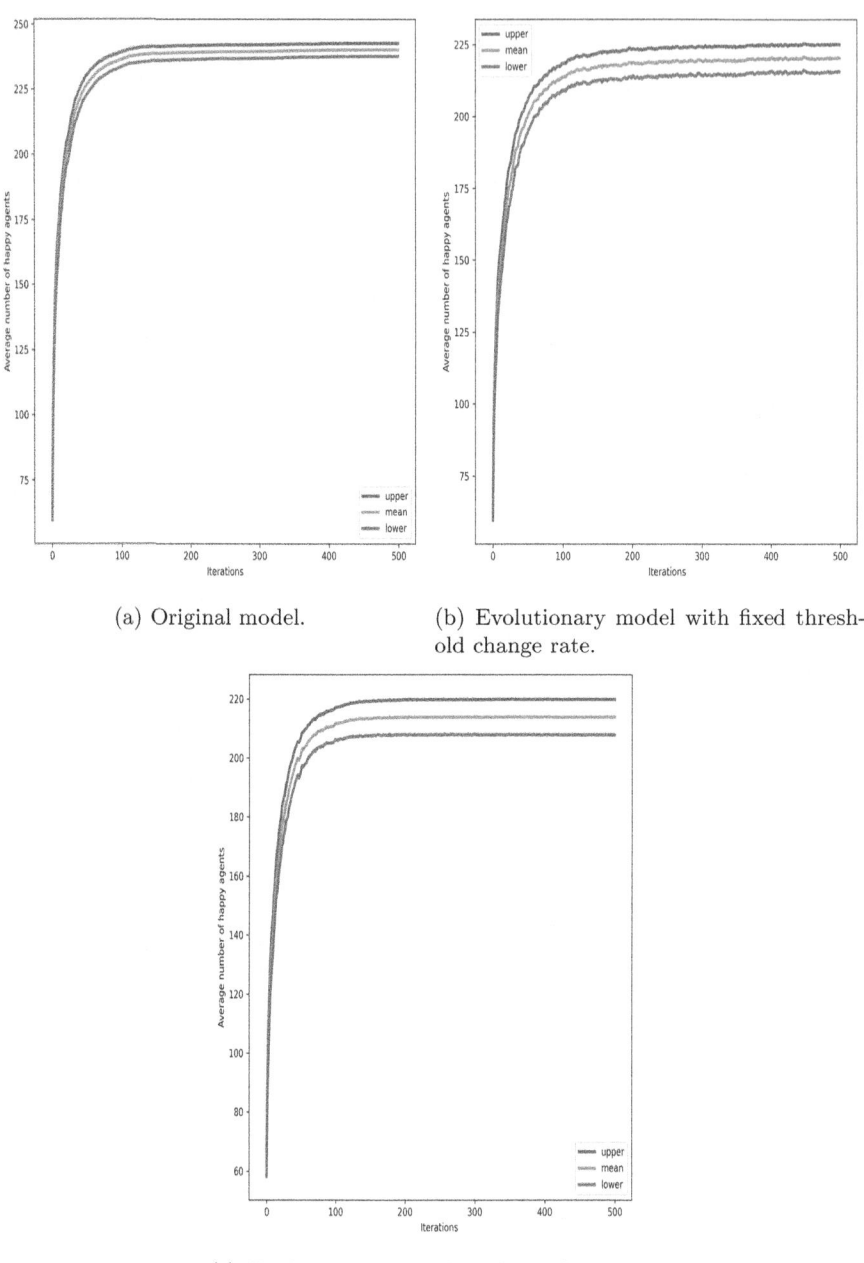

(a) Original model.

(b) Evolutionary model with fixed threshold change rate.

(c) Evolutionary model with randomized threshold change rate.

Fig. 2. Change of the number of happy agents for both models with density 0.6 and threshold 0.5.

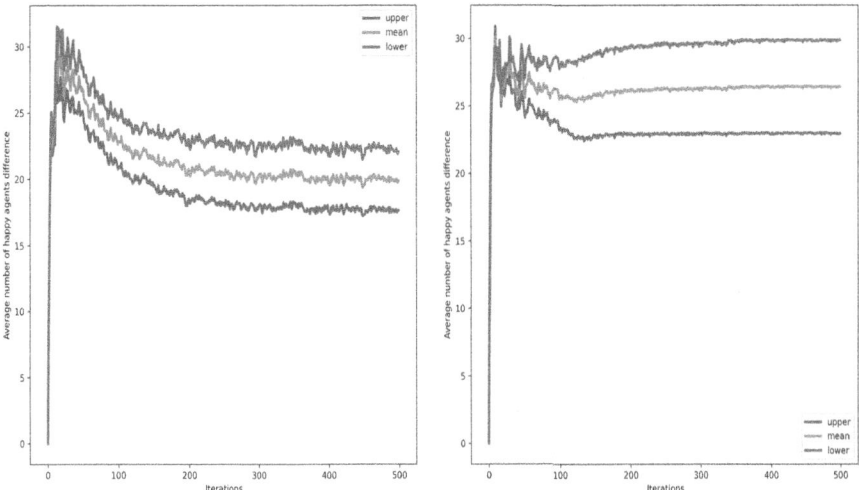

(a) Evolutionary model with fixed thresh- (b) Evolutionary model with randomized
old change rate.   threshold change rate.

Fig. 3. Difference in the number of happy agents between the original and evolutionary models for density of 0.6 and threshold of 0.5.

threshold values to compare the evolutionary models with the original model. We simulated the original and evolutionary models with the same seed numbers 30 times and plotted 95% confidence intervals to determine how the evolutionary component influences the two outputs (which are given in Sect. 2) of Schelling's model across iterations. We also provide snapshots of the simulations for both models at the 500th iteration in order to visually compare the segregation effect (see Fig. 1 and 4).

We conducted studies with the following threshold values and densities: 0.4, 0.5, and 0.6; and 0.5, 0.6, and 0.7; respectively. In both models, segregation does not occur at density of 0.5 for all provided threshold values because these threshold values cannot be achieved and agents continue to move without ever settling down in a grid cell. For other densities and threshold values, the original model and evolutionary model yield comparable outcomes. Due to space limitation, we provide results for two scenarios, shown in Figs. 1 through 8. We further provide a summary of the findings by discussing the observed outcomes as follows:

– The first point to note is that the variability of the evolutionary model is greater than that of the original model, as shown in Figs. 2, 3, 5, 6, 7 and 8. In the evolutionary model, the threshold value is modified based on the current position of each agent through simulations, making the position of each agent in each step more influential in the movement decision than in the original model. In this way, the evolutionary model has become more complex than the original model. Also, the variability of the evolutionary

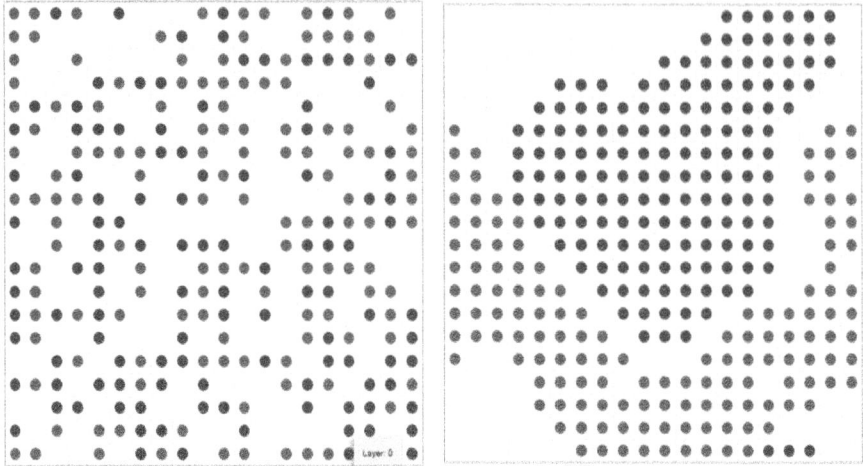

(a) Initial position of agents at time step 0. (b) Positions of agents at time step 500 in the original model.

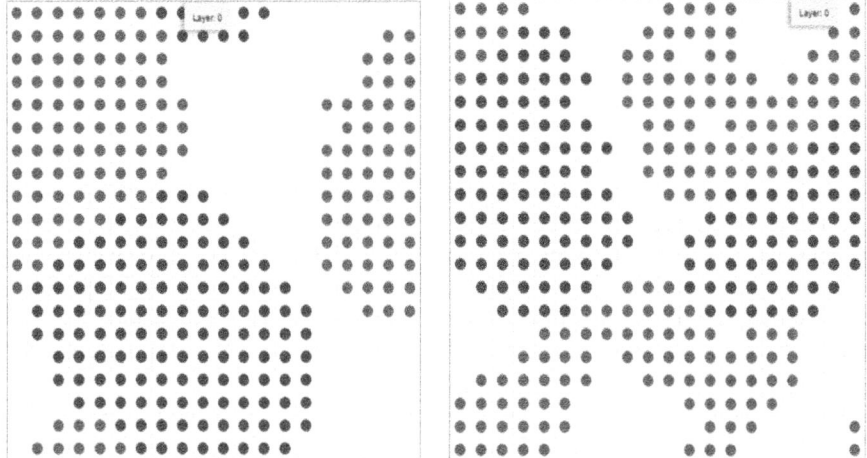

(c) Positions of agents at time step 500 in the evolutionary model (for the fixed threshold change rate).

(d) Positions of agents at time step 500 in the evolutionary model (for the randomized threshold change rate).

Fig. 4. Occurrence of segregation in both model with density 0.7 for threshold 0.5

model increases more with the uniformly distributed threshold change rate because it increases the heterogeneity of agents' behaviors.

– The second issue is that some agents in the evolutionary model are unable to settle due to their evolved individual threshold value. As a result, the model requires additional steps to converge or cannot converge at all. (see Fig. 1 and 4). In addition, in the vast majority of iterations, the average SI of the original model is greater than that of the evolutionary model (see Fig. 7). This also

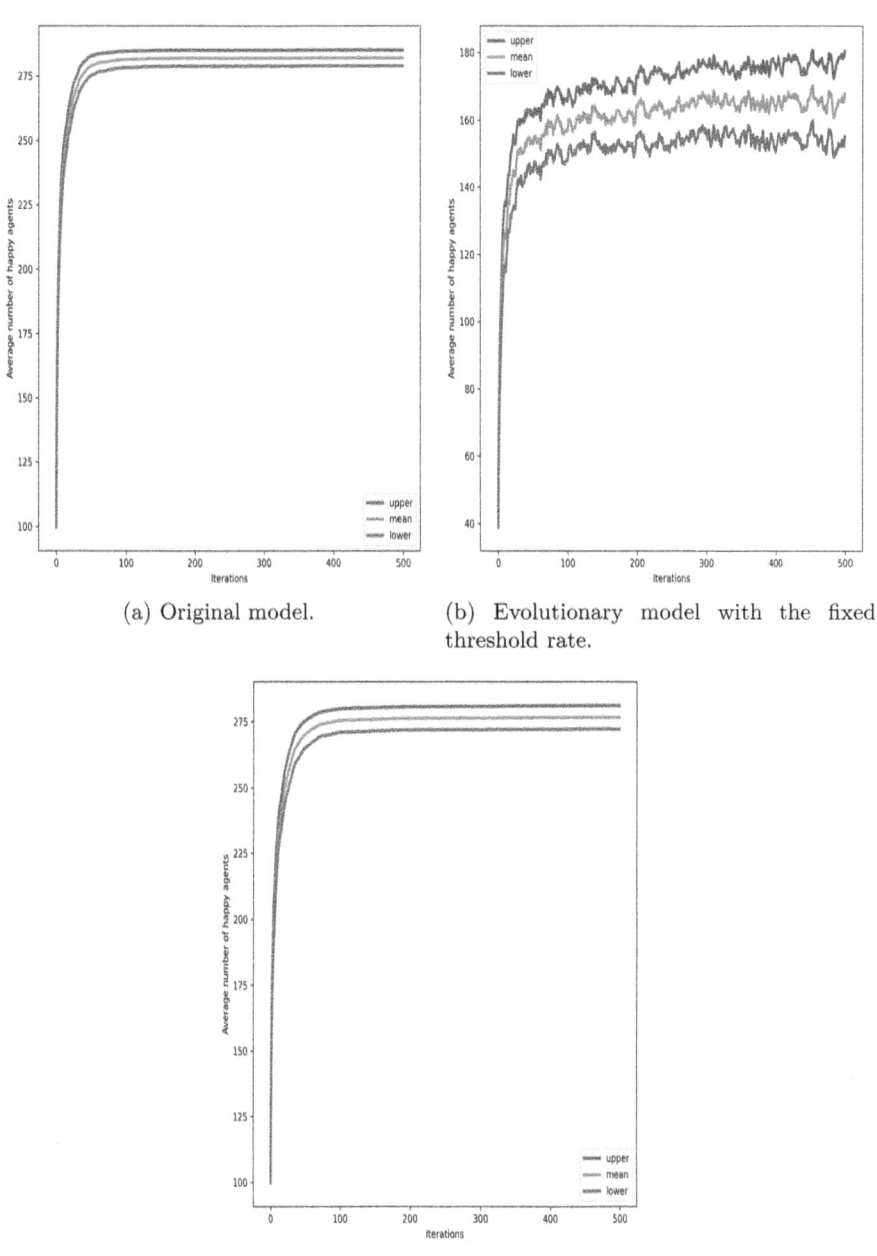

(a) Original model.

(b) Evolutionary model with the fixed threshold rate.

(c) Evolutionary model with randomized threshold change rate.

Fig. 5. Change of the number of happy agents for both models with density 0.7 for threshold 0.5.

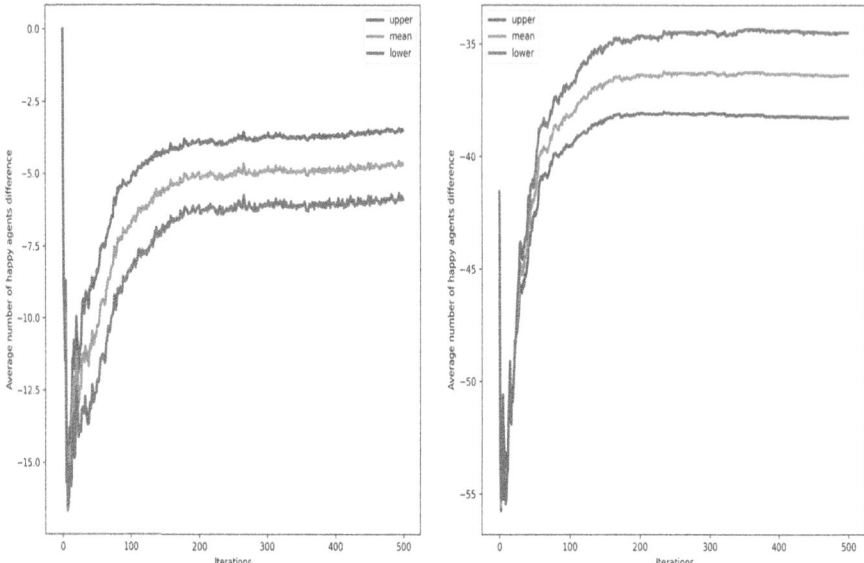

(a) Evolutionary model through iterations (b) Evolutionary model through iterations
with fixed threshold change rate.   with randomized threshold change rate.

Fig. 6. The difference in the number of happy agents between the original model and
evolutionary models for density 0.7 and threshold 0.5.

emphasizes that the mild preferences of agents in the original model accelerate
the process of segregation. In addition, Fig. 8 displays the average difference
in the number of moves between the original model and the evolutionary
model. We may observe that the difference in the number of moves between
two models increases with each iteration. This also suggests that the original
model converges more rapidly than the evolutionary model.

– Thirdly, the evolutionary model has more subgroups than the original model.
We believe that this is due to the fact that the evolution of threshold values causes certain agents to have identical threshold values, resulting in the
formation of subgroups within groups (see Fig. 1 and 4). We noted that the
threshold values of the agents who had not yet settled down by the end of
the simulation were higher than those of the settled down agents because
of their interaction over the simulation's iterations. This also demonstrates
how agents' preferences change over the course of the simulation, even if they
started off with the same ones. Moreover, The evolutionary model with the
randomized threshold change rate results in more subgroups than the fixed
threshold change rate because of the increase in the heterogeneity of agents'
behaviors.

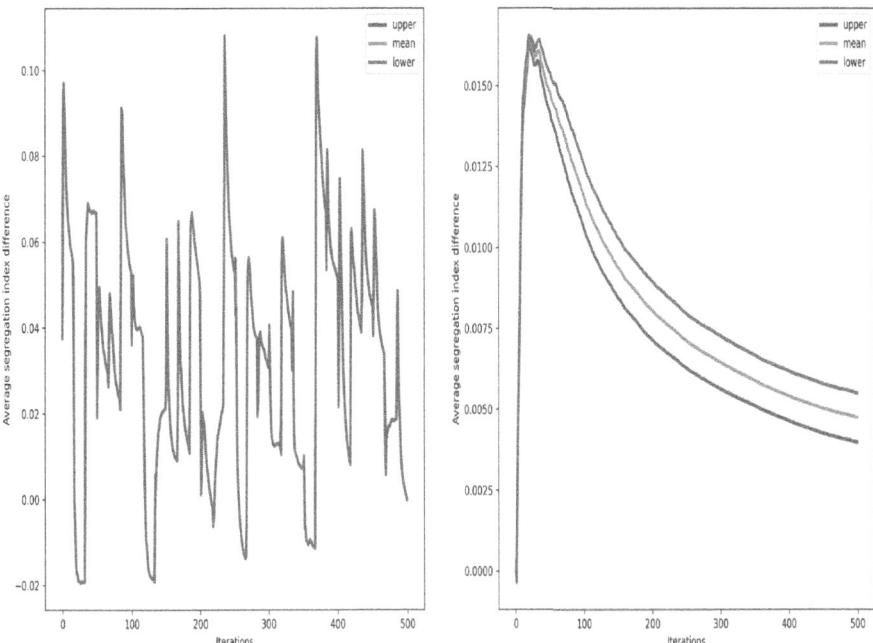

(a) For fixed threshold change rate, since upper, mean, and lower values are almost equal, only lower values (in green) are visible.

(b) For randomized threshold change rate.

Fig. 7. The difference in average segregation index between the original model and evolutionary model. (Color figure online)

- Fourth, we examined the effect of the fixed size of r_t on the model's outputs. We experimented with several numbers ranging from 0.000001 to 0.1 by multiplying them by 10 to reach the next r_t value. For smaller values of r_t, the number of happy agents reduces while the variability of evolutionary model outcomes increases; conversely, for larger values of r_t, the number of happy agents increases while the variability of evolutionary model outcomes decreases.
- The last one is the visual distinction between both model outcomes. Models generate distinct segregation patterns under the same initial conditions. This variation in pattern has a peculiar source. However, we believe it to be crucial from an emergent behavior perspective. Patterns evolve differently in the two models.

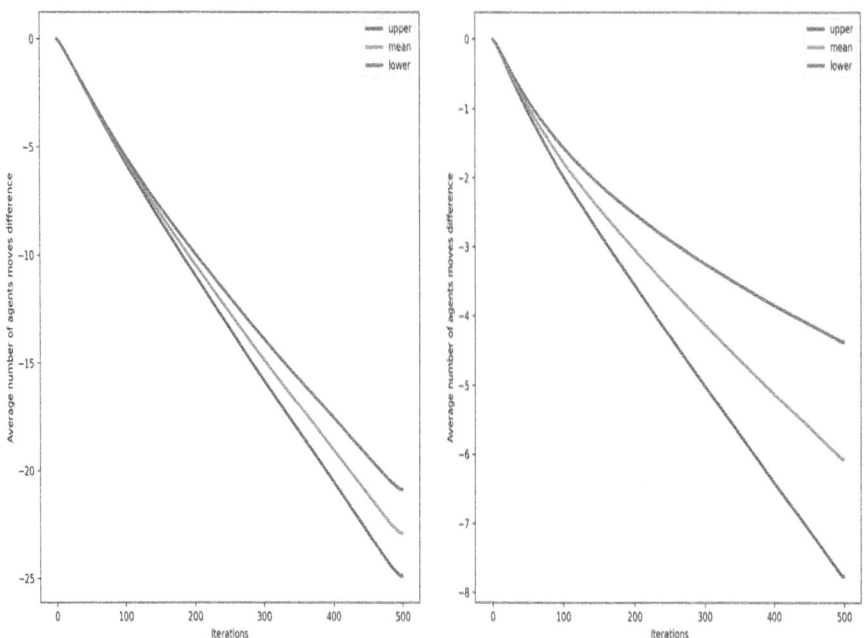

(a) For the fixed threshold change rate. (b) For randomized threshold change rate.

Fig. 8. The difference in the average number of agents' moves between the original model and evolutionary model (for the fixed threshold change rate).

5   Conclusion

In this study, we aimed to relax one of the main assumptions of the Schelling's original model, namely the static movement behavior pattern of agents over iterations, and analyze how this influences the dynamics of segregation. The reasoning behind this is that this evolving behaviour may be closer to reality. For this, we made the threshold parameter of the original model interaction-dependent and conducted tests to determine its effect on the dynamics of segregation. The results indicate that the outcomes of the evolutionary (dynamic) model are more variable than those of the original model. This variability modifies segregation patterns, causing the evolutionary model to converge later and to have more subgroups than the original model. Moreover, this variability implies that the evolutionary model is more complex than the original model.

Understanding the impact of real-world dynamics on segregation dynamics was a driving force for the study. Specifically, we were interested in the potential impact of evolutionary behavior on the dynamics of segregation. In real life, a person's preferences are influenced by their interactions with others. There cannot be distinct separations between human groups in the real world for a variety of reasons, the evolution of human preferences over time being one of them. Individual preferences evolve as a result of a person's exposure to new experiences and social interactions. Therefore, we may achieve more credible segregation patterns by incorporating these interaction-driven evolutionary aspects in the segregation model.

All other model parameters are kept constant to remain faithful to Schelling's original work and isolate the effect of the interaction-driven evolution. For instance, we conducted every experiment with two groups as in the original model. In future research, we intend to examine the effect of evolutionary behavior on the existence of more than two groups. In addition, the change in threshold resulting from interactions may vary for different groups; for instance, it may increase the threshold value for one group while decreasing it for another. So, we aim to investigate these scenarios in order to analyze the effect of evolutionary behavior on segregation dynamics with greater precision. Furthermore, we intend to improve the concept of evolutionary behavior in segregation dynamics by extending the original model, such as by incorporating the educational background of agents into their movement decisions and by incorporating political regulations that influence the movement decisions of agents.

References

1. Abella, D., San Miguel, M., Ramasco, J.J.: Aging effects in Schelling segregation model. Sci. Rep. **12**(1), 19376 (2022)
2. Ardestani, B.M., O'Sullivan, D., Davis, P.: A multi-scaled agent-based model of residential segregation applied to a real metropolitan area. Comput. Environ. Urban Syst. **69**, 1–16 (2018)
3. Banos, A.: Network effects in Schelling's model of segregation: new evidence from agent-based simulation. Environ. Plann. B. Plann. Des. **39**(2), 393–405 (2012)
4. Benito, J.M., Brañas-Garza, P., Sanchis, J.A., et al.: Strategic behavior in Schelling dynamics: a new result and experimental evidence (2011)
5. Flaig, J., Houy, N.: Altruism and fairness in Schelling's segregation model. Phys. A **527**, 121298 (2019)
6. Funder, D.C., Colvin, C.R.: Explorations in behavioral consistency: properties of persons, situations, and behaviors. J. Pers. Soc. Psychol. **60**(5), 773 (1991)
7. Ingold, T.: Becoming persons: consciousness and sociality in human evolution. Cult. Dyn. **4**(3), 355–378 (1991)
8. Jani, A.: An extension of Schelling's segregation model: modeling the impact of individuals' intolerance in the presence of resource scarcity. Commun. Nonlinear Sci. Numer. Simul. **85**, 105202 (2020)
9. Kazil, J., Masad, D., Crooks, A.: Utilizing python for agent-based modeling: the mesa framework. In: Thomson, R., Bisgin, H., Dancy, C., Hyder, A., Hussain, M. (eds.) SBP-BRiMS 2020. LNCS, vol. 12268, pp. 308–317. Springer, Cham (2020). https://doi.org/10.1007/978-3-030-61255-9_30
10. Liu, Z., Li, X., Khojandi, A., Lazarova-Molnar, S.: On the extension of Schelling's segregation model. In: 2019 Winter Simulation Conference (WSC), pp. 285–296. IEEE (2019)
11. Logan, J.R., Burdick-Will, J.: School segregation and disparities in urban, suburban, and rural areas. Ann. Am. Acad. Pol. Soc. Sci. **674**(1), 199–216 (2017)

12. Opp, S.M.: The forgotten pillar: a definition for the measurement of social sustainability in American cities. Local Environ. **22**(3), 286–305 (2017)
13. Ortega, D., Rodríguez-Laguna, J., Korutcheva, E.: A Schelling model with a variable threshold in a closed city segregation mode. Analysis of the universality classes. Phys. A: Stat. Mech. Appl. **574**, 126010 (2021)
14. Park, Y.M., Kwan, M.P.: Beyond residential segregation: a spatiotemporal approach to examining multi-contextual segregation. Comput. Environ. Urban Syst. **71**, 98–108 (2018)
15. Schelling, T.C.: Dynamic models of segregation. J. Math. Sociol. **1**(2), 143–186 (1971)
16. Stoica, V.I., Flache, A.: From Schelling to schools: a comparison of a model of residential segregation with a model of school segregation. J. Artif. Soc. Soc. Simul. **17**(1), 5 (2014)
17. Wang, Y.: Beyond preference: modelling segregation under regulation. Comput. Environ. Urban Syst. **54**, 388–396 (2015)
18. Washburn, S.L., Lancaster, G.: The evolution of hunting. In: Man the Hunter, pp. 293–303. Routledge (2017)

Identification of Hybrid Systems by Fuzzy C-Regression Clustering

Sašo Blažič$^{(\boxtimes)}$ (ID) and Igor Škrjanc (ID)

Faculty of Electrical Engineering, University of Ljubljana, Ljubljana, Slovenia
{saso.blazic,igor.skrjanc}@fe.uni-lj.si

Abstract. This paper introduces a method for identifying hybrid systems using fuzzy C-regression clustering. A distinctive aspect of this approach is the use of hyperplanes as cluster prototypes in the input space, with shared parameters between the antecedent and consequent parts. The algorithm features a small number of design parameters, making it easy to implement and tune. The effectiveness of the proposed method is demonstrated on three simulated examples.

Keywords: Fuzzy C-regression clustering · Identification · Hybrid systems

1 Introduction

We are in the era of data. Data are being generated everywhere. There is a great need to extract knowledge from these data. These streaming data need to be processed online. Many different areas of science and technology are looking at machine learning algorithms. Systems and control engineers search for the relationships between data that can be described using models. Since these underlying models are mostly nonlinear, we need to apply appropriate techniques. Online identification of nonlinear systems based on fuzzy systems and neuro-fuzzy networks is covered in many publications [3,4,6,11,12,15,17,19,21,22,30–32,34,36,37]. Streaming data are inherently non-stationary and not only the parameters but also the structure of the models needs to adapt over time, leading to the so-called evolving systems [1,2,5,13,14,16,23,28,35]. A common method for identifying a nonlinear model, such as a fuzzy model, involves partitioning the input space and estimating the parameters of the consequent part. Evolving systems do these two tasks automatically and in parallel. But so many degrees of freedom make these algorithms vulnerable to the creation of incorrect local models, local models that are not accurate enough, models based on old data, etc. Several supervisory mechanisms have been proposed to avoid these problems.

Hybrid systems include both continuous and discrete dynamics, which makes their analysis and control a challenging task. An alternative approach to describing nonlinear systems is to use hybrid models in which the system dynamics

This work has been supported by Slovenian Research Agency (ARRS) with the Research Program P2-0219.

© The Author(s), under exclusive license to Springer Nature Switzerland AG 2024
M. Mujica Mota and P. Scala (Eds.): EUROSIM 2023, CCIS 2033, pp. 259–273, 2024.
https://doi.org/10.1007/978-3-031-68438-8_19

change abruptly and switch between different modes of operation [20]. The switchings can occur due to the partitioning of the state space or as a result of an external excitation. Typical natural examples of hybrid systems include systems with multiple modes of operation, systems with actuator nonlinearities (saturation, backlash, dead zone, etc.). Very often local models are affine, leading to the so-called piecewise affine systems. Similar to the fuzzy models, hybrid systems are universal approximators. Hybrid systems identification is not a novel technique, with some methods (under various names) traceable back several decades [29]. In recent years, this field has seen a surge in research activity [7–9, 20, 24–27, 33].

In this paper, we propose adapting the incremental fuzzy C-regression clustering method [10] for hybrid system identification. This approach is based on the evolving principle, allowing models to be updated online and new local models to be created as needed. The robustness of the approach is improved by the possibility to keep the measurements in the buffer – the actual decision whether to keep or discard the data can be made at a later stage. The key features of the proposed approach include the straightforward tuning of a few design parameters and the simplicity of its implementation.

2   Hybrid System Represented as a Local Model Network

Numerous approaches to nonlinear systems modelling (including those referenced in the introduction) assume that the system output y can be represented by a local model network:

$$y = \sum_{j=1}^{m} \mu_j(\mathbf{u}_p) y_j(\mathbf{u}) \tag{1}$$

Here $\mathbf{u}_p \in \mathbb{R}^q$ and $\mathbf{u} \in \mathbb{R}^r$ while $\mu_j \in [0,1]$ denotes the validity of the corresponding local model output y_j, where $\mu_j(\mathbf{u}_p)$ provides information about the regions in $\mathbf{u}_p$ space where the local models are valid. The validity function μ_j corresponds to the partition of unity within the convex set $\mathcal{C}$ that encompasses the entire region of interest of $\mathbf{u}_p$. In the framework of Takagi-Sugeno fuzzy models [4, 15], $\mathbf{u}_p$ represents antecedent variables, $\mathbf{u}$ represents consequent variables, $\mu_j(\cdot)$ denotes membership functions, and m stands for the number of rules in the rule base. Equation (1) can be interpreted as a mapping from $\mathbf{u}_p$ and $\mathbf{u}$ to y.

The functions $y_j(\mathbf{u})$ can adopt nearly any form, although linear or affine functions are frequently chosen for their simplicity. In our approach, we will employ affine functions:

$$y = \sum_{j=1}^{m} \mu_j(\mathbf{u}_p) \mathbf{u}_e^T \boldsymbol{\theta}_j \tag{2}$$

with $\mathbf{u}_e^T = [1, u_1, u_2, \ldots, u_r]$ and $\boldsymbol{\theta}_j^T = [\theta_{j0}, \theta_{j1}, \ldots, \theta_{j,r}]$.

The concept behind hybrid systems remains fundamentally similar. They also use multiple local models. However, the difference between hybrid systems and the aforementioned local model networks is that in a hybrid system, at any given moment only one local model is active. To be consistent with the notation in (2), the output of the hybrid system can be defined as follows:

$$y = \boldsymbol{u}_e^T \boldsymbol{\theta}_j \quad 1 \leq j \leq m \tag{3}$$

All models are indexed by integers j. Any (abrupt) transition of the operating mode is reflected by a sudden change of the model index j in (3). The transition can occur due to three distinct scenarios:

- The operating mode changes due to the change of an internal state. The corresponding state variable should be contained in the vector $\boldsymbol{u}_p$. The operating mode is determined by a particular partition of the domain of $\boldsymbol{u}_p$, ensuring a unique mapping from $\boldsymbol{u}_p$ to j. Equation (2) remains valid, although only binary validities (0 or 1) are allowed.
- The operating mode changes as a result of some (external) quantity(s) or signal(s). If these quantities are known, they can be incorporated into the vector $\boldsymbol{u}_p$, and again Eq. still (2) holds.
- The operating mode changes based on time (schedule). Strictly speaking, this transformation makes the system a linear (or affine) time-varying system, which is a departure from the classical definition of a hybrid system.

As demonstrated, hybrid systems represent a subset of the broader class of nonlinear systems modelled by Eq. (2). This paper deals with the task of identifying hybrid systems by adapting an approach [10] originally developed for local model networks to the requirements of hybrid systems.

3 Incremental C-Regression for Identification of Hybrid Systems

Incremental fuzzy C-regression [10] is an evolving technique that generates a fuzzy model from streaming data. Unlike most existing approaches that partition the input space $\boldsymbol{u}_p$ based on the finite compact clusters, this approach assumes that the cluster prototypes are hyperplanes. Additionally, for the sake of model simplification and parameter number reduction, we assume that the hyperplane in the antecedent part of the model and a function found in the consequent part are identical. The main difference between the proposed approach and that of [10] is that here only one local model is updated.

The construction of the model can either start from scratch or we can construct some initial local models if a priori information is available. In either case, the model can add additional local models as needed. This approach does not provide techniques for merging, deleting, or splitting the existing clusters, although existing methods from the literature could be implemented.

When a new measurement (or information in some other form) arrives, the distances to the existing m clusters (hyperplanes) are calculated:

$$d_{jk}^2 = \left(y_k - \boldsymbol{u}_{ek}^T \boldsymbol{\theta}_j \right)^2 \quad j = 1, \ldots, m \tag{4}$$

Then we determine the nearest cluster, which we call the winning one and denote it by the index w ($w = \arg\min_{1 \leq j \leq m} d^2_{jk}$). A decision must be made as to how to proceed. We have several options:

1. If the nearest cluster is *close* enough, we update its parameters. The distance measure is determined by examining the previous samples associated with that particular cluster.
2. If the nearest cluster is *far* but not very far away, we keep the last measurement in the buffer. The contents of the buffer will be re-evaluated after the arrival of future data.
3. If the nearest cluster is *very far* away, the new measurement is discarded based on the assessment that it is a bad measurement or an outlier.

3.1   The Distance Measure

The distance measure that determines which of the three options above you proceed with should be relative in nature. After assigning a new measurement to a specific winning cluster, the most recent distance d^2_{wk} is compared to the average distance of previous samples assigned to that cluster, denoted as $\overline{d^2}_w$. The choice between the three options depends on the following conditions:

1. The nearest cluster is considered *close* if $d^2_{wk} < \kappa_{min}\overline{d^2}_{w,k-1}$.
2. The nearest cluster is considered *far* if $\kappa_{min}\overline{d^2}_{w,k-1} \leq d^2_{wk} \leq \kappa_{max}\overline{d^2}_{w,k-1}$.
3. The nearest cluster is considered *very far* if $d^2_{wk} > \kappa_{max}\overline{d^2}_{w,k-1}$.

Here we introduce two dimensionless parameters κ_{min} and κ_{max}. While inherently relative, κ_{min} and κ_{max} still require tuning to yield satisfactory results (typically κ_{min} is chosen from the interval $(1, 100)$). A specific choice of κ_{min} allows a simple trade-off between the size of the modelling error and the number of local models. The parameter κ_{max} is easier to tune because we can put it in a simple relation to κ_{min}, e.g. $\kappa_{max} = 2\kappa_{min}$.

If the first condition is met, the average $\overline{d^2}_w$ of the winning cluster must be updated:

$$\overline{d^2}_{wk} = \frac{k_w-1}{k_w}\overline{d^2}_{w,k-1} + \frac{1}{k_w}d^2_{wk} \tag{5}$$

where k_w is the number of samples in the w-th local model, and the indices $k-1$ and k of $\overline{d^2}_w$ in Eq. (5) represent the time dependence. At time $k = 0$ all local models are initialised to 0: $\overline{d^2}_{j,0} = 0$.

3.2   Update of the Local Cluster Parameters

If the first condition above is satisfied, the parameters of the w-th local model are updated. This is done by a weighted recursive least squares algorithm, which consists of the following steps [18]:

1. The error e_{wk} at time k, representing the difference between the current output y_k and the weighted model output based on the previous parameter estimate $\boldsymbol{u}^T_{ek}\boldsymbol{\theta}_{w,k-1}$, is computed first:

$$e_{wk} = y_k - \boldsymbol{u}^T_{ek}\boldsymbol{\theta}_{j,k-1} \tag{6}$$

2. Next, the vector of innovation gain $\boldsymbol{K}_{wk}$ at k is obtained:

$$\boldsymbol{K}_{wk} = \boldsymbol{P}_{w,k-1} \boldsymbol{u}_{ek} \left(\gamma + \boldsymbol{u}_{ek}^T \boldsymbol{P}_{w,k-1} \boldsymbol{u}_{ek} \right)^{-1} \tag{7}$$

where $\boldsymbol{P}_{w,k-1}$ stands for the covariance matrix of the estimation error at time $(k-1)$, and a positive constant $\gamma \leq 1$ is the user-defined forgetting factor (typically between 0.95 and 1).

3. The covariance matrix of the estimation error $\boldsymbol{P}_{wk}$ is then computed:

$$\boldsymbol{P}_{wk} = \frac{1}{\gamma} \left(\boldsymbol{I} - \boldsymbol{K}_{wk} \boldsymbol{u}_{ek}^T \right) \boldsymbol{P}_{w,k-1} \tag{8}$$

4. The parameters of the model $\boldsymbol{\theta}_w$ are finally updated:

$$\boldsymbol{\theta}_{wk} = \boldsymbol{\theta}_{w,k-1} + \boldsymbol{K}_{wk} e_{wk} \tag{9}$$

The general approach outlined in this paper is not exclusively confined to the online parameter estimation algorithm described above. Other algorithms based on recursive least squares can also be employed.

3.3   Temporary Buffer for Recent Data

It is important not to initiate a new local model as soon as it is determined that the current measurement does not match any existing local models. This precautionary measure helps to prevent the formation of erroneous local models, which can occur due to outliers or insufficient information content in the data. Instead, temporary data is stored in a buffer. Only when it is determined that the data in the buffer can form a new local model will a new model be initialised based on this data. Various criteria can be used for this purpose. Typically, it is strongly recommended to ensure that the data in the buffer span a distinct hyperplane. In addition, for practical reasons, it is advisable not to create a new model based on non-consecutive data patches.

With the decision to construct a new local model, the number of local models m increases by 1, which requires the initialisation of all parameters defining the new local cluster. The parameter vector of the local affine model defining the hyperplane can be derived from the data in the buffer using the established least squares method for affine systems. Initialisation of the other parameters such as $\boldsymbol{P}_m, k_m, \overline{d_m^2}$ is straightforward.

4   Examples

The approach in this paper has been tested on two simple systems that are clear representatives of hybrid systems. Even from such simple systems, some interesting conclusions can be drawn. It is very important to emphasise that it is relatively easy to tune the design parameters of this algorithm. Only a minimal set of tuning parameters is required and often default values are sufficient. In fact, throughout all the experiments described in this paper, all the parameters remain constant, i.e. $\gamma = 0.98$, $\kappa_{min} = 5$, $\kappa_{max} = 2\kappa_{min} = 10$.

4.1   Tank with Two Cross-Sections

The process discussed in this section (see Fig. 1) can be classified as a piece-wise affine system with two distinct dynamics – partitioning is induced in the state space. In Example 2 the same system is used but another partitioning is introduced by the input saturation (as a result of the actuator limitation). The dynamics of the system will therefore change depending on the input and/or state.

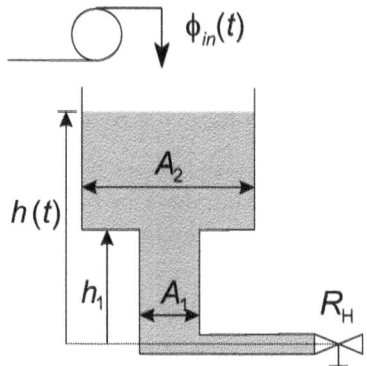

Fig. 1. The illustration of the tank used in Example 1. The tank has two different cross-sections, being wide above a certain water level and narrow below. The system also consists of an inlet pump and an outlet valve with a linear static curve.

Example 1. The tank shown in Fig. 1 has a variable cross-section, with the area changing with the water level h:

$$A(h) = \begin{cases} A_1 & \text{if } h < h_1 \\ A_2 & \text{otherwise.} \end{cases} \tag{10}$$

It is very easy to obtain the continuous-time model of this dynamic system:

$$A(h)\dot{h} = \phi_{in} - \frac{1}{R_H}h \tag{11}$$

where the discharge is modelled based on the assumption of a linear hydraulic resistance R_H of the discharge valve. This system can be viewed as a continuous-time hybrid system with two modes of operation ('low level' and 'high level'), each of which exhibits affine dynamics. The change of dynamics is governed by the system state (water level h).

The simulations are conducted using the continuous-time model of the process outlined in Eq. (11). However, the identification is performed using the sampled measurements. To avoid numerical problems, a relatively slow sampling period of $T = 10$ is used. The input varies sinusoidally between $0.5 \cdot 10^{-3}$ and $2.5 \cdot 10^{-3}$ with linearly increasing frequency (from $1.2 \cdot 10^{-3}$ to $2.8 \cdot 10^{-3}$). The process parameters are: $h_1 = 0.5$, $A_1 = 0.1$, $A_2 = 0.4$, $R_H = 400$. A zero-mean measurement noise with a standard deviation of 0.001 was added to the measured output. *Note that all units of measurement are omitted for brevity.*

The local model in an operating point follows the discrete-time representation:

$$h_{k+1} = \theta_1 \phi_{1k} + \theta_2 h_k + \theta_3 \tag{12}$$

where θ_1, θ_2 and θ_3 are unknown parameters that could be obtained if the process parameters and sampling time were known. However, this mapping also depends on assumptions about the behaviour of the system between samples. In this paper, these parameters are estimated using system identification.

Figure 2 shows the data used by the algorithm – different colours indicate different modes of operation. Figure 3 shows how the data were associated with a particular local model during identification. We can see that some data were skipped because they could not be reliably associated with a particular local model. We can also see some errors due to incorrect associations. Figure 4 shows how the identified parameters evolve over time. The red colour is associated with a few samples at the beginning. The resulting "red" model is not good enough and could be deleted as it is not used after the beginning. The other two models correctly describe the model in the upper and lower parts of the tank. The dotted lines show the "ideal" parameters, but note that although the original system (11) is piecewise affine, these "ideal" model parameters for (12) are not completely unique. There are at least two reasons for this: the input $\phi_{in}(t)$ changes between samples; and the change between two modes occurs at a certain time between two samples.

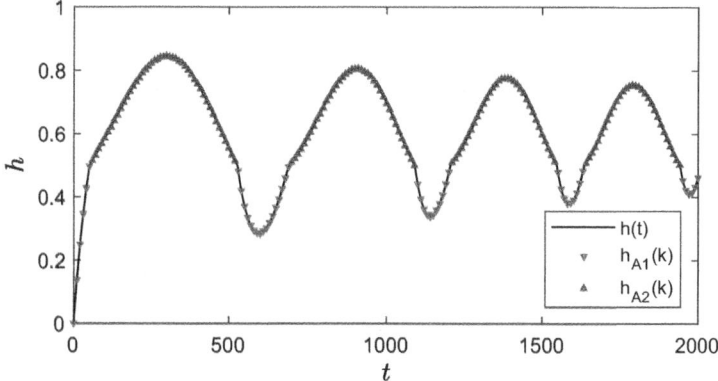

Fig. 2. Example 1 measurements: the actual water level in the tank (black), noisy measurements (red signs in the upper part, blue signs in the lower part). (Color figure online)

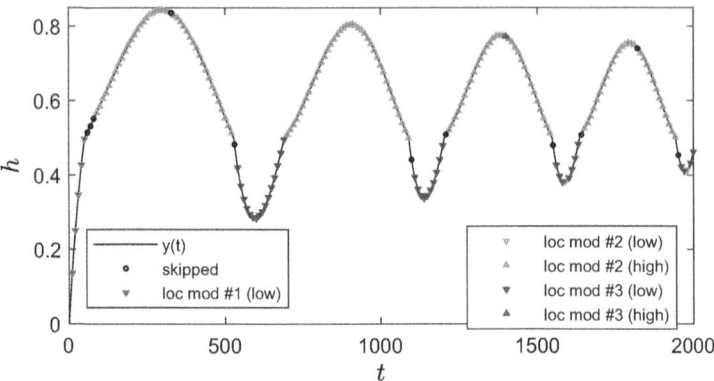

Fig. 3. Example 1 classification results: red signs depict measurements associated with local model #1, green with #2, and blue with #3. (Color figure online)

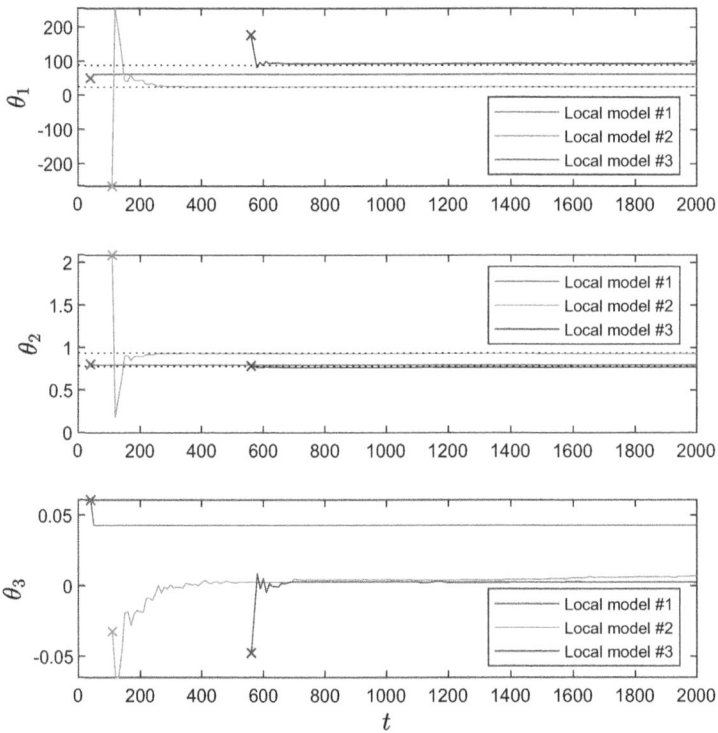

Fig. 4. Example 1 parameter evolution: initialisation (shown with crosses) and adaptation of individual parameters.

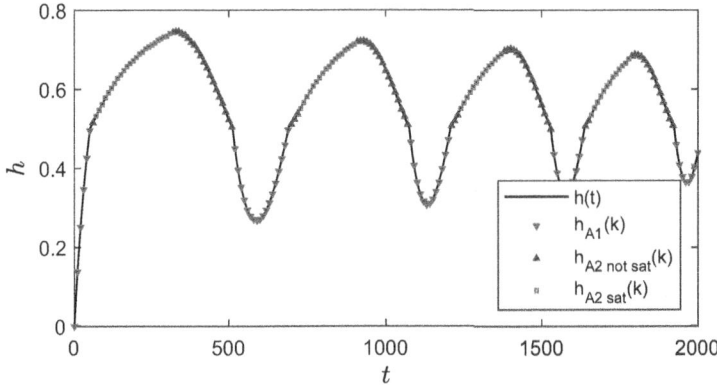

Fig. 5. Example 2 measurements: the actual water level in the tank (black), noisy measurements (red signs for the upper part, magenta squares for the saturation, blue signs for the lower part). (Color figure online)

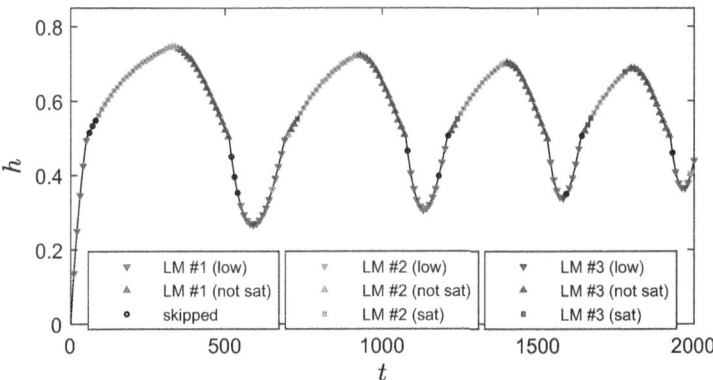

Fig. 6. Example 2 classification results: red signs depict measurements associated with local model #1, green with #2, and blue with #3. (Color figure online)

Example 2. This example deals with the same system as the previous one, with an additional partition introduced due to the saturation of the input (at $2 \cdot 10^{-3}$). Most often, the closed-loop system is analysed after the saturation of the input, but it is more transparent to analyse it in the open-loop. The effect of saturation is that high inputs are limited in the process, while this information is not considered in the identification algorithm, i.e. the non-saturated input is processed. Figure 5 shows the collected data with different colours indicating different modes. Figure 6 shows how these data are associated to local models during identification. Some errors due to noise and problems with "ideal" parameters can be seen. Figure 7 shows the identified parameters.

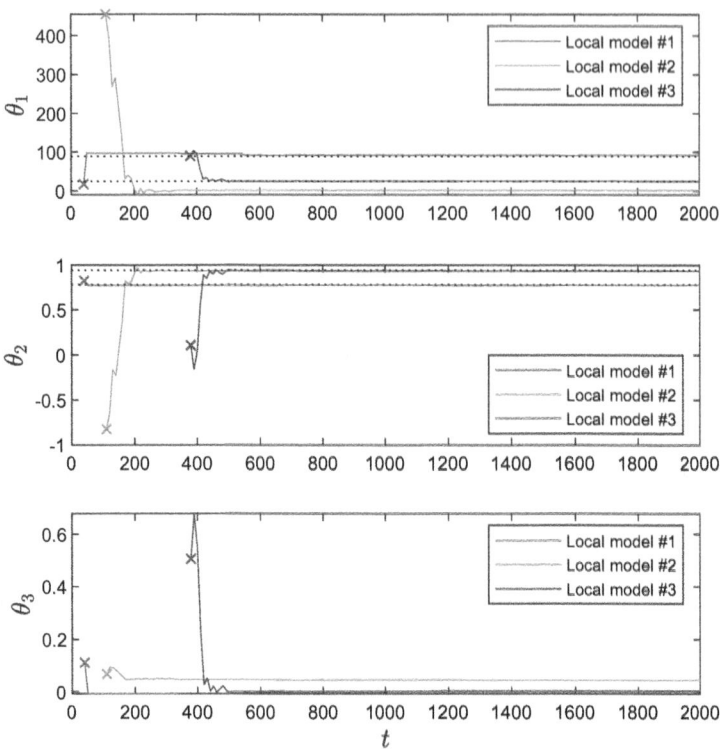

Fig. 7. Example 2 parameter evolution: initialisation (shown with crosses) and adaptation of individual parameters.

4.2 Control System

Example 3. Controllers with multiple control modes are typical examples of hybrid systems. In this case, an on-off controller with hysteresis (the width is 0.2) controls the process with a transfer function $G(s) = \frac{1}{(s+1)^2}$. Our objective is to identify the control system that already exists. This control system is a continuous system, which means that the time of the change of the control mode is completely independent of the sampling time, which complicates the identification problem. A zero-mean normally distributed disturbance is added to the system state (not the output).

We assume the following local discrete-time model:

$$y_{k+2} = \theta_1 r_k + \theta_2 y_{k+1} + \theta_3 y_k \tag{13}$$

where the reference r_k provides the desired value for the process output y_k.

Figures 8 and 9 show the measured data. The effect of the disturbance is significant. Figure 10 shows the identified parameters.

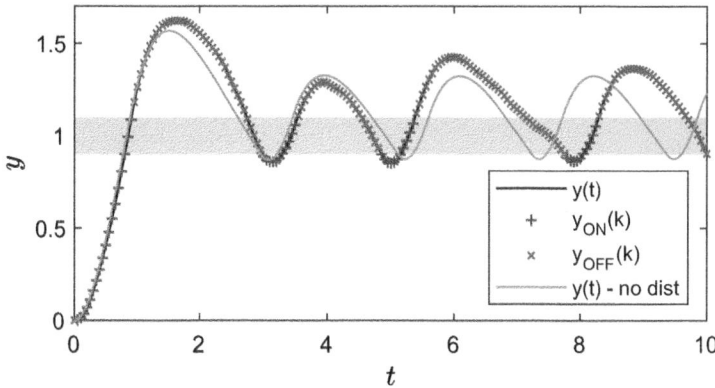

Fig. 8. Example 3 measurements: green – output (without disturbance), black – output (with disturbance), blue – controller in ON state, red – controller in OFF state. (Color figure online)

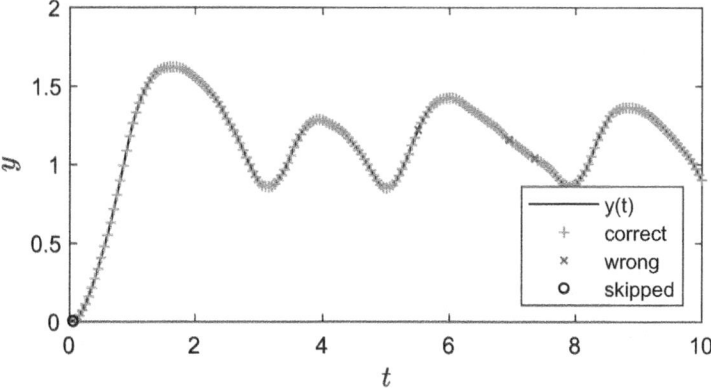

Fig. 9. Example 3 online data association during identification: green pluses represent measurements correctly assigned to a local model, red crosses denote incorrect assignments, and black circles indicate skipped measurements. (Color figure online)

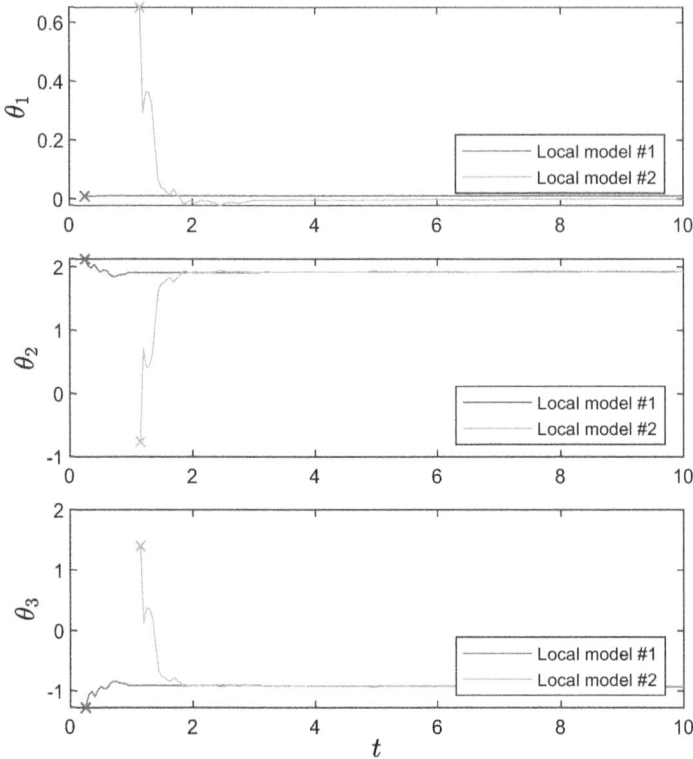

Fig. 10. Example 3 parameter evolution: initialisation (shown with crosses) and adaptation of individual parameters, red – controller in ON state, green – controller in OFF state. (Color figure online)

5   Conclusion

In this paper we present an adaptation of the incremental fuzzy C-regression clustering method [10] for hybrid system identification. The original approach, developed for Takagi-Sugeno fuzzy model identification, operates on the evolving principle, updating models online and creating new local models if needed. The method proposed in this paper retains the evolving principle, but with the constraint that only one model is updated at each instant.

This online identification approach for hybrid systems is very promising. It offers ease of implementation, requiring the selection of only a few design parameters, and high interpretability. Only one parameter (κ_{min}) significantly influences the behaviour of the identification process, allowing a trade-off between the accuracy (the size of the modelling error) and the complexity (the number of local models). One potential problem that may arise is the inadvertent creation of incorrect local models. However, this is a common challenge in evolving systems, and additional mechanisms are needed to retain only true local models, discard obsolete or erroneous models, merge similar models, and so on.

References

1. Andonovski, G., Angelov, P.P., Blažič, S., Škrjanc, I.: Robust evolving cloud-based controller (RECCo). In: Škrjanc, I., Blažič, S. (eds.) 2017 Evolving and Adaptive Intelligent Systems, EAIS 2017, Ljubljana, Slovenia, 31 May–2 June 2017, pp. 1–6. IEEE (2017)
2. Andonovski, G., Costa, B.S.J., Blažič, S., Škrjanc, I.: Robust evolving controller for simulated surge tank and for real two-tank plant. Automatisierungstechnik **66**(9), 725–734 (2018)
3. Andonovski, G., Mušič, G., Blažič, S., Škrjanc, I.: Evolving model identification for process monitoring and prediction of non-linear systems. Eng. Appl. Artif. Intell. **68**, 214–221 (2018)
4. Angelov, P.P., Filev, D.P.: An approach to online identification of Takagi-Sugeno fuzzy models. IEEE Trans. Syst. Man Cyber. Part B **34**(1), 484–497 (2004)
5. Angelov, P.P., Škrjanc, I., Blažič, S.: Robust evolving cloud-based controller for a hydraulic plant. In: Proceedings of the 2013 IEEE Conference on Evolving and Adaptive Intelligent Systems, EAIS 2013. IEEE Symposium Series on Computational Intelligence (SSCI), Singapore, 16–19 April 2013, pp. 1–8. IEEE (2013)
6. Azeem, M.F., Hanmandlu, M., Ahmad, N.: Structure identification of generalized adaptive neuro-fuzzy inference systems. IEEE Trans. Fuzzy Syst. **11**(5), 666–681 (2003)
7. Bako, L.: Identification of switched linear systems via sparse optimization. Automatica **47**(4), 668–677 (2011)
8. Bemporad, A., Garulli, A., Paoletti, S., Vicino, A.: A bounded-error approach to piecewise affine system identification. IEEE Trans. Autom. Control **50**(10), 1567–1580 (2005)
9. Bianchi, F., Prandini, M., Piroddi, L.: A randomized two-stage iterative method for switched nonlinear systems identification. Nonlinear Anal. Hybrid Syst **35**, 100818 (2020)
10. Blažič, S., Škrjanc, I.: Incremental fuzzy C-regression clustering from streaming data for local-model-network identification. IEEE Trans. Fuzzy Syst. **28**(4), 758–767 (2020)
11. Blažič, S., Dovžan, D., Škrjanc, I.: Cloud-based identification of an evolving system with supervisory mechanisms. In: 2014 IEEE International Symposium on Intelligent Control, ISIC 2014, Juan Les Pins, France, 8–10 October 2014, pp. 1906–1911. IEEE (2014)
12. Borlea, I.D., Precup, R.E., Borlea, A.B., Iercan, D.: A unified form of fuzzy C-means and K-means algorithms and its partitional implementation. Knowl.-Based Syst. **214**, 106731 (2021)
13. Dovžan, D., Škrjanc, I.: Recursive clustering based on a Gustafson-Kessel algorithm. Evol. Syst. **2**, 15–24 (2011)
14. Dovžan, D., Škrjanc, I.: Possible use of evolving C-regression clustering for energy consumption profiles classification. In: 2015 IEEE International Conference on Evolving and Adaptive Intelligent Systems, EAIS 2015, Douai, France, 1–3 December 2015, pp. 1–6. IEEE (2015)

15. Dovžan, D., Škrjanc, I.: Fuzzy space partitioning based on hyperplanes defined by eigenvectors for Takagi-Sugeno fuzzy model identification. IEEE Trans. Ind. Electron. **67**(6), 5144–5153 (2020)
16. Filev, D., Georgieva, O.: An extended version of the Gustafson-Kessel algorithm for evolving data stream clustering, chap. 12, pp. 273–299. Wiley (2010)
17. Garcia, C., Esmin, A., Leite, D.F., Škrjanc, I.: Evolvable fuzzy systems from data streams with missing values: with application to temporal pattern recognition and cryptocurrency prediction. Pattern Recogn. Lett. **128**, 278–282 (2019)
18. Haykin, S.S.: Adaptive Filter Theory, 4th edn. Prentice Hall, Upper Saddle River (2002)
19. Johanyak, Z.: Fuzzy rule interpolation based model for student result prediction. J. Intell. Fuzzy Syst. **36**(2), 999–1008 (2019)
20. Lauer, F., Bloch, G.: Hybrid System Identification. Springer, Cham (2019). https://doi.org/10.1007/978-3-030-00193-3
21. Leite, D.F., Andonovski, G., Škrjanc, I., Gomide, F.: Optimal rule-based granular systems from data streams. IEEE Trans. Fuzzy Syst. **28**(3), 583–596 (2020)
22. Lin, C.T.: A neural fuzzy control system with structure and parameter learning. Fuzzy Sets Syst. **70**, 183–212 (1995)
23. Lughofer, E., Pratama, M., Škrjanc, I.: Incremental rule splitting in generalized evolving fuzzy regression models. In: Škrjanc, I., Blažič, S. (eds.) 2017 Evolving and Adaptive Intelligent Systems, EAIS 2017, Ljubljana, Slovenia, 31 May–2 June 2017, pp. 1–8. IEEE (2017)
24. Nelles, O.: Nonlinear System Identification. From Classical Approaches to Neural Networks and Fuzzy Models. Springer, Berlin (2001). https://doi.org/10.1007/978-3-662-04323-3
25. Ohlsson, H., Ljung, L.: Identification of switched linear regression models using sum-of-norms regularization. Automatica **49**(4), 1045–1050 (2013)
26. Ozay, N., Sznaier, M., Lagoa, C.M., Camps, O.I.: A sparsification approach to set membership identification of switched affine systems. IEEE Trans. Autom. Control **57**(3), 634–648 (2012)
27. Pillonetto, G.: A new kernel-based approach to hybrid system identification. Automatica **70**, 21–31 (2016)
28. Precup, R.E., Bojan-Dragos, C.A., Hedrea, E.L., Roman, R.C., Petriu, E.M.: Evolving fuzzy models of shape memory alloy wire actuators. Rom. J. Inf. Sci. Technol. **24**(4), 353–365 (2021)
29. Quandt, R.E.: The estimation of the parameters of a linear regression system obeying two separate regimes. J. Am. Stat. Assoc. **53**(284), 873–880 (1958)
30. Rong, H.J., Sundararajan, N., Huang, G.B., Saratchandran, P.: Sequential adaptive fuzzy inference system (SAFIS) for nonlinear system identification and prediction. Fuzzy Sets Syst. **157**(9), 1260–1275 (2006)
31. Tzafestas, S.G., Zikidis, K.C.: NeuroFAST: on-line neuro-fuzzy art-based structure and parameter learning tsk model. IEEE Trans. Syst. Man Cyber. Part B **31**(5), 797–802 (2001)
32. Černe, G., Dovžan, D., Škrjanc, I.: Short-term load forecasting by separating daily profiles and using a single fuzzy model across the entire domain. IEEE Trans. Ind. Electron. **65**(9), 7406–7415 (2018)
33. Vo Tan, P., Millérioux, G., Daafouz, J.: A contribution to the identification of switched dynamical systems over finite fields. In: 49th IEEE Conference on Decision and Control (CDC), pp. 4429–4434 (2010)

34. Škrjanc, I., Blažič, S., Lughofer, E., Dovžan, D.: Inner matrix norms in evolving cauchy possibilistic clustering for classification and regression from data streams. Inf. Sci. **478**, 540–563 (2019)
35. Škrjanc, I., Dovžan, D., Gomide, F.A.C.: Evolving fuzzy-madel-based on C-regression clustering. In: 2014 IEEE Conference on Evolving and Adaptive Intelligent Systems, EAIS 2014, Linz, Austria, 2–4 June 2014, pp. 1–7. IEEE (2014)
36. Zamfirache, I.A., Precup, R.E., Roman, R.C., Petriu, E.M.: Reinforcement learning-based control using Q-learning and gravitational search algorithm with experimental validation on a nonlinear servo system. Inf. Sci. **583**, 99–120 (2022)
37. Škrjanc, I., Iglesias, J.A., Sanchis, A., Leite, D., Lughofer, E., Gomide, F.: Evolving fuzzy and neuro-fuzzy approaches in clustering, regression, identification, and classification: a survey. Inf. Sci. **490**, 344–368 (2019)

Closed-Loop Workload Input-Output Control of Production Systems: A Hybrid Simulation Study

G. Mušič[1]([⊠])[iD] and J. K. Sagawa[2]

[1] Faculty of Electrical Engineering, University of Ljubljana, Ljubljana, Slovenia
gasper.music@fe.uni-lj.si
[2] Production Engineering Department, Federal University of São Carlos,
Rodovia Washington Luís, km 235, São Carlos, SP, Brazil
juliana@dep.ufscar.br

Abstract. Workload Control (WLC) is a method of adjusting the overall workload in the production system by controlling the input of production orders and the output of produced items. It contributes to the predictability of lead times and more accurate delivery date commitments in make-to-order manufacturing. Input control relates to order release and output control to capacity adjustment. Recently, a novel approach to simultaneous and integrated input-output control has been proposed: a dynamic closed-loop model where automatic feedback control is added and where order release and capacity decisions are based on the observed shop floor state. The control was applied to an abstracted continuous dynamic model of a shop with unidirectional flow and three control rules were tested in scenarios with unbalanced work in process (WIP) and demand fluctuations. In this paper, the approach was further investigated in a more realistic hybrid simulation setting, combining discrete-time controllers and a discrete-event model of a manufacturing shop. The results show that simultaneous input-output control is effective and enables the system to maintain WIP balance, absorb demand fluctuations and effectively reject disturbances, especially when global information is provided to the controller.

Keywords: Production · Workload control · Closed-loop control systems · Industry 4.0 · Simulation

1 Introduction

Workload Control (WLC) is a shop-floor managing concept adapted to the specifics of make-to-order (MTO) companies [3,7,9]. MTO companies can reap many of the benefits of lean production planning and control by balancing demand and production over time when work is not standardised and when it is not possible to synchronise production flows. WLC is an important solution for production planning and control in small to medium-sized MTO manufacturing companies. One of the biggest challenges for these companies is finding

© The Author(s), under exclusive license to Springer Nature Switzerland AG 2024
M. Mujica Mota and P. Scala (Eds.): EUROSIM 2023, CCIS 2033, pp. 274–288, 2024.
https://doi.org/10.1007/978-3-031-68438-8_20

a balance between incoming orders and production capacity to ensure that the operation stays busy while delivering confirmed orders on time. By applying input-output control, work-in-process (WIP) can be stabilised and reduced. The main benefits of WLC for small and medium-sized MTO companies are maintaining short, predictable and achievable lead times, controlling and effectively utilising capacity, and controlling WIP and inventory for lean production.

This paper presents a simulation study of a recently introduced approach to workload control, in which the workload control of manufacturing systems is represented as a dynamic model [5,7]. The approach has primarily been studied through a continuous dynamic representation of the manufacturing system. In this paper, we attempt to verify the developed concept in a hybrid setting combining discrete event simulation (DES) with discrete-time approximation of continuous controllers. The novel implementation of the simulation model in Matlab Simulink/SimEvents is described, implementation challenges are highlighted, and a number of simulation scenarios are presented with corresponding conclusions. The approach is novel because a feedback-controlled DES model is proposed (i.e., a DES production model with closed-loop controllers). Typically, DES models are open-loop, i.e., not coupled to controllers.

2 Problem Setting

The majority of existing studies on WLC typically treat the parameterisation and application of order release and of capacity adjustment as separate entities. For instance, the study on the parameterisation of order release can be found in [3], while the parameterisation of capacity adjustment can be found in [9]. In these studies, different parameter combinations are evaluated through experiments on DES models, followed by an analysis of the results (typically throughput times) to help in selecting the optimal parameters. The simulation runs in an open loop, as no feedback information is used within the simulation to modify the system parameters. Although the parameters are affected by the production load, they remain constant throughout the simulation experiment. This approach follows the logic of static optimisation and is referred to as the "existing approach" in Fig. 1.

A novel approach, as proposed in [5,7], uses differential equations to model the WLC system. The resulting dynamic model facilitates the use of feedback control. The controller outputs change continuously, simultaneously defining the order release rate and capacity adjustments. Consequently, a reactive model is obtained with the time-dependent parameters of the WLC instead of being a fixed increment as in the existing approaches. For instance, the capacity adjustment level is subject to a continuous change over time, as calculated by the controllers. The workshop workload is reflected by the current work in progress (WIP) in the system buffers, which are monitored as a set of state variables. The manipulated variables are the job pool release rate (U_{pool}) and the set of workstation capacity adjustments (U_i), where i denotes the i-th workstation. The simulation is performed by feeding back information to the system in order to change the system parameters over time. This is achieved through the use of a closed-loop simulation. In the event of machine failure, fluctuations in demand or order rate, the corresponding control system will automatically respond by dynamically adjusting the WLC parameters. Furthermore, the WLC input and output control are performed simultaneously. The aforementioned approach is illustrated in Fig. 2.

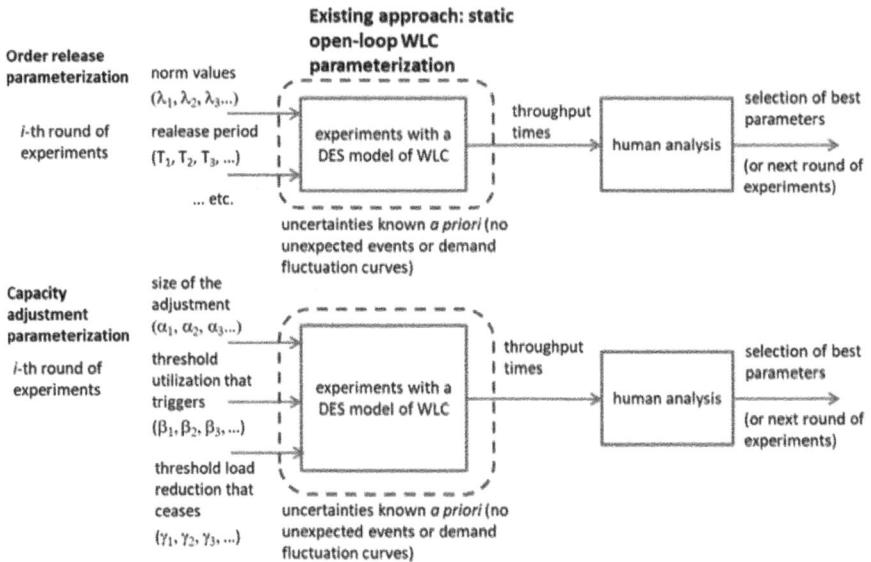

Fig. 1. Traditional WLC parameterisation approach (open-loop DES experiments)

3   Model of a Production System

A real production system was considered for WLC modelling and simulation. The system deals with metal parts manufacturing and consists of four workstations. It is fed by a single order and material source, $S_{01\text{-}pool}$, which represents the arrival of production orders (Fig. 3). The workstations are composed of parallel machines, which are represented as a single entity in the model. In addition to the four workstations, the model represents the job pool, i.e. the buffer of waiting jobs/orders, in another workstation. The system exhibits a unidirectional order flow [1], and comprises a variety of product types, each with a distinct production route. However, the routings are directed from the initial stage to the final workstation, with no re-entry flows. As described in [4], this is referred to as a general flow shop, which is a common configuration observed in various industrial settings. Table 1 presents data on the families, processing routes and demand of the manufacturing system under consideration. The data are used to calculate the proportions of material flow into different routes of the system (Fig. 3) in effort to achieve the required production mix (see right column of Table 1).

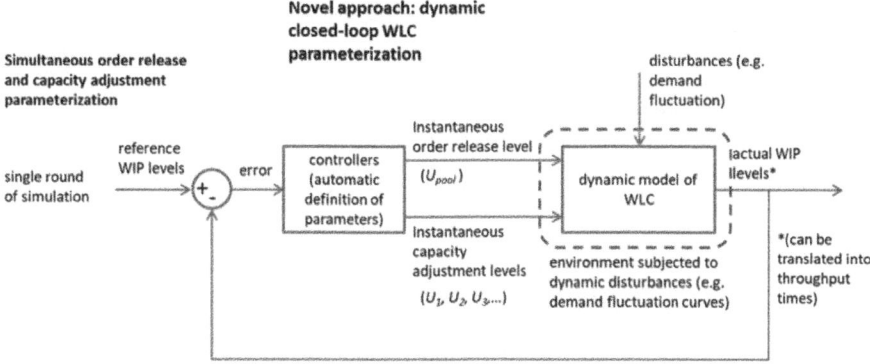

Fig. 2. Proposed WLC parameterisation approach (feedback control approach)

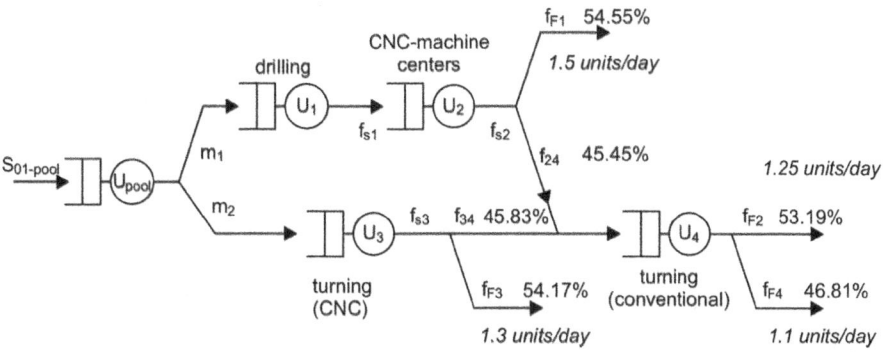

Fig. 3. Schematic representation of the modelled production system

The job pool and the workstations represent the controlled components in the developed model. The manipulated variables in this model are the release rate, represented by U_{pool}, and the workstations processing frequencies, represented by U_i (where $i = 1, 2, 3$ and 4). The variables that are subject to control are the quantity of orders awaiting processing in the pool, represented by q_{pool}, and the work in progress (WIP) in the buffers, represented by q_i ($i = 1, 2, 3$ and 4). The control inputs affect the flows in different branches of the manufacturing system (represented by f in Fig. 3). The list of variables involved in control is the following:

Table 1. Routings and demand data for the studied production system

	Station 1 (drilling)	Station 2 (CNC machine centers)	Station 3 (CNC turning)	Station 4 (convent. turning)	Demand rate (units/day)	Mix (%)
Family 1	X	X			1.5	29.1%
Family 2	X	X		X	1.25	24.3%
Family 3			X		1.3	25.2%
Family 4			X	X	1.1	21.4%

- $\dot{q}_{pool}$: The rate at which the quantity of orders or jobs in the pool changes. This quantity is the difference between the input and output flows in the pool;
- $\dot{q}_i$: The rate of material consumption or storage in buffer i, $i = 1, 2, 3, 4$;
- q_{pool}: Quantity of orders in the pool, waiting for release;
- q_i: Quantity of material held in each buffer, with $i = 1, 2, 3, 4$;
- U_{01pool}: Frequency of the orders arriving to the job pool. It represents the demand;
- U_{pool}: Frequency with which the orders are released to the production;
- U_i: Frequency of workstation i processing, with $i = 1, 2, 3, 4$.

The specific features of the system include the material flow proportions, determined by the required mix of products, and the product family demand rates, which can be changed to represent fluctuating demand.

Using the Bond graph-based modelling method (see [6] and [7] for details), the following dynamic model of the system is derived:

$$
\begin{bmatrix} \dot{q}_{pool} \\ \dot{q}_1 \\ \dot{q}_2 \\ \dot{q}_3 \\ \dot{q}_4 \end{bmatrix} = \begin{bmatrix} -1 & 0 & 0 & 0 & 0 & 1 \\ m_1 & -1 & 0 & 0 & 0 & 0 \\ 0 & 1 & -1 & 0 & 0 & 0 \\ m_2 & 0 & 0 & -1 & 0 & 0 \\ 0 & 0 & 0.4545 & 0.4583 & -1 & 0 \end{bmatrix} \begin{bmatrix} U_{pool}min(1, q_{pool}) \\ U_1 min(1, q_1) \\ U_2 min(1, q_2) \\ U_3 min(1, q_3) \\ U_4 min(1, q_4) \\ U_{01pool} \end{bmatrix} \tag{1}
$$

Parameters m_1 and m_2 provide additional degree of freedom in the order release strategy. In the considered case the parameters were fixed at $m_1 = 0.53398$ and $m_2 = 0.46602$.

The manufacturing system under consideration produces a number of products. To facilitate the development of the continuous dynamic model, the products have been grouped into families. This approximation is justified by the fact that products belonging to the same family have similar processing times. This aggregated information is also sufficient for planning purposes. Furthermore, this approach provides an integrated understanding of WLC dynamics when input and output control are performed simultaneously, which is sufficient for proof of concept. Following the feasibility proof, the approach can be extended and tested in more detailed and disaggregated scenarios. One possible avenue for further investigation is to apply these adjustments to a more detailed DES model that incorporates the company's individual orders and products. This will allow an examination of whether the generic rules deduced from this study (see below) also apply to the more detailed scenario.

4   Input and Output Control

The proposed WLC concept includes both control of input and control of output. Input control is concerned with the release of orders. The suggested model introduces an order release workstation, and input control is defined as the adjustments to the processing frequency of this workstation, denoted U_{pool}. In contrast, the output control refers to the capacity adjustments. In the suggested model, the processing capacity of the workstations is adjusted by changing the processing frequency of each workstation, i.e. U_1 to U_4. The approach has been evaluated by testing specific combinations of input and output control, formulated as control rules described below.

The proposed order release strategy is determined by the observed shop floor workload and the pool workload. The shop floor workload is quantified by measuring the buffer levels at workstations 1 and 3. These are the workstations immediately following the order release workstation and all four different product production flows start at one of these two workstations (Fig. 3, Table 1). The workload of the pool is quantified by the level of its order buffer, q_{pool}. The same order release rule (input control) is used in all simulations.

The proposed capacity adjustments, i.e. output control of a particular workstation, are formulated as control rules A, B and C. Rule A states that processing frequencies should change in relation to the amount of load in the own buffer of each workstation. Consequently, the control of workstations is based on local measurements. Rule B represents a more global perspective of the system, as the workstation control considers not only its own buffer, but also the buffers of the workstations immediately preceding it. Rule C (Fig. 4) represents the most global of the three implemented rules. The workstation control takes into account not only the buffers of the workstations, but also the job pool, which represents the work waiting to be released. The dashed lines in Fig. 4 represent measurement signals, i.e. information sent to the controller of a particular workstation, and control signals, i.e. adjustments to the processing frequency. For instance, the pool controller receives data from buffers 1 and 3 and from the buffer of waiting orders, and subsequently modifies the release rate of the orders.

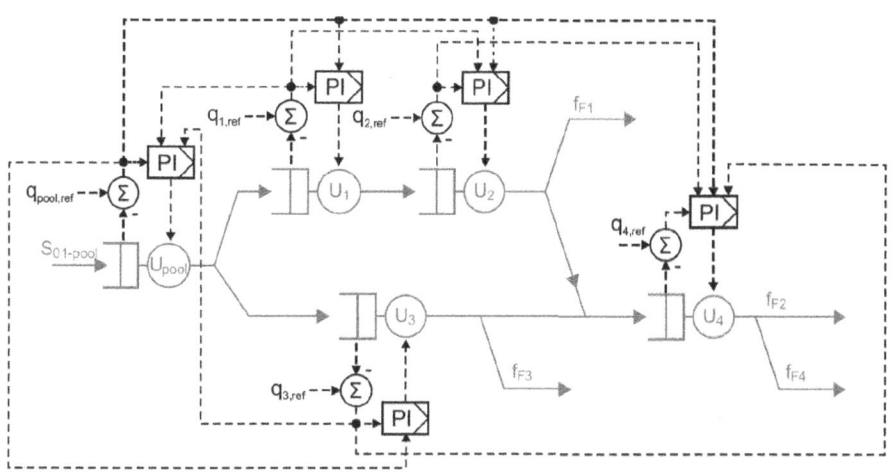

Fig. 4. Control rule C signal flow

5   Simulation in Simulink/SimEvents

This section describes the hybrid simulation of the proposed WLC concept and control rules A, B and C. It also provides some specific Simulink/SimEvents implementation details. In particular, it presents a novel implementation of the discrete-event workstation model, which is based on a server with externally controlled service time and preemption.

5.1   Hybrid WLC System Simulation Model

The control design and simulation of the WLC system in [7] employs the state space model (Eq. (1)). The implementation of the corresponding simulation model is straightforward as Simulink allows integration and vector-matrix calculations.

This paper investigates a different, hybrid implementation of the simulation model (Fig. 5). The model is hybrid in the sense of a combined discrete-time/discrete-event simulation. The production system is simulated in SimEvents as a discrete-event system of queues and servers (Fig. 6), which is similar to the structure of the system in Fig. 3. The WLC control part is simulated as a discrete-time approximation of the PI controllers, as indicated by the shaded region of the simulation scheme in Fig. 5. The control structure of rule C, based on the signal flow of Fig. 4, is depicted. The parameters of the controller are determined in accordance with the methodology outlined in [7].

The discrete-time setting was chosen because a continuous adjustment of the processing frequencies, i.e. the capacity of the production system, is unrealistic. Instead, a suitable sampling time could correspond to a repeated periodic adjustment of the main production parameters, e.g. at the beginning of a working day. Therefore, a uniform sampling time $T_s = 1$ day was chosen. A similar approach was used in [6]. The discrete-time setting of the controllers (left part of Fig. 5) and the discrete-event setting of the production system (Fig. 6) make this model different from the existing ones [5,7].

The two sub-models interact with the monitoring logic that sets the simulation scenarios. More importantly, they interact during the simulation run, where the buffer levels are monitored by the control section at the specified sampling rate and the control section outputs set the nominal order release and workstation processing frequencies during each sampling period. The actual order inter-release times and workstation processing times are then varied with respect to a lognormal distribution with mean μ being the reciprocal of the nominal processing frequency.

The lognormal distribution is additionally parameterised with the squared coefficient of variation $c_v^2 = \sigma^2/\mu^2$, where σ^2 is the variance of the lognormal distribution. More precisely, different degrees of variability of processing times can be tested, e.g. $c_v^2 = 0.25$ stands for a low variability of processing times, while $c_v^2 = 1$ gives a variability of processing times comparable to the exponential distribution [10].

A large variation in the parameters for the probability distribution of the processing times during the simulation is challenging and requires a modification of the standard approach for modelling SimEvents queues and servers. While the SimEvents Server block provides a fairly simple means of adjusting the service time during the simulation run, a conceptual problem must be solved to deal with situations where the processing frequency approaches zero. Simply setting a large processing time will slow down or almost stop processing, but any increase in processing frequency can then only come into effect after the processing time has elapsed. If the workstation stops, the processing time tends towards infinity and a later resumption of workstation processing cannot be realised in this way.

This is solved by allowing pre-emption of the Server block [8]. A change in the processing time distribution parameters triggers the creation of a single high-priority entity with a service time of zero, which immediately preempts the server's current processing. Apart from removing the currently processed entity from the server, this high priority entity has no impact and is later discarded. The remaining processing time T_{rem} of the removed entity is recorded and the entity is re-entered into the server with a new processing time T_p defined according to

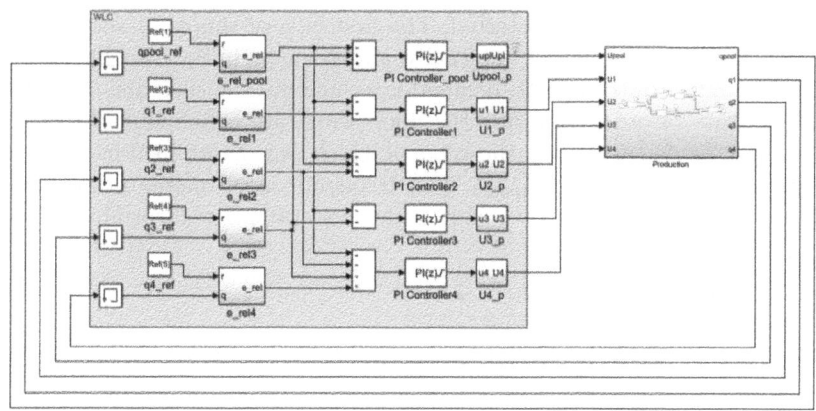

Fig. 5. Hybrid implementation of the simulation model

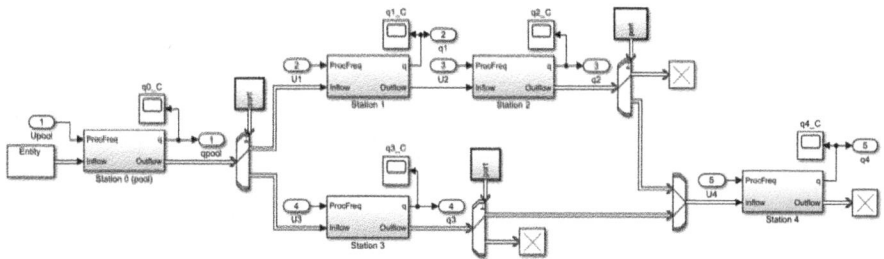

Fig. 6. Production system simulation in SimEvents

the new parameters and the ratio between remaining and previous processing time $T_{p,old}$.

$$T_p = T_{p,new} \frac{T_{rem}}{T_{p,old}} \tag{2}$$

This corresponds to a change in processing frequency between the processing of a batch of products. The corresponding implementation of the workstation submodel in SimEvents is shown in Fig. 7.

The Simulink function t=lognorm() is called on each server entry to generate a random value that defines the service time, i.e. the processing time $T_{p,new}$. The mean value of the processing time is adjusted to the reciprocal of the current processing frequency specified by the corresponding controller. Additional attributes are used to distinguish between entities that are recirculated and entities that come directly from the workstation buffer. Only the processing time of a recirculated entity is rescaled by Eq. (2).

6   Tested Scenarios

Three different scenarios were tested. The first is very simple; it served to validate the implemented model internally and to gain initial insights into the system dynamics. In scenario 1, demand is assumed to be constant, i.e. there is a constant rate of orders entering the pool. This rate corresponds to the constant source frequency U_{01pool}, the value of which was determined by the steady state processing frequency $U_{01poolp}$. This frequency is calculated from the data in Table 1 and corresponds to a sum of demand rates for all product families

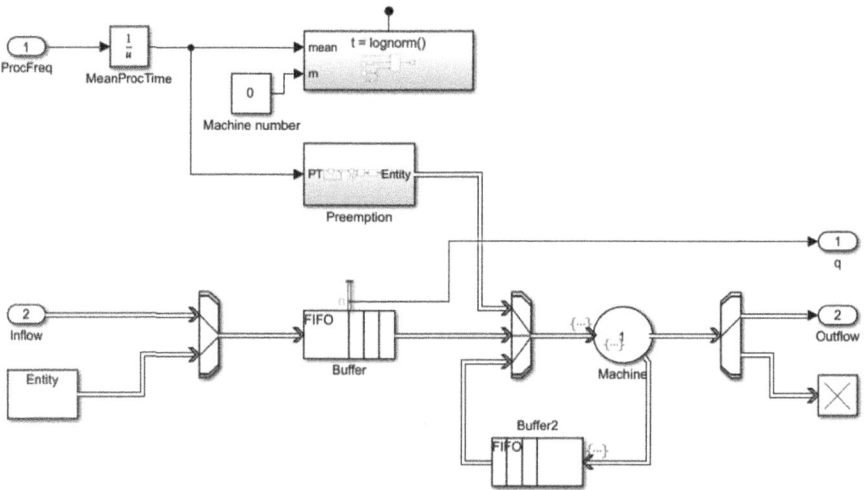

Fig. 7. Manufacturing workstation with preemption at the change of the production rate

($U_{01poolp} = 5.15$). Scenarios 2 and 3 were designed to investigate the response of the WLC system with concurrent input and output control to fluctuations in demand. In scenario 2, the rate of order arrivals over time has a trapezoidal shape. In Scenario 3, the order arrival rate follows a sinusoidal signal to test the system's response to periodic fluctuations in demand.

Control rules A, B and C are first simulated under Scenario 1. The initial conditions for the buffer levels were set above or below the reference values in order to ensure that the initial WIP levels were not balanced. Table 1 illustrates the selected initial conditions. The objective was to determine whether the controllers would be capable of rebalancing the system and whether the buffer levels would stabilise at the reference values, despite the fact that the controllers had to deal with composite error signals with components of opposing sign. The initial levels of the buffers for Scenarios 2 and 3 were set precisely to the corresponding reference values (the middle column of Table 2). In these scenarios, there are significant fluctuations in demand, yet the initial WIP levels are balanced, i.e. they are at the desired level. The objective of these scenarios is to ascertain whether the controllers are capable of maintaining the WIP balance (and buffer levels at the reference values) despite fluctuations in demand.

7   Results and Discussions

Three types of graph are used to present the results. The first shows the relative order release rate and relative processing frequency of the workstations over time. The second graph shows the status of the job pool and the third shows the buffer levels over time. From a production management point of view, the first graph shows the capacity adjustments of the workstations and the release or orders. The second graph shows the quantity of pending orders and the third graph corresponds to the WIP on the shop floor.

Note that the relative processing frequency is the ratio between the actual processing frequency and the steady state value. Therefore, the value 1 corresponds to a steady-state processing frequency required for stable system operation at the nominal rate of incoming orders $U_{01poolp}$. Details on the calculation of this steady-state frequency can be found in [6].

Scenario 1 is very simple and was primarily developed to validate the implemented model and to get a first insight into the dynamics of the closed loop [7]. Due to space constraints, we therefore focus mainly on the simulation results of Scenario 2.

7.1   Scenario 2 Simulation Results

Figure 8 shows the simulation results without varying processing time ($c_v^2 = 0$). In the figure, diagrams a) show the open loop response with constant processing frequencies (including the order release frequency U_{pool}), while diagrams b) and c) show the closed loop responses with control rules A and C respectively.

It can be observed that as demand increases (indicated by the change in U_{01pool}), a large job pool backlog builds up when the order release frequency U_{pool} is constant, while both control rules effectively keep the quantity of waiting orders in the job pool close to the reference value. Control rule C performs slightly better in this respect, but at the cost of larger fluctuations in processing frequency (cf. Fig. 8 b and c, top row). The performance of rules A and C regarding the buffer levels (Fig. 8 b and c, bottom row) is very similar.

The variability of the q_{pool}-level and the q_i-buffer level in Fig. 8 is entirely due to the random nature of the material flow distribution (i.e. the type of the products that are queued up, according to the 4 different families, as in Table 1). This was implemented by making random decisions during the simulation, adjusting the probabilities to the percentages of material flow in the continuous model. Although the overall ratio of material flows is constant, large momentary deviations from this overall ratio occur and cause significant fluctuations in the input flows of buffers 1, 3 and 4.

Table 2. Reference levels and initial buffer levels for Scenario 1

Element	Reference level	Initial level
Job Pool	7 units	6 units
Buffer 1	7 units	8 units
Buffer 2	7 units	8 units
Buffer 3	7 units	6 units
Buffer 4	6 units	5 units

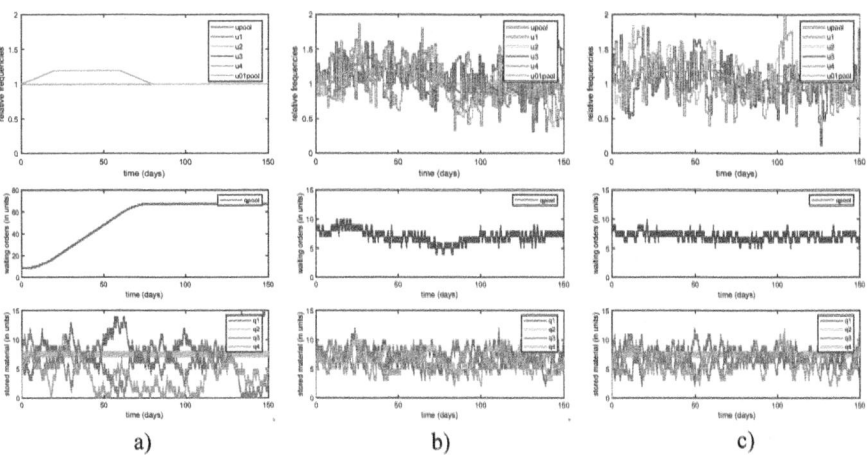

a)   b)   c)

Fig. 8. Relative order release rate and relative processing frequencies of workstations (top), pending orders (middle), workstation buffer levels (bottom) in Scenario 2: a) open-loop case; b) rule A; c) rule C

In [7] a continuous model for workload control of this production system is presented. The same features and parameters are used in the present work to allow comparison. To validate the continuous model in [7], a discrete-event simulation was performed a posteriori. The control inputs of the discrete-event system were replaced by discretised control input trajectories from the continuous simulation, but the discrete-event simulation operated in open loop. Furthermore, only the DES system without external disturbances was simulated.

In contrast, the work presented in this paper uses a closed-loop hybrid simulation where changing buffer levels directly influence the processing frequencies through feedback loops. Note that the results in Fig. 8 are similar to those in [7], although derived in a conceptually different way. The results are similar when there are no disturbances, showing that the assumptions used in [7] were correct.

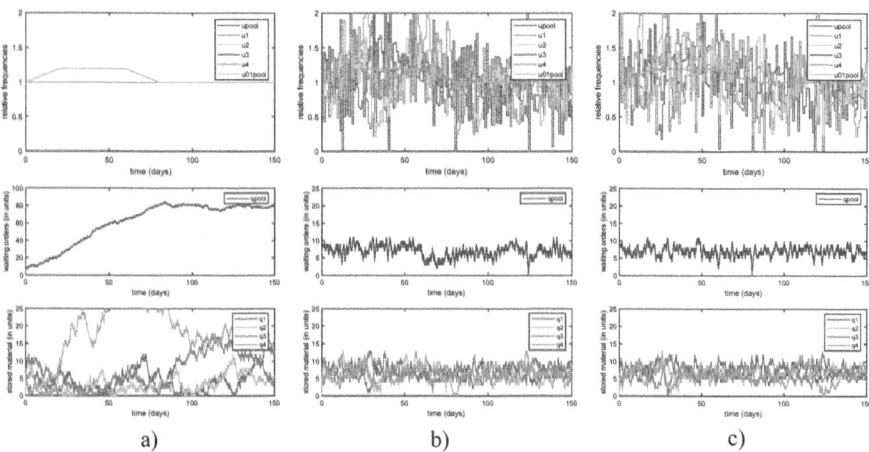

Fig. 9. Relative order release rate and relative processing frequencies of workstations (top), pending orders (middle), workstation buffer levels (bottom) in Scenario 2 with processing time variability: a) open-loop case; b) rule A; c) rule C

7.2   Additional Experiments

The presented hybrid simulation model allows for various extensions of the simulated scenarios. The first extension is to take into account the variability of processing times. A low variability of processing times was chosen, represented by $c_v^2 = 0.25$ [10], the corresponding results are shown in Fig. 9.

The closed loop still performs much better compared to the open loop when q_{pool} is considered, but the performance of control rules A and C is quite similar. Compared to Fig. 8, a much larger variation of the buffer levels can also be observed. This is an expression of one of the fundamental properties of variability in production systems, i.e. regardless of its source, any variability in a production system is buffered [2].

The implementation of the simulation model in this work allows to go one step further: as the control is based on the actual feedback of the buffer states, different disturbances can be simulated.

As an example, another scenario (Scenario 4) is presented where a temporary failure of workstation 1 has been simulated. Also, as in the previous case, a variability of processing time with $c_v^2 = 0.25$ is assumed. The results are shown in Fig. 10. The vertical dashed lines indicate the interval of failure of workstation 1 (between day 20 and 25).

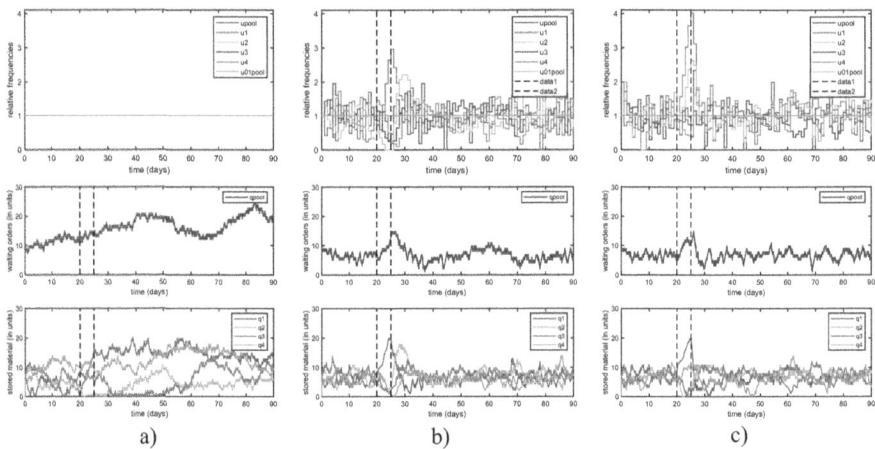

Fig. 10. Relative order release rate and relative processing frequencies of workstations (top), pending orders (middle), workstation buffer levels (bottom) in Scenario 4 with workstation 1 failure: a) open-loop case; b) rule A; c) rule C

In the case of the open loop, the effect of the workstation failure on q_{pool} is masked by the fluctuations caused by the variability of the flow splitting. However, a gradual increase of q_1 can be observed during the time of the workstation failure. On the other hand, diagrams b) and c) in the figures clearly show how control loop strategies A and C react to the failure. The controllers detect the increase in waiting orders and in the WIP of buffer 1, and react by increasing the processing frequency of workstation 1. This has no effect until the workstation is restarted, and, after that, the buffer levels quickly return to normal. Both control strategies are effective in this sense, with control rule C responding faster and having a shorter settling time, but again at the cost of a greater divergence from the steady-state processing frequency.

8   Conclusions

In the simulation study presented, a novel approach to workload control system synthesis through dynamic modelling and application of control theory was addressed. The synthesised continuous controllers enable automatic and cooperative control of order release and manufacturing workstation capacity adjustments. The approach was evaluated through a hybrid simulation study modelling the workload control of a 4-workstation manufacturing facility with unidirectional order flow.

The simulation results confirm the validity of the continuous simulations previously reported in the literature, and additional scenarios confirm the responsiveness of the WLC control even in case of processing time variability and workstation operation disturbances. The closed-loop WLC is able to maintain the order backlog and the WIP level in the nominal operating range and is therefore a promising alternative to existing WLC approaches.

Moreover, the hybrid simulation model proposed here, i.e., the discrete-time setting of the PI controllers coupled with the DES model of the production system, allows the direct modelling and simulation of closed-loop DES systems. This is a spin-off of this study that should be emphasized. To the best of our knowledge, DES is employed in production research only to generate open-loop systems, i.e., systems without feedback loops (with the term "feedback" employed strictly in the sense of control theory). DES is descriptive and, in the operations management domain, is usually coupled with design of experiments, heuristics or optimization, but not directly with feedback controllers. One existing (and restricted) branch of research explores continuous closed-loop systems for shop floor representation, but not DES closed-loop systems. Thus, the proposed approach opens up a relevant research avenue, since it could be applied to a variety of different production systems (flow shops, job shops, open shops) and to a variety of production control systems (push, pull, conwip and base stock systems, among others). For many types of production systems, the discrete closed-loop representation is much more accurate than the existing continuous one, and, the closed-loop feature of the models allows the proposition of smart production control systems, reactive to real time states of the system, instead of descriptive (static) systems.

Acknowledgements. The first author acknowledges financial support from the Slovenian Research Agency (research core funding No. P2-0219). The second author also acknowledges the financial support from the National Council for Scientific and Technological Development under process #310812/2021-6.

References

1. Duffie, N.A., Shi, L.: Dynamics of WIP regulation in large production networks of autonomous work systems. IEEE Trans. Autom. Sci. Eng. **7**(3), 665–670 (2010). https://doi.org/10.1109/TASE.2009.2036374
2. Hopp, W.J., Spearman, M.L.: To pull or not to pull: what is the question? Manuf. Serv. Oper. Manag. **6**(2), 133–148 (2004). https://doi.org/10.1287/msom.1030.0028
3. Land, M.J.: Parameters and sensitivity in workload control. Int. J. Prod. Econ. **104**(2), 625–638 (2006). https://doi.org/10.1016/j.ijpe.2005.03.001

4. Oosterman, B., Land, M., Gaalman, G.: Influence of shop characteristics on work-load control. Int. J. Prod. Econ. **68**(1), 107–119 (2000). https://doi.org/10.1016/S0925-5273(99)00141-3

5. Sagawa, J.K., Land, M.J.: Representing workload control of manufacturing systems as a dynamic model. IFAC PapersOnLine **51**(2), 825–830 (2018). https://doi.org/10.1016/j.ifacol.2018.04.016

6. Sagawa, J.K., Mušič, G.: Towards the use of bond graphs for manufacturing control: design of controllers. Int. J. Prod. Econ. **214**, 53–72 (2019). https://doi.org/10.1016/j.ijpe.2019.03.017

7. Sagawa, J.K., Oliveira, A.F., Mušič, G., Land, M.J., Maluf, A.S.: Smart workload input-output control of production systems: a proof of concept. Eur. J. Oper. Res. **309**, 286–305 (2023). https://doi.org/10.1016/j.ejor.2022.12.034

8. SimEvents. https://www.mathworks.com/products/simevents.html. Accessed 15 Feb 2023

9. Thürer, M., Stevenson, M., Land, M.J.: On the integration of input and output control: Workload control order release. Int. J. Prod. Econ. **174**, 43–53 (2016). https://doi.org/10.1016/j.ijpe.2016.01.005

10. Thürer, M., Fernandes, N.O., Stevenson, M., Qu, T., Li, C.D.: Centralised vs. decentralised control decision in card-based control systems: comparing Kanban systems and COBACABANA. Int. J. Prod. Res. **57**(2), 322–337 (2019). https://doi.org/10.1080/00207543.2018.1425018

Nonlinear Control of a Helio-Crane Laboratory Device

Goran Andonovski[(✉)], Martin Porenta, and Igor Škrjanc

Faculty of Electrical Engineering, University of Ljubljana, Ljubljana, Slovenia
{goran.andonovski,igor.skrjanc}@fe.uni-lj.si, mp0251@student.uni-lj.si

Abstract. When designing control loops, we often encounter non-linear systems that have much more complex dynamics than linear systems. This means that with classical control algorithms we cannot guarantee that the control loop works sufficiently well over the entire operating range. Therefore, in this paper we focus on controlling a nonlinear dynamic system with more advanced nonlinear controllers. All experiments were performed on a helio-crane laboratory device. The device exhibits nonlinear operation and rather oscillatory behaviour and is therefore particularly well suited for testing developed nonlinear controllers. In this work we have developed and tested three controllers: PID controller with optimised parameters, fuzzy cloud-based predictive functional controller (FCPFC) and NARMA L2 controller. We first present the results for each controller, which we compare to determine the best type of controller for our nonlinear problem. The goal is not only to find the best controller in terms of efficiency, but we also consider the time required to develop the control solution. In this way, each of the controllers has its own advantages and disadvantages.

Keywords: Fuzzy predictive functional controller · Neural NARMA L2 controller · nonlinear control

1 Introduction

In practise, we often deal with non-linear dynamic systems, which are characterised by much more complex dynamics than linear systems. The classical approach to controlling such systems is to drive the system to a certain operating point and tune the controller around this point, e.g. a PID controller [1]. Such a controller only works well near the chosen operating point; outside the operating point, the performance of such a control system is usually worse.

In this paper we have implemented three different controllers to control a non-linear dynamic system. The first controller is a Fuzzy Predictive Functional Controller (FPFC) based on the principle of predictive control. The second controller is the NARMA L2, which is based on modelling and controlling the system with a pair of neural networks. Both controllers are advanced control methods

© The Author(s), under exclusive license to Springer Nature Switzerland AG 2024
M. Mujica Mota and P. Scala (Eds.): EUROSIM 2023, CCIS 2033, pp. 289–300, 2024.
https://doi.org/10.1007/978-3-031-68438-8_21

that are not widely used in practise. Therefore, we compare their control performance with a classical PID controller with optimised parameters (function *fminsearch* in MATLAB with a given criterion function as root mean square error). Given the significant differences between these three controllers, we provide an overview of the theory and basic principles of each controller. Overall, the aim of this paper is to investigate the effectiveness of advanced control techniques such as FPFC and NARMA-L2 controllers in controlling a non-linear dynamic system and compare their performance with that of a classical PID controller. Through our analysis, we aim to highlight the strengths and limitations of each controller and provide insight into their practical applicability.

The first controller is a fuzzy predictive functional controller (FPFC), which uses a fuzzy model of the controlled process to predict the output signal at a certain time in the future [3–6]. The algorithm [2] used in this paper combines the fuzzy cloud-based model presented in [8] with the well-known model based predictive control [7]. The data clouds are used to calculate the membership degree of the current data point to all clouds (fuzzy rules). In addition, the parameters of the local models are calculated using the recursive weighted least squares (rWLS) method. The identified cloud-based fuzzy model is used to predict the future behaviour of the process over a certain prediction horizon. The control law is defined to minimise the criterion, usually the difference between the predicted signal and the desired reference trajectory.

The Narma L2 Controller [9,10,20] is a type of model-based controller based on a neural network. The controller is designed to control the output of a system based on a reference signal by learning from past control inputs and system responses. The Narma L2 controller uses a non-linear autoregressive moving average (NARMA) model trained with the backpropagation through time algorithm. The controller's inputs include the system's reference signal and previous outputs, while its output is the control signal that regulates the system's behaviour. The Narma L2 controller is highly adaptable and can be applied to various control problems, such as robotics [11], industrial processes [12] and mechanical systems [13]. Its ability to learn from past data makes it a powerful tool for control systems that exhibit non-linearity and uncertainty.

All experiments were performed on a helio-crane laboratory device and its mathematical model [14]. The device consists of a pendulum that is pivoted in the middle. On the left side of the pendulum is a counterweight and on the right side is an electric motor with a propeller. The system receives the voltage at the electric motor as input and returns the offset angle as output. The device shows non-linear operation and rather oscillatory behaviour, so it is particularly well suited for testing developed non-linear controllers. It has been shown that a good knowledge of the properties of the system is crucial for the successful development of experiments and control algorithms. We first developed all the control algorithms on the mathematical model of the device, as it is much faster to collect measurements and identify the model, and it is also much easier to look for errors in the programme code. Then we developed the controls on the real system and observed what changes we had to make to make a more complex control work successfully on the real system.

The comparison results show that the Fuzzy Predictive Functional Controller (FPFC) performs the best but require quite good model of the controlled system. The Narma L2 model shows good and comparable performance on the mathematical model of the process, but very poor performance on the real helio-crane system. As expected, the optimised PID controller shows satisfactory results with some oscillation behaviour at some points of the operating range of the real plant.

The following paper is divided into six sections. After the introduction, in the second section we briefly review the theoretical foundations of FCPFC and Narma L2 controllers and derive the expressions for both controllers. As we encounter more advanced nonlinear controllers based on a nonlinear system model, we introduce the theory of nonlinear system model identification. In the third section, we introduce the helio-crane laboratory device. In the fourth section, we present experimental results and discuss them. Finally, in the last section, we draw conclusions based on our findings.

2   Nonlinear Control Systems

In the introduction we mentioned that the controllers used are based on the motel of the controlled system, which can be developed through different approaches. These approaches include theoretical modelling, experimental modelling or identification and a combination of these approaches. In this paper, both controllers are based on experimental modelling of the controlled system. In experimental modelling or identification, experiments are performed on a real system to obtain data to train a model. The theoretical basis of the identification process is explained in more detail in the following subsections.

2.1   Fuzzy Predictive Functional Controller (FPFC)

The Fuzzy Predictive Functional Controller (FPFC) [2,3] uses a fuzzy model of the controlled process to predict the output signal at a certain future horizon. By building a dynamic model of the process based on past knowledge, the FPFC predicts the controlled signal and determines the control signal that minimises the difference between the output signal and the reference signal at a certain prediction horizon H. The FPFC is intended for non-linear processes and uses fuzzy models to divide the process space into linear regions.

Identification of the Fuzzy Cloud-Based Model. In this section, we give a brief overview of the identification process for a cloud-based fuzzy model. A more detailed explanation of the identification process can be found in [8]. The aim of the identification process is to develop a fuzzy model that can accurately reflect the dynamics of the system. The proposed fuzzy model takes the form of a cloud-based model, which is characterised by its ability to deal with uncertainty and non-linearity [15]. The model is capable of representing complex systems and its structure is designed to capture the relationships between input and output variables. The fuzzy cloud-based model for the ith rule has the following form:

$$\mathcal{R}^i : \textbf{IF} \quad (\boldsymbol{x}_f \sim X^i) \quad \textbf{THEN} \quad y^i(k) = f^i(\boldsymbol{x}_f), \quad where \quad i = 1, \ldots, c \quad (1)$$

where X^i is i^{th} data cloud (fuzzy rule); c is number of data clouds; $\boldsymbol{x}_f(k) = [y(k-1), \ldots, y(k-n_a), u(k-1), \ldots, u(k-n_b)]$ is input vector; and n_a and n_b define the dynamics of the process. The model 1 consists of two parts: the conditional (IF) part and the consequent (THEN) part. In the conditional part, the membership of the current data to existing clouds is determined using a self-evolving model common to all [8]. The consequent part is defined by a nonlinear autoregressive exogenous (NARX) model. In addition, a partial local model is defined as part of the NARX model:

$$y_k^i = f_k^i(\boldsymbol{x}_f) = \boldsymbol{\theta}_k^{i\,T} \boldsymbol{\psi}_k \quad (2)$$

where $\boldsymbol{\psi}_k = [\boldsymbol{x}_f, 1]^T$ is the regression vector and $\boldsymbol{\theta}_k^i$ is a vector of unknown parameters $\boldsymbol{\theta}_k^{i\,T} = [a_1^i, \ldots, a_{n_a}^i, b_1^i, \ldots, b_{n_b}^i, r^i]$. These parameters are calculated using Recursive Fuzzy Weighted Least Squares (RFWLS) method [16].

Finally, the model output $\hat{y}(k)$ is calculated as follows:

$$\hat{y}(k) = \sum_{j=1}^{c} \beta^j(\boldsymbol{x}_f) \boldsymbol{\theta}_k^{j\,T} \boldsymbol{\psi}_k \quad (3)$$

where $\beta^j(\boldsymbol{x}_f)$ is the membership function (normalized relative density [8]) of the data point $\boldsymbol{x}_f$ to the cloud X^j.

The fuzzy rules $(i = 1, \ldots, c)$ are obtained by evolving mechanism explained in detail in [2,8].

Control Law. In the previous subsections we established a principle for identifying the process model $\hat{y}(k)$. The subsequent part of the predictive functional control approach based on fuzzy clouds is to identify the output signal of the controllers $u(k)$. The predictive control law is formulated to minimise the criterion, typically the deviation between the reference trajectory $y_r(k)$ and the predicted controlled signal $\hat{y}(k)$ over a certain predictive horizon H [5]. The desired closed loop trajectory is determined by the first order reference model in the following manner:

$$y_r(k+1) = a_r y_r(k) + (1 - a_r) r(k), \quad 0 < a_r < 1 \quad (4)$$

In the same manner as in (4) we can choose $n_a = 1$ and $n_b = 1$ and the Eq. (2) can be rewritten as follows:

$$y_m^i(k+1) = a_m^i y_m(k) + b_m^i u(k) + r_m^i, \quad i = 1, \ldots, c \quad (5)$$

where a_m^i, b_m^i and r_m^i are unknown parameters of the theta vector $\boldsymbol{\theta}_k^i$ and are determined using Recursive Fuzzy Weighted Least Squares (RFWLS) method [16]. The final output of the process model is calculated as:

$$y_m(k+1) = \hat{a}_m y_m(k) + \hat{b}_m u(k) + \hat{r}_m \quad (6)$$

where

$$\hat{a}_m = \sum_{j=1}^{c} \beta^j a_m^j; \qquad \hat{b}_m = \sum_{j=1}^{c} \beta^j b_m^j; \qquad \hat{r}_m = \sum_{j=1}^{c} \beta^j r_m^j. \tag{7}$$

where β^j is the membership function (normalized relative density [8]) of the data point x_f to the cloud X^j.

In the end, the control signal is determined by minimising a certain cost function. For a more detailed explanation of the analytical derivation of the control law, refer to [5]. The calculation of the control signal is carried out as follows:

$$u(k) = \frac{(1 - a_r^H)(r(k) - y_p(k))}{\dfrac{\hat{b}_m}{1 - \hat{a}_m}(1 - \hat{a}_m^H)} + \frac{y_m(k)}{\dfrac{\hat{b}_m}{1 - \hat{a}_m}} - \frac{\hat{r}_m}{\hat{b}_m} \tag{8}$$

where H is a prediction horizon and should be chosen within the interval:

$$N \leq H \leq \frac{\tau}{2T_s} \tag{9}$$

where N is the process order, τ and T_s are time constant and sampling time, respectively.

2.2   Narma L2 Controller

There are several control approaches that use neural networks, such as model-referenced control, model-predictive control, adaptive control and others. For this particular task, we chose the NARMA L2 controller. This involves modelling the system using the NARMA L2 model, which takes into account both the current and previous inputs and outputs of the system to estimate the current output.

Identification of the Narma L2 Model. The NARMA L2 controller is a control approach for a non-linear process. First, the dynamic system is modelled using pairs of neural networks to obtain the NARMA L2 model. This model is then inverted to obtain the control law. The control law is then derived analytically. This eliminates the need for optimisation during system control and makes the algorithm work faster.

The NARMA L2 algorithm was presented in [20] and is based on the NARMA model, which stands for "Nonlinear autoregressive moving average model". For a more detailed theory of identifying dynamic systems with neural networks, readers can refer to [17–19]. The NARMA L2 controller is derived from this NARMA model as defined in the following expression:

$$\hat{y}(k + d) = F[\varphi_0(k)] \tag{10}$$

where $\hat{y}$ is the output of the NARMA model, F is a nonliner function, $\varphi_0(k) = [y(k), \ldots, y(k - n_a + 1), u(k), \ldots, u(k - n_b + 1)]$ is regression vector and d is the relative system order.

Furthermore, for the NARMA model (10) an approximation is introduced named NARMA L2 model [20,21], defined as follows:

$$\hat{y}(k + d) = N_1 \left[\boldsymbol{\varphi_2}(k)\right] + N_2 \left[\boldsymbol{\varphi_2}(k)\right] u(k) \tag{11}$$

where N_1 and N_2 are outputs of two neural networks in a form of Multi Layer Perceptron (MLP), $\hat{y}$ is the model output, d is the relative process order, $u(k)$ is the process input and $\boldsymbol{\varphi_2} = [y(k), \ldots, y(k - n_a), u(k - 1), \ldots, u(k - n_b)]^T$ is the regressor of NARMA L2 model.

To implement the NARMA L2 controller, we use a pair of neural networks N_1 and N_2 that are trained simultaneously. The training process involves iterating over the training dataset and optimising the networks. The optimisation is done with gradient-based methods and backpropagation of the error.

During training, the neural networks are adjusted to minimise the difference between the predicted output and the actual output. In this process, the error is calculated in the output layer and backpropagated through the network to adjust the weights of the connections between the neurons.

When we train the neural networks using this method, we obtain a well-tuned NARMA L2 model that accurately reflects the dynamics of the system. This allows us to develop a control law that can effectively control the system based on past and current inputs and outputs.

Control Law. Once the identification of the system is completed we can control the system. From Eq. (11), we express the current input to the system $u(k)$ and assume that the output of the model $\hat{y}(k+d)$ is equal to the prescribed reference trajectory $y_r(k + d)$:

$$\hat{y}(k + d) = y_r(k + d) \tag{12}$$

Thus, we get the regulation law of the NARMA L2 controller, which is shown by the equation:

$$u(k) = \frac{y_r(k + d) - N_1[\boldsymbol{\varphi_2}(k)]}{N_2[\boldsymbol{\varphi_2}(k)]} \tag{13}$$

The control law given by Eq. (13) is designed to ensure that the system follows the reference trajectory y_r. Similar to the FPFC controller, the trajectory of y_r is planned and depends on the particular system to be controlled. The goal is to achieve the desired behaviour by controlling the input $u(k)$ to the system based on the current and past inputs and outputs of the system.

3   Helio-Crane Laboratory Device

In this paper we will perform all the experiments on the helio-crane system. The system consists of a pendulum that is clamped in the middle so that it can rotate ("pivot"). On the left side of the pendulum we have a counterweight and on the right side an electric motor with a propeller. When the motor rotates, it generates a buoyancy force F_m and the whole system is displaced by a deflection angle ϕ. When the motor is not activated, the end of the rod in which the motor is located remains in the down position due to gravity. The motor operates in a single direction, resulting in a fixed thrust direction relative to the motor. When you activate the motor, the rod can only be lifted and then passively lowered by gravity. Additional weights are attached to the main rod, which influence the behaviour of the system. The detailed mathematical model and explanation of the system can be found in [22] and Fig. 1 shows a schematic representation of the helio-crane system.

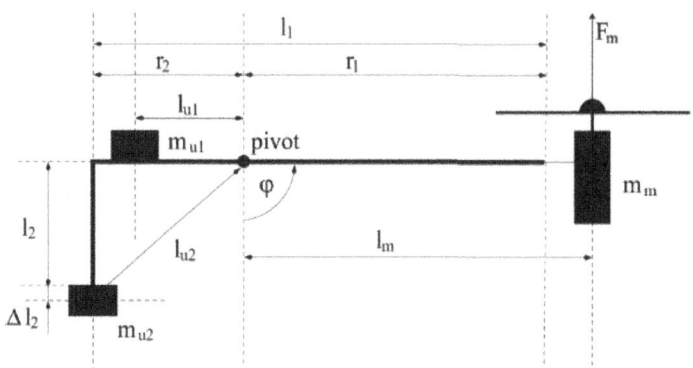

Fig. 1. Schematic representation of the helio-crane system [22].

4   Experimental Results

In this section, the results of the previously presented non-linear advanced controllers are presented. As mentioned earlier, the challenge in implementing advanced control algorithms is to transfer the developed algorithm from the simulation environment, where the control is performed on a model of the real system, to the physical world (real helio-crane laboratory device). The aim of this section is to compare the performance of all the algorithms used and evaluate which algorithm is more suitable for our control problem.

In this section we will compare the results of the control algorithms on the system model and then on the real system. We will evaluate the quality of the control algorithms using two criterion functions: Mean Squared Error - MSE (14) and Sum of Absolute Input Differences - $SADU$ (15):

$$MSE = \frac{1}{N} \sum_{k=1}^{N} (y_p(k) - y_r(k))^2 \qquad (14)$$

$$SADU = \sum_{k=2}^{N} \left| (u(k) - u(k-1)) \right| \qquad (15)$$

where N is the number of data points, $y_p(k)$ is the true value of the system, and $y_r(k)$ is the reference or desired value.

4.1   Control Results Comparison on Mathematical Model

In this subsection we first present the performance of the controllers used to control the mathematical model of the system. For our reference signal, we have chosen a series of step signals that cover the entire operating range of the system. Figure 2 shows a comparison between the responses of the controlled system and Fig. 3 shows a comparison between the control signals for each controller.

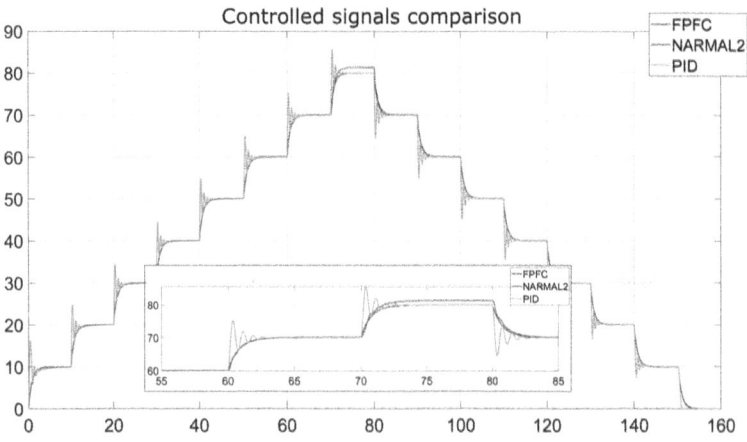

Fig. 2. Comparison between the controlled signals on the mathematical model of the system.

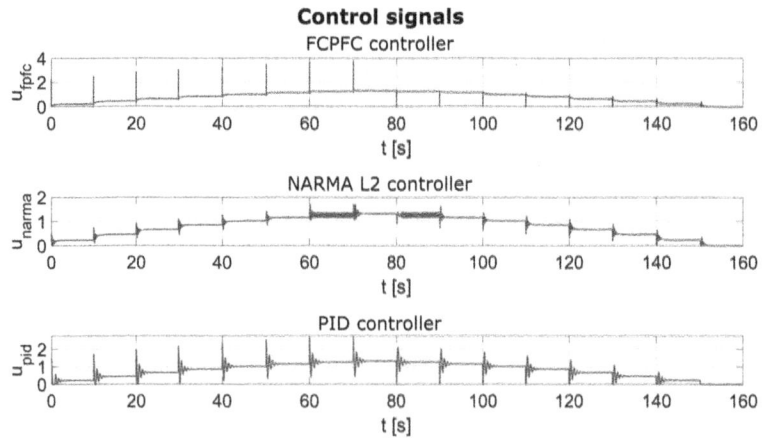

Fig. 3. Comparison between the control signals on the mathematical model of the system.

Analysis of the results (see Figs. 2 and 3) shows that the FCPFC and NARMA L2 controllers follow the reference trajectory closely without showing significant oscillations, unlike the PID controller which produces an oscillatory response. A closer look at Fig. 2 shows that the NARMA L2 controller produces a smaller steady-state error at the highest reference point. The MSE criterion function values in Table 1 show that the best performance is achieved with the FPFC controller, followed by the NARMA L2 and PID controllers.

However, it should be noted that although the FCPFC controller has the lowest MSE value, it has larger jumps in the controlled variable compared to the NARMA L2 and PID controllers, which is reflected in the higher SADU criterion function value.

Table 1. Values of criterion functions MSE and SADU for an individual controller on a mathematical model.

Controller type	MSE	SADU
FCPFC	0.063	1041.14
NARMA L2	0.108	268.502
PID optimized	4.995	89.1607

4.2 Control Results Comparison on Real Plant

In this subsection we compare the performance of the control algorithms with a real system. Unfortunately, we did not obtain satisfactory results in the learning phase of the NARMA L2 model. Therefore, we exclude it from the comparison in this section and only compare the results of the FCPFC and the PID controller.

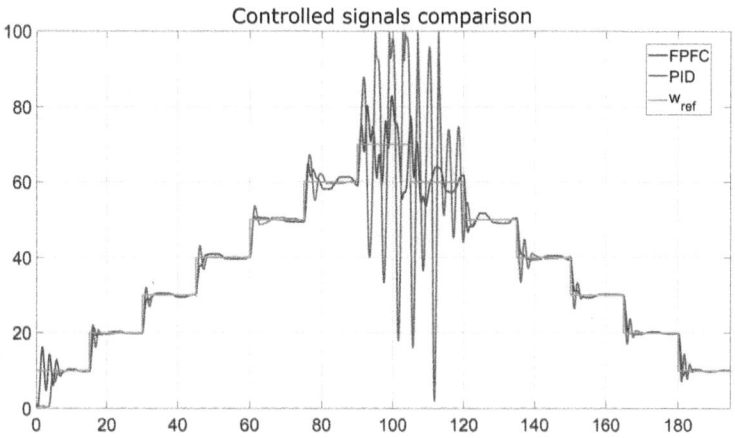

Fig. 4. Comparison between the controlled signals on the real system.

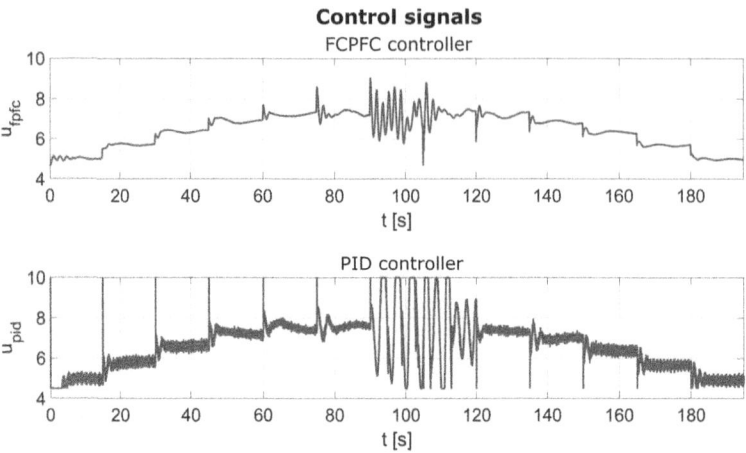

Fig. 5. Comparison between the control signals on the real system.

Figure 4 illustrates a comparison of the system responses when operating with both controllers. Figure 5 shows a comparison of the control signals of the two controllers.

From the results in Figs. 4 and 5 and Table 2, we can see that we get better control results with the FCPFC than with the PID controller. We also notice that the response becomes quite oscillatory at larger offset angles (e.g. $70°$), especially with the PID controller. The reason for the oscillations at larger offset angles is that the system is much more non-linear in this range than at smaller offset angles of the system. In this case, the PID controller performs worse than the FCPFC because it is a linear controller, which in principle works best near the operating point to which the PID controller is optimised. However, where the system is clearly non-linear, the PID controller gives much worse results. The FCPFC also exhibits oscillations at an angle of $70°$ because we performed the identification of the soft model in the range of $0°$–$50°$, because at larger deviations the system becomes unstable and we cannot perform an identification, which means that the model is not determined in this range of device operation. From the SADU criteria and the control signals, we see that the FCPFC has better performance than the PID controller. The output of the PID controller is much more oscillatory and at some points the amplitude of the control signal reaches the maximum excitation voltage of the system.

Table 2. Values of criterion functions MSE and SADU for an individual controller on a real system.

Controller type	MSE	SADU
FCPFC	4.425	78.365
PID optimized	62.54	309.50

5   Conclusion

In this paper, we have presented and compared the performance of three control algorithms on helio-crane laboratory device. We have presented fuzzy cloud-based predictive functional controller, Narma L2 controller and PID controller with optimised parameters. We first present the results for each controller, which we compare in the last subsection of section four to determine the best controller type for our problem. It turns out that the best control results are obtained with an FCPFC controller, but the FCPFC is more time-consuming to develop than a PID controller. The PID controller works well at the beginning of the operating range (smaller offset angles of the helicopter), but at higher offsets the system response becomes quite oscillatory. The FCPFC shows much less oscillation than the PID controller, but at steady state it performs worse than the PID controller in removing disturbances. The NARMA L2 controller performs quite well when running on a mathematical model of the system, but we were not able to train the controller on a real system. When we run the NARMA L2 controller on a real system, we get an unstable system response and never manage to get good performance.

References

1. Kansha, Y., Jia, L., Chiu, M.S.: Self-tuning PID controllers based on the Lyapunov approach. Chem. Eng. Sci. **63**(10), 2732–2740 (2008)
2. Andonovski, G., Lughofer, E., Škrjanc, I.: A comparison of RECCo and FCPFC controller on nonlinear chemical reactor. Model. Identification Control 214–221 (2017)
3. Dovžan, D., Škrjanc, I.: Predictive functional control based on an adaptive fuzzy model of a hybrid semi-batch reactor. Control. Eng. Pract. **18**(8), 979–989 (2010)
4. Karer, G., Škrjanc, I., Zupančič, B.: Self-adaptive predictive functional control of the temperature in an exothermic batch reactor. Chem. Eng. Process. **47**(12), 2379–2385 (2008)
5. Škrjanc, I., Matko, D.: Predictive functional control based on fuzzy model for heat-exchanger pilot plant. IEEE Trans. Fuzzy Syst. **8**(6), 705–712 (2000)
6. Lepetič, M., Škrjanc, I., Chiacchiarini, H.G., Matko, D.: Predictive control based on fuzzy model: a case study. In: 10th IEEE International Conference on Fuzzy Systems, vol. 2, pp. 868–871 (2001)
7. Richalet, J.: Industrial applications of model based predictive control. Automatica **29**(5), 1251–1274 (1993)
8. Andonovski, G., Mušič, G., Blažič, S., Škrjanc, I.: Evolving fuzzy model based performance identification for production control. In: IEEE International Conference on Evolving and Adaptive Intelligent Systems (EAIS), pp. 85–91 (2016)
9. Pukrittayakamee, A., De Jesus, O., Hagan, M.T.: Smoothing the control action for NARMA-L2 controllers. In: The 2002 45th Midwest Symposium on Circuits and Systems, MWSCAS-2002, Tulsa, OK, USA, p. III (2002). https://doi.org/10.1109/MWSCAS.2002.1186964
10. Awwad, A., Abu-Rub, H., Toliyat, H.A.: Nonlinear autoregressive moving average (NARMA-L2) controller for advanced ac motor control. In: 34th Annual Conference of IEEE Industrial Electronics. Orlando, FL, USA, pp. 1287–1292 (2008). https://doi.org/10.1109/IECON.2008.4758140

11. Sahbani, A.: NARMA-L2 Neuro controller for speed regulation of an intelligent vehicle based on image processing techniques. In: 2018 21st Saudi Computer Society National Computer Conference. NCC), Riyadh, Saudi Arabia (2018)

12. Sambariya, D.K., Nath, V.: Application of NARMA L2 controller for load frequency control of multi-area power system. In: 2016 10th International Conference on Intelligent Systems and Control (ISCO), Coimbatore, India (2016)

13. Valluru, S.K., Singh, M., Kumar, N.: Implementation of NARMA-L2 Neuro controller for speed regulation of series connected DC motor. In: 2012 IEEE 5th India International Conference on Power Electronics (IICPE), Delhi, India (2012)

14. Zdešar, A., Cerman, O., Dovžan, D., Hušek, P., Škrjanc, I.: Fuzzy control of a helio-crane: comparison of two control approaches. J. Intell. Robot. Syst. **72**(3–4), 497–515 (2013)

15. Angelov, P., Yager, R.: Simplified fuzzy rule-based systems using non-parametric antecedents and relative data density. In: Symposium Series on Computational Intelligence (IEEE SSCI 2011) - IEEE Workshop on Evolving and Adaptive Intelligent Systems (EAIS 2011), pp. 62–69 (2011)

16. Lughofer, E.: Evolving Fuzzy Systems - Methodologies, Advanced Concepts and Applications. STUDFUZZ, Springer, Heidelberg (2011). https://doi.org/10.1007/978-3-642-18087-3

17. Narendra, K.S., Parthasarathy, K.: Identification and control of dynamical systems using neural networks. IEEE Trans. Neural Netw. **1**(1), 4–27 (1990)

18. Levin, A.U., Narendra, K.S.: Control of nonlinear dynamical systems using neural networks. II. Observability, identification, and control. IEEE Trans. Neural Netw. **7**(1), 30–42 (1996)

19. Narendra, K.S.: Neural networks for control theory and practice. Proc. IEEE **84**(10), 1385–1406 (1996)

20. Narendra, K.S., Mukhopadhyay, S.: Adaptive control using neural networks and approximate models. IEEE Trans. Neural Netw. **8**(3), 475–485 (1997)

21. Yang, Y., Xiang, C., Gao, S., Lee, T.H.: Data-driven identification and control of nonlinear systems using multiple NARMA-L2 models. Int. J. Robust Nonlinear Control **28**, 3806–3833 (2018)

22. Zdešar, A., Cerman, O., Dovžan, D., Hušek, P., Škrjanc, I.: Fuzzy control of a helio-crane: comparison of two control approaches. J. Intell. Robot. Syst. Theory Appl. **72**(3–4), 497–515 (2013)

Evolving Neuro-Fuzzy Design of Experiments: A Novel Approach of Nonlinear Process Identification

Miha Ožbot$^{(\boxtimes)}$ ⓘ and Igor Škrjanc ⓘ

Faculty of Electrical Engineering, University of Ljubljana, Ljubljana, Slovenia
{miha.ozbot,igor.skrjanc}@fe.uni-lj.si

Abstract. This paper presents a novel model-oriented sequential online design of experiments for model identification. An evolving neuro-fuzzy model identification of nonlinear dynamical systems is combined with two approaches to experiment design. First, a sequential design based on the maximin space-filling criterion was used to select an input space with the least model coverage. Second, an optimal experiment design method was used to select the optimal input signal online near the operating point selected with the first step. The use of an evolving model allows online generation of a model-based optimal input signal for identification of nonlinear dynamical processes without the need for multiple iterations or an initial model. The method was validated on a multiple-input single-output theoretical model of a plate heat exchanger.

Keywords: Evolving neuro-fuzzy · Space-filling · Optimal design of experiments · Plate heat exchanger · Filtered recursive least squares

1 Introduction

The goal of Design of Experiments (DoE) is to optimally select values of independent variables for data-driven identification with minimal effort. The DoE method used depends on the application of the experiment, e.g., dynamical or static model identification, analysis of the significance of inputs, design of controls, etc. Space-filling designs that select input signals in a particular pattern are used to identify the static characteristics of a system, while optimal designs can also be used to design input signals with better frequency information for dynamical processes. A review of space-filling designs, e.g., Minimax, Maximin, and Latih hypercube, was presented in [11] and a review of the history of optimal experimental designs, e.g., D-optimal and A-optimal, was given in [3]. Another type of experimental designs are the sequential designs that select new measurements based on the previously identified model [2,14]. A DoE for systems with ARX structure with non-Gaussian noise is presented in [13]. A D-optimal

The authors acknowledge the financial support from the Slovenian Research Agency ARRS (Ph.D. grant for Miha Ožbot, funding number P2-0219).

© The Author(s), under exclusive license to Springer Nature Switzerland AG 2024
M. Mujica Mota and P. Scala (Eds.): EUROSIM 2023, CCIS 2033, pp. 301–316, 2024.
https://doi.org/10.1007/978-3-031-68438-8_22

design was used in [12] for medical experiments and in [5] using local model networks with external dynamics (which is structurally identical to a neuro-fuzzy dynamical model) for the identification of an engine.

Online DoE for the identification of dynamical systems requires computationally efficient methods and has recently become possible due to the advancement of online methods and the availability of computing power. A common property of classical excitation signals is random perturbation based on the persistent excitation condition [16], which cannot always be applied to industrial processes. The main advantage of the online approach over offline methods is that the input signal is optimized for the target system, allowing a reduction in the number of samples required. In [15], a sample-efficient online design for nonparametric linear time-invariant systems based on Hankel matrices was proposed. In [14], an incremental Latin hypercube additive design (ILHAD) method based on Gaussian process regression for static model identification is presented and used to identify an industrial bitumen furnace. This online DoE is able to expand the dimension of the design space used for identification in each iteration with an additive model structure, where each submodel uses a subset of the available data. The DoE method combined with interpretable fuzzy systems can improve understanding of how the process works and identify opportunities to improve quality, reduce costs, and increase efficiency. We propose an online experimental design method based on an evolving neuro-fuzzy system, the Evolving Design of Experiments (eDoE). Evolving systems are characterized by their ability to incrementally adapt the structure of the model and identify the consequent parameters online with the last available sample in one-pass manner and immediately discard the sample [21]. An evolving design of experiment method for static nonlinear systems was proposed in [18]. The region with the worst local model is divided into two halves and a Maximin space filling method is used to design measurements in this region. In [17], a full factorial DoE based on the ANOVA analysis for the popular ANFIS [6] neuro-fuzzy systems and the EFuNN [7] evolving neuro-fuzzy system were presented. A comprehensive survey of evolving fuzzy and neuro-fuzzy systems was presented in [21]. The main contributions of this work are:

- An online single-pass DoE based on an evolving neuro-fuzzy identification method for nonlinear dynamical systems.
- A hybrid model-based space-filling and optimal DoE method. The first method allows identification of static nonlinearities without a priori knowledge, while the second design maximizes the information obtained from each measurement.

The proposed hybrid design consists of a model-based sequential design used to select an appropriate identification region and an optimal design method based on the Fisher information matrix that stimulates the target system with informative input signals to determine local linear models (LLM). Both DoE methods were formulated so that it is not necessary to store previous samples. The sequential space-filling method uses the evolving model antecedent structure of the model and the optimal design uses the covariance matrix and the identified parameters of the consequent LLMs.

The remainder of the paper is organized as follows. Section 2 describes the methodology of the evolving neuro-fuzzy model. Sections 3 and 4 present the sequential space-filling design and the optimal experimental design, respectively. The simulation experiments are performed in Sect. 5 and the results are discussed in Sect. 6. Conclusions and future work are presented in Sect. 7.

2   Evolving Neuro-Fuzzy system

The evolving neuro-fuzzy model structure consists of neurons in a network structure based on fuzzy Takagi-Sugeno rules with a single cluster-based membership function as antecedent and an affine function in the consequent part of the rule [6]. A neuro-fuzzy Takagi-Sugeno rule is described as follows

$$R_i : \quad \text{if} \quad \big(\underline{z}(k) \quad \text{is} \quad \mathcal{Z}_i\big) \quad \text{then} \quad y_i(k) = \underline{\varphi}^\top(k)\underline{\theta}_i, \tag{1}$$

where $i = 1, 2, \ldots, c$ is the index of the fuzzy rule, k is the discrete time step, $\underline{z}(k) \in \mathbb{R}^{n_z}$ is the antecedent input vector, $\mathcal{Z}_i$ is the antecedent fuzzy set, $y_i(k) \in \mathbb{R}$ is the output, and $\underline{\theta}_i \in \mathbb{R}^{n_\varphi}$ is the parameter vector of the consequent local linear model (LLM) associated with the ith fuzzy rule, and $\underline{\varphi}^\top(k) = [u^1(k-1), u^1(k-2), \ldots, u^1(k-m), \ldots, u^2(k-1), \ldots, u^{n_u}(k-m), \ldots, y(k-1), \ldots, y(k-n), 1] \in \mathbb{R}^{n_\varphi}$ is the consequent regression vector. The antecedent fuzzy sets $\mathcal{Z}_i$ are usually represented by Gaussian clusters, since they can approximate a variety of data distributions [19]. In addition, the multivariate Gaussian distribution can be used to describe axis-oblique clusters that can be rotated arbitrarily to define correlations between variables (as opposed to axis-parallel representation)

$$\gamma_i(k) = \exp\Big(-\big(\underline{z}(k) - \underline{\mu}_i\big)^\top \underline{\Sigma}_i^{-1}\big(\underline{z}(k) - \underline{\mu}_i\big)\Big), \tag{2}$$

where $\gamma_i \in [0, 1]$ is the non-normalized Gaussian membership function of the ith fuzzy rule, $\underline{\mu}_i \in \mathbb{R}^{n_z}$ and $\underline{\Sigma}_i \in \mathbb{R}^{n_z \times n_z}$ are the center and the covariance matrix of the cluster associated with the ith fuzzy rule. The membership functions are normalized as $\phi_i(k) = \gamma_i(k) / \sum_i^c \gamma_i(k) \in [0, 1]$, so that a unit partition is obtained $\sum_{i=1}^c \phi_i(k) = 1$. The output of the evolving neuro-fuzzy model is aggregated from all activated rules as $y(k) = \sum_{i=1}^c \phi_i(k) y_i(k) \in \mathbb{R}$. In this study, the evolving neuro-fuzzy identification method presented in [10,20] was used because it is able to identify Nonlinear Output Error (NOE) dynamical systems online without the need to compute membership functions at every time step, by assuming that consecutive samples belong to the same rule. This reduces computational burden by avoiding the computation of the inverse covariance matrices for each rule for every sample in the data stream.

2.1   Evolving Additive Mechanism

In the proposed approach, a new fuzzy rule is added when a new region is selected with the space-filling DoE. A new rule is added by adding a Gaussian cluster in the antecedent and an associated consequent Output Error Local Linear Model (OE-LLM). The new c^{th} cluster is initialized by setting the cluster center to the last sample $\underline{\mu}_c = \underline{z}(k) \in \mathbb{R}^{n_z}$, the covariance matrix to $\underline{S}_c = \underline{0} \in \mathbb{R}^{n_z \times n_z}$, and the number of samples to $n_c = 1$. The consequent model is defined by the parameter vector which is set to zero $\hat{\underline{\theta}}_c = \underline{0} \in \mathbb{R}^{n_\varphi}$. The information matrix of the RLS method is also reset as $\underline{P} = \alpha_P \underline{I} \in \mathbb{R}^{n_\varphi \times n_\varphi}$. The parameter of the RLS method $\alpha_P \in \mathbb{R}$ is usually large enough to allow a fast initial estimate of the parameters.

2.2 Incremental Clustering

The centers $\underline{\mu}_i$ and Covariance matrices $\underline{\Sigma}_i$ of the Gaussian clusters in the antecedent are incrementally clustered based on the online algorithm for computing variance [20]

$$\underline{e}_i(k) = \underline{z}(k) - \underline{\mu}_i(n_i), \tag{3}$$

$$\underline{\mu}_i(n_i + 1) = \underline{\mu}_i(n_i) + \underline{e}_i(k)/(n_i + 1), \tag{4}$$

$$\underline{S}_i(n_i + 1) = \underline{S}_i(n_i) + \underline{e}_i(k)\big(\underline{z}(k) - \underline{\mu}_i(n_i + 1)\big)^\top, \tag{5}$$

where the number of samples can be adapted as $n_i = n_i + 1$. The covariance matrix is computed only as it is required as $\underline{\Sigma}_i = \underline{S}_i/n_i$ to reduce computational demand.

2.3 Evolving Merging Mechanism

Data from the stream can coalesce over time, resulting in overlapping clusters with similar consecutive models [9]. In this case, merging is used to simplify the evolving neuro-fuzzy model structure. Rules were merged by combining the antecedent clusters $(\underline{\mu}_{pq}, \underline{\Sigma}_{pq}, n_{pq})$ and consequent models $\hat{\underline{\theta}}_{pq}$. Valid candidates for merging were first selected based on the degree of overlapping computed with the Bhattacharyya distance for multi-variate Gaussian distributions [9] as

$$d_B(p,q) = \frac{1}{8}\big(\underline{\mu}_p - \underline{\mu}_q\big)^\top \left(\frac{\underline{\Sigma}_p + \underline{\Sigma}_q}{2}\right)^{-1}\big(\underline{\mu}_p - \underline{\mu}_q\big) + \frac{1}{2}\ln\left(\frac{\det(\frac{1}{2}(\underline{\Sigma}_p + \underline{\Sigma}_q))}{\sqrt{\det\underline{\Sigma}_p \det\underline{\Sigma}_q}}\right), \tag{6}$$

where $p = 1, 2, \ldots, c$ and $q = 1, 2, \ldots, c$ are the indexes of the two compared clusters for $p \neq q$. A useful feature of this measure is the symmetry $d_B(p,q) = d_B(q,p)$. The pth and q^{th} clusters are considered to be overlapping if (6) is lower than a threshold κ_V. The candidates selected using the overlap measure (6) were further compared based on the dissimilarity of the consequent models. This was simplified by comparing the dc-gain and the bias of the transfer functions of the consequent models as proposed in [10, 20]

$$\left|\hat{\underline{G}}_p(1) - \hat{\underline{G}}_q(1)\right| < \underline{\kappa}_K \in \mathbb{R}_0^+, \tag{7}$$

$$\left|\hat{\underline{\theta}}_{p,n_\varphi} - \hat{\underline{\theta}}_{q,n_\varphi}\right| < \kappa_r \in \mathbb{R}_0^+, \tag{8}$$

where $\hat{G}_p(1) = \lim_{z \to 1} (\hat{B}_p(z)/\hat{A}_p(z))$ is the dc-gain of the pth model transfer function, which is constructed from $\underline{\hat{\theta}}_p$ for each input-output combination, and $\hat{\theta}_{p,n_\varphi}$ is the static bias associated with the affine constant. The choice of these dissimilarity parameters depends on the desired simplicity/accuracy of the model. This approach can be used to merge clusters that are close to each other, while avoiding merging clusters that have similar consequent LLMs but are far apart. The region between clusters could contain nonlinear features that have not yet been sampled. Conversely, these models can be merged later when more data are available.

The clusters are merged by estimating the new covariance matrix $\underline{\Sigma}_{pq}(k)$ based on $\underline{Z}_{pq}^\top \underline{Z}_{pq} = \underline{Z}_p^\top \underline{Z}_p + \underline{Z}_q^\top \underline{Z}_q$ and $\underline{\Sigma}_{pq} = \frac{1}{(n_{pq}-1)}(\underline{Z}_{pq}^\top \underline{Z}_{pq} - n_{pq}\underline{\mu}_{pq}\underline{\mu}_{pq}^\top)$, where $\underline{Z}_{pq}^\top = [\underline{Z}_p^\top, \underline{Z}_q^\top] = [\underline{z}_p(1), \dots, \underline{z}_p(n_p), \underline{z}_q(1), \dots, \underline{z}_q(n_q)] \in \mathbb{R}^{(n_p+n_q) \times n_z}$ is the data matrix containing all clustering samples. This can be formulated in a way that avoids the data matrices as [19]

$$n_{pq}(k) = n_p(k) + n_q(k), \tag{9}$$

$$\underline{\mu}_{pq}(k) = \big(n_p(k)\underline{\mu}_p(k) + n_q(k)\underline{\mu}_q(k)\big)/n_{pq}(k), \tag{10}$$

$$\underline{\Sigma}_{pq}(k) = \frac{1}{n_{pq}-1}\Big((n_p-1)\underline{\Sigma}_p + (n_q-1)\underline{\Sigma}_q + $$
$$+ \underline{M}_p^\top \underline{E}_p^\top \underline{E}_p \underline{M}_p + \underline{M}_q^\top \underline{E}_q^\top \underline{E}_q \underline{M}_q - \underline{M}_{pq}^\top \underline{E}_{pq}^\top \underline{E}_{pq} \underline{M}_{pq}\Big), \tag{11}$$

where $\underline{M}_p = \underline{\mu}_p^\top I \in \mathbb{R}^{n_z \times n_z}$ is a diagonal matrix containing the cluster center variables on the diagonal, $\underline{E}_p \in \mathbb{R}^{n_p \times n_z}$ is a matrix in which all elements equal 1, likewise for q and pq. The consequent parameters $\underline{\hat{\theta}}_{pq}$ are merged based on a weighted average similar to (10).

2.4   Filtered Recursive Least Squares Optimization

The NOE model structure assumes an output with additive white Gaussian noise $v \sim \mathcal{N}(0, \sigma^2)$, which is common in real processes but requires a specialized parameter identification method. The consecutive OE-LLMs are linear-in-parameters and can be identified with the Filtered Recursive Least Squares (FRLS) method [10,20]. It is a modified version of the Recursive Least Squares (RLS) method that is commonly used in the identification of fuzzy models. The identification of OE-LLM is enabled by leveraging the identified system transfer function denominator as a filter to reshape the prediction error formulation and uncorrelate the input signals $\underline{u}(k)$ with the affine constant 1 in the regression vector

$$\underline{\hat{\theta}}(k) = \underline{\hat{\theta}}(k-1) + \underline{P}(k)\underline{\varphi}_f(k)\big(y_f(k) - \underline{\varphi}_f^\top(k)\underline{\hat{\theta}}\big), \tag{12}$$

$$\underline{P}(k) = \underline{P}(k-1) - \frac{\underline{P}(k-1)\underline{\varphi}_f(k)\underline{\varphi}_f^\top(k)\underline{P}(k-1)}{\underline{\varphi}_f^\top(k)\underline{P}(k-1)\underline{\varphi}_f(k) + 1}, \tag{13}$$

where the regression signals and output signal are filtered as $\underline{\varphi}_f(k) = \frac{1}{A_f(z)}\underline{\varphi}(k)$. The filter is adapted online during the identification when the normalized confidence interval of the identified LLM falls under a threshold value [10]

$$A_f(z) = \begin{cases} \dfrac{\hat{A}(z)}{\lim_{z \to 1} \hat{A}(z)}, & \text{if } \underline{\varphi}^\top(k)\underline{P}(k)\underline{\varphi}(k) \leq \kappa_f \\ & \text{and } |z_j| < 1 \, \forall j \\ A_f(z), & \text{otherwise} \end{cases} \tag{14}$$

where $\hat{A}(z)$ is the denominator of the transfer function of the identified model, z_j are the poles of the denominator $\hat{A}(z_j) = 0$, and $\kappa_f \in [0,1]$ is a threshold parameter. The condition $|z_j| < 1$ ensures the stability of the filter. This avoids the initial instability of the optimization after a new rule is added. Each sample is used to identify the consecutive parameters of only the cluster with the highest activation.

3 Sequential Space-Filling Design of Experiments

The proposed design modifies the Maximin space-filling design to select the input region that is the least covered with the antecedent structure of the evolving neuro-fuzzy model. The classical Maximin space-filling design aims to maximize the distance between samples by finding the largest distance between the new sample and the closest existing sample [18]. Conversely, in this approach the previous samples are substituted by the evolving antecedent clusters to increase the computation speed and reduce storage requirement. The Maximin problem for the optimal input vector $\underline{v}^* \in \mathbb{R}^{n_u}$ can be formulated as

$$\begin{aligned} \text{minimize} \quad & \max_{i \in \{1,\dots,c\}} \left(-d_i^2(\underline{v}) \right), \\ \text{subject to} \quad & \underline{u}_{\min} \leq \underline{v} \leq \underline{u}_{\max}, \end{aligned} \tag{15}$$

where $d_i^2(\underline{v}) = (\underline{v} - \underline{\mu}_i^u)^\top (\underline{\Sigma}_i^u)^{-1}(\underline{v} - \underline{\mu}_i^u) \in \mathbb{R}$ is the Mahalanobis distance from the point $\underline{v}$ to the i-th cluster in the model antecedent, $\underline{u}_{\min}$ and $\underline{u}_{\max}$ are the lower and upper limits of the input signals, respectively. The superscript ^{u} means that only the input components are used since the corresponding outputs are not available.

In [10] a particle swarm optimization (PSO) [8] method was used to compute $\underline{v}^*$, but it can be too computationally expensive for some real time applications. The gradient of the Maximin function is not continuous but it can be computed for the current value $\underline{v}$ as

$$\frac{\partial \max_i \left(-d_i^2(\underline{v}) \right)}{\partial \underline{v}} = -2(\underline{v} - \underline{\mu}_j^u)^\top (\underline{\Sigma}_j^u)^{-1}, \tag{16}$$

where $\underline{\mu}_j^u$ and $\underline{\Sigma}_j^u$ are the center and covariance matrix of the closest cluster to the independent variable $\underline{v}$, respectively, i.e. the cluster with index

$j = \text{argmax}_{i \in \{1,\ldots,c\}}\, d_i^2(\underline{v})$. An example with three clusters is shown in Fig. 1. The optimal input vector $\underline{v}^*$ is used to select a hypercube region $[\underline{v}^* - \underline{\Delta u}, \underline{v}^* + \underline{\Delta u}]$ that is used as the design space for the local optimal experiment, where $\underline{\Delta u} \in \mathbb{R}^{n_u}$ is a parameter of the method. This design is not computed at every time step, but only after a rule was identified in the previous region, after k_h time steps.

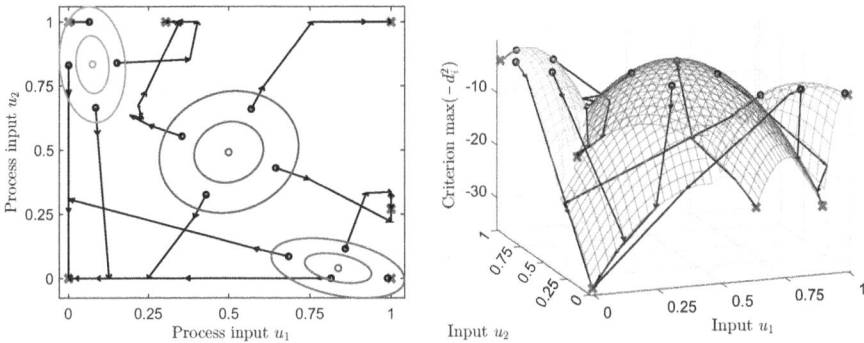

Fig. 1. The Sequential Quadratic Programming method was employed for space-filling optimization. In this approach, the initial optimization points (depicted as black dots) were chosen based on the eigenvectors of the clusters represented by 1σ and 2σ elliptic contours, as illustrated in the left figure. The Maximin criterion is displayed in the right figure.

4   Optimal Design of Experiments

An optimal design is used to select the most informative input signal for the identification of the consequent parameters in the region selected with the space-filling design. Optimal experiment design methods search for an optimal input signal that maximizes a measure of information gained from each sample. We examined the D-optimal and A-optimal criterion since our objective is to maximize the accuracy of the identified parameters. Both designs are based on the Fisher information matrix $\underline{F}\big(\hat{\underline{\Psi}}(k+1)\big) \in \mathbb{R}^{n_\varphi \times n_\varphi}$ [5]

$$\hat{\underline{\psi}}^\top(k+1) = \frac{\partial \hat{y}(k+1)}{\partial \hat{\underline{\theta}}} = \frac{\partial\big(\hat{\underline{\varphi}}^\top(k+1)\hat{\underline{\theta}}\big)}{\partial \hat{\underline{\theta}}} = \hat{\underline{\varphi}}^\top(k+1), \tag{17}$$

$$\hat{\underline{\Psi}}(k+1) = [\hat{\underline{\psi}}(1),\ldots,\hat{\underline{\psi}}(k+1)]^\top = [\hat{\underline{\varphi}}(1),\ldots,\hat{\underline{\varphi}}(k+1)]^\top, \tag{18}$$

$$\underline{F}\big(\hat{\underline{\Psi}}(k+1)\big) = \frac{1}{\sigma^2}\hat{\underline{\Psi}}^\top(k+1)\hat{\underline{\Psi}}(k+1) \simeq \frac{1}{\sigma^2}\hat{\underline{P}}^{-1}(k+1), \tag{19}$$

where $\hat{\underline{\Psi}}(k+1) \in \mathbb{R}^{(k+1)\times n_\varphi}$ is the sensitivity matrix of the LLM output with respect to the parameters of the model, σ^2 is the estimated white noise variance, and $\underline{P}^{-1}(k+1) \in \mathbb{R}^{n_\varphi \times n_\varphi}$ is the symmetric information matrix used in the recursive identification of the parameters of the LLM. Note, that the numerator layout convention for matrix differentiation is used in this study. Since we assume a winner-takes-all optimization strategy for the consecutive parameters, we optimize for a single LLM in the vicinity of the operating point $\underline{v}^*$ that was selected with the space-filling experimental design method (15). Computational effort is reduced because only a subset of the problem space is identified at a time [14]. The main idea of the proposed approach is that the covariance matrix of the RLS method $\underline{P}^{-1}(k+1)$ can be used to recursively compute the Fisher information matrix without storing any samples [3].

4.1   D-Optimal Criterion

The D-optimal design minimizes the area of the confidence interval of the estimated parameters [12] by maximizing the determinant of the inverse of the covariance matrix. The determinant of a covariance matrix is proportional to the area of the hyper-ellipsoid that describes the covariance matrix [19]. The D-optimal design is defined as

$$I_D\big(\hat{\underline{\Psi}}(k+1)\big) = \det\Big(\underline{F}\big(\hat{\underline{\Psi}}(k+1)\big)\Big) \Rightarrow \underline{u}^*(k) = \mathrm{argmax}_{\underline{u}(k)}\Big(I_D\big(\hat{\underline{\Psi}}(k+1)\big)\Big), \quad (20)$$

The analytical derivative of the criterion $I_D\big(\hat{\underline{\Psi}}(k+1)\big) \in \mathbb{R}$ w.r.t. the input signal $u_j(k)$ is required for efficient optimization. The chain rule for scalar-by-scalar derivation with matrices involved can be expressed as [5]

$$\frac{dI_D\big(\hat{\underline{\Psi}}(k+1)\big)}{du_j(k)} = \mathrm{tr}\left(\frac{dI_D\big(\hat{\underline{\Psi}}(k+1)\big)}{d\hat{\underline{\Psi}}(k+1)} \frac{d\hat{\underline{\Psi}}(k+1)}{du_j(k)}\right) \in \mathbb{R}, \quad (21)$$

where $\mathrm{tr}(\cdot)$ is the trace of a square matrix. The first term $\frac{dI_D(\hat{\underline{\Psi}}(k+1))}{d\hat{\underline{\Psi}}(k+1)} \in \mathbb{R}^{n_\varphi \times (k+1)}$ can be expanded with the identity $\frac{\partial}{\partial \underline{X}}(\det(\underline{X}^\top \underline{X})) = 2\det(\underline{X}^\top \underline{X}) (\underline{X}^\top \underline{X})^{-1}\underline{X}^\top$ as

$$\frac{dI_D\big(\hat{\underline{\Psi}}(k+1)\big)}{d\hat{\underline{\Psi}}(k+1)} = 2I_D\big(\hat{\underline{\Psi}}(k+1)\big)\big(\hat{\underline{\Psi}}^\top(k+1)\hat{\underline{\Psi}}(k+1)\big)^{-1}\hat{\underline{\Psi}}^\top(k+1), \quad (22)$$

The second term in (21) is equal to

$$\frac{d\hat{\underline{\Psi}}(k+1)}{du_j(k)} = \left[\underline{0}, .., \frac{d\varphi(k+1)}{du_j(k)}\right]^\top = [\underline{0}, .., \underline{\iota}_{jk}]^\top \in \mathbb{R}^{(k+1)\times n_\varphi}, \quad (23)$$

where $\underline{\iota}_{jk} = [\delta_{1j}\delta_{kk}, \delta_{1j}\delta_{(k-1)k}, \ldots, \delta_{n_u j}\delta_{(k-n)k}, 0, \ldots, 0]^\top = [0, \ldots, 0, \delta_{jj}\delta_{kk}, 0, \ldots, 0]^\top \in \mathbb{R}^{n_\varphi}$ is a vector with a single nonzero element, and $\delta_{ij} = [i = j]$ is the Kronecker delta function with two variables. The last term in (22) can be combined with (23) and simplified as

$$\hat{\underline{\Psi}}^\top(k+1)\frac{d\hat{\underline{\Psi}}(k+1)}{du_j(k)} = [\underline{\varphi}(1), \ldots, \hat{\underline{\varphi}}(k+1)][\underline{0}, .., \underline{\iota}_{jk}]^\top = \hat{\underline{\varphi}}(k+1)\underline{\iota}_{jk}^\top \in \mathbb{R}^{n_\varphi \times n_\varphi}. \quad (24)$$

This eliminates the need to store previous samples from the data matrix $\hat{\underline{\Psi}}^\top(k+1)$. Finally, the derivative of the D-optimal criterion with respect to the input signal (21) is equal to

$$\frac{\mathrm{d}I_\mathrm{D}(\hat{\underline{\Psi}}(k+1))}{\mathrm{d}u_j(k)} = \frac{2}{\sigma^2} \det(\hat{\underline{P}}^{-1}(k+1)) \mathrm{tr}(\hat{\underline{P}}(k+1)\hat{\underline{\varphi}}(k+1)\underline{\iota}_{jk}^\top), \qquad (25)$$

4.2   A-Optimal Criterion

A drawback of using the D-optimal criterion is that it requires the inverse of the covariance matrix, which is computationally expensive and it has to be non-singular. With the proposed online approach and the A-optimal criterion the inverse is not required in the criterion or it's derivative. The A-optimal criterion optimizes the average variance of the regression parameters [3] and is defined as

$$I_A(\hat{\underline{\Psi}}(k+1)) = \mathrm{tr}\left(\underline{F}^{-1}(\hat{\underline{\Psi}}(k+1))\right) \Rightarrow \underline{u}^*(k) = \mathrm{argmin}_{\underline{u}(k)}\left(I_A(\hat{\underline{\Psi}}(k+1))\right). \ (26)$$

Note that with the A-optimal criterion $I_A(\hat{\underline{\Psi}}(k+1)) \in \mathbb{R}$ a minimum is searched as opposed to a maximum with the D-optimal criterion (20). The derivative of the design w.r.t. the input signal can be computed analytically with the chain rule as

$$\frac{\mathrm{d}I_A(\hat{\underline{\Psi}}(k+1))}{\mathrm{d}u_j(k)} = \mathrm{tr}\left(\frac{\mathrm{d}I_A(\hat{\underline{\Psi}}(k+1))}{\mathrm{d}\hat{\underline{\Psi}}(k+1)} \frac{\mathrm{d}\hat{\underline{\Psi}}(k+1)}{\mathrm{d}u_j(k)}\right) \in \mathbb{R}, \qquad (27)$$

The first term in (27) can be expanded with the identity $\frac{\partial}{\partial \underline{X}} \mathrm{tr}((\underline{X}^\top \underline{X})^{-1}) = -2(\underline{X}^\top \underline{X})^{-1}(\underline{X}^\top \underline{X})^{-1}\underline{X}^\top$ as

$$\frac{\mathrm{d}I_A(\hat{\underline{\Psi}}(k+1))}{\mathrm{d}\hat{\underline{\Psi}}(k+1)} = -\frac{2}{\sigma^2}(\hat{\underline{\Psi}}^\top(k+1)\hat{\underline{\Psi}}(k+1))^{-1}(\hat{\underline{\Psi}}^\top(k+1)\hat{\underline{\Psi}}(k+1))^{-1}\hat{\underline{\Psi}}^\top(k+1). \quad (28)$$

Similar to the D-optimal design, the rightmost data matrix in (28) and the derivative of the data matrix (23) are combined in (27) which results in (24). Finally the derivative of the A-optimal design w.r.t. the input signal (27) equals

$$\frac{\mathrm{d}I_A(\hat{\underline{\Psi}}(k+1))}{\mathrm{d}u_j(k)} = -\frac{2}{\sigma^2} \mathrm{tr}(\hat{\underline{P}}(k+1)\hat{\underline{P}}(k+1)\hat{\underline{\varphi}}(k+1)\underline{\iota}_{jk}^\top). \qquad (29)$$

The one-step-ahead model output prediction $\hat{y}(k) = \underline{\varphi}^\top(k)\hat{\underline{\theta}}$ is used to compute the optimal $\underline{u}^*(k)$ while also taking into account the filter $A_f(z)$ used to identify the parameters of the OE-LLM with the FRLS method. The future information matrix $\hat{\underline{P}}(k+1)$ that depends on the input signal $\underline{u}(k)$ is obtained by constructing the future regression vector $\hat{\underline{\varphi}}_f^\top(k+1) = [u_{1f}(k), u_{2f}(k-1), \ldots,$ $u_{mf}(k-n), \hat{y}_f(k), y_f(k-1), \ldots, y_f(k-n), 1]$, where $\underline{u}_f(k) = \frac{A_f(1)}{A_f(z)}\underline{u}(k)$ and

$\hat{y}_f(k) = \frac{A_f(1)}{A_f(z)}\hat{y}(k)$ are the filtered input signal and one-step-ahead model prediction, respectively. The information matrix $\hat{\underline{P}}(k+1)$ is then computed as (13).
To obtain the D-optimal input signals (20) and the A-optimal input signals (26),
we used Sequential Quadratic Programming (SQP) [4] based on the derivatives
(25)/(29) and the estimated information matrix $\hat{\underline{P}}(k+1)$. Importantly, since the
FRLS optimization method is used the derivative of the D-optimal design (25)
and A-optimal design (29) the derivatives must also be multiplied by $A_f(1)$.
This is a necessary step as we are optimizing for $u(k)$ and not $u_f(k)$ in (23).
The noise variance σ^2 is a scalar that has no effect on the minimum of the optimal criterion and was therefore omitted from the calculation. An example of the
proposed hybrid eDoE is shown in Fig. 2.

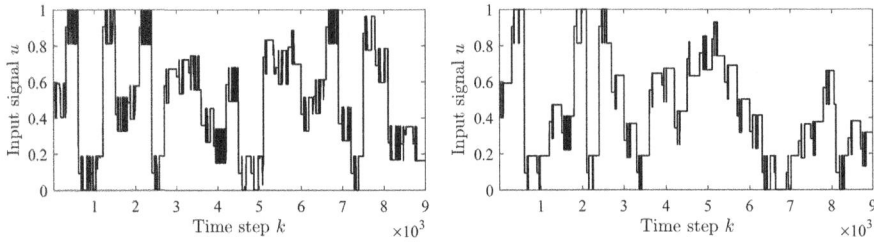

Fig. 2. An example of the proposed Hybrid DoE. The signal is sequentially generated
online – regions of interest are selected with the space-filling design (15) while local
perturbations are optimized with the D-optimal DoE (20)(left) and the A-optimal DoE
(26)(right).

5 Experimentation

The combined evolving neuro-fuzzy hybrid algorithm for experimental design is
presented in Algorithm 1. The parameters of the algorithm can be divided into
three groups, namely, the merging thresholds of the evolving method ($\underline{\kappa}_K, \kappa_r$,
and κ_V), the parameter identification method (α_P and κ_f), and the DoE ($\underline{\Delta}u$
and k_h). We propose the following ranges for these parameters. The merging
thresholds of the consequent OE-LLM $\underline{\kappa}_K$ and $\underline{\kappa}_r$ depend on the application of
the model, but one-tenth of the estimated range of the parameters is a good
first choice. A good range for the overlap threshold is $\kappa_V \in [1,3]$, with a higher
number leading to more clusters detected as overlapping. The FRLS method
covariance matrix must be initialized with a value of $\alpha_P \in [10^3, 10^5]$ to allow
fast initial convergence, and the filter adaptation threshold in the range $\kappa_f \in
[10^{-2}, 10^{-1}]$, with a smaller value resulting in higher confidence being required
before filter adaptation can begin. The DoE parameter $\underline{\Delta}u$ has a significant
impact on the accuracy/simplicity of the identified model, as it affects the size
of the clusters in the model's antecedent structure. For a simple model, one-tenth
of the input range might be appropriate, while a smaller value would improve

---

Algorithm 1: Evolving neuro-fuzzy model identification with Hybrid DoE

Parameters: $\underline{\Delta u}$, α_P, $\underline{\kappa}_K$, κ_r, κ_V, κ_f, k_h

Initialize: $\underline{S}_1 \leftarrow \underline{0}$, $n_1 \leftarrow 1$, $\underline{\mu}_1 \leftarrow \underline{z}(1)$, $\hat{\underline{\theta}}_1 \leftarrow \underline{0}$, $\underline{P} \leftarrow \alpha_P \underline{I}$, $\underline{u}(1) \leftarrow \underline{v}^*$, $y(1)$, $\underline{\varphi}(1)$

repeat

   | $k \leftarrow k+1$

   | **if** $\underline{\varphi}(k-1)\underline{P}(k-1)\underline{\varphi}^\top(k-1) \leq \kappa_f$ **then**

   | | Compute optimal $u(k) \in [\underline{v}^* - \underline{\Delta u}, \underline{v}^* + \underline{\Delta u}]$ with (20, 25) or (26, 29)

   | | Filter adaptation $A_f(z) \leftarrow A_c(z)$ (14).

   | **else**

   | | Use previous inputs $\underline{u}(k) \leftarrow \underline{u}(k-1)$

   | System measurement $z(k)$

   | Incremental Gaussian clustering (3, 4, 5)

   | Recursive parameter identification (12, 13).

   | **if** *Local model identified* **then**

   | | **repeat**

   | | | **for** $i \leftarrow 1$ **to** c **do**

   | | | | Find overlapping candidates (6).

   | | | | Compute dissimilarity of LLMs (7, 8).

   | | | | Cluster merging mechanism (9, 10, 11).

   | | **until** *no merge is possible*

   | | Compute $\underline{v}^*$ with space-filling DoE (15, 16).

   | | Cluster addition $c \leftarrow c+1$, $\underline{S}_c \leftarrow \underline{0}$, $n_c \leftarrow 1$, $\underline{\mu}_c \leftarrow \underline{z}(k)$, $\hat{\underline{\theta}}_c \leftarrow \underline{0}$, $\underline{P} \leftarrow \alpha_P \underline{I}$.

until *termination condition is met*

Output:

---

the accuracy of the model but would require a larger number of experiments. The holding time k_h should be chosen to be 4 to 5 times the dominant time constant of the process.

The proposed eDoE method was used to identify a theoretical model of a plate heat exchanger (PHE) in a computer simulation experiment where the parameters of the model were optimized with data from a real pilot plant [1, 10, 22]. Heat exchangers are used to transfer thermal energy from one medium to another and are essential components of cars, servers, buildings, chemical reactors, etc. A PHE uses thermally conductive surfaces between which fluids of different temperatures flow alternately, increasing the surface area available to transfer energy without mixing the fluids. The schematic of the PHE device under consideration is shown in Fig. 3. The purpose of this device is to transfer thermal energy from the heated hot water in the reservoir $d(t) = T_{ec}(t) \in \mathbb{R}$ in the primary loop (solid line) to the colder water $T_{ep}(t) \in \mathbb{R}$ in the secondary loop (dashed line), resulting in the output $y(t) = T_{sp}(t) \in \mathbb{R}$. Two motor-driven valves control the hot water flow $F_c(t) \in [0, 1]$ and the cold water flow $F_p(t) \in [0, 1]$. The first principle theoretical model of the PHE is defined as [22]

$$\gamma(t) = \frac{1 + \mathrm{k}_c \left(\frac{1}{F_c(t)}\right)^{m_c}}{1 + \mathrm{k}_c \left(\left(\frac{1}{F_c(t)}\right)^{m_c} + \left(\frac{1}{F_p(t)}\right)^{m_c}\right)}, \tag{30}$$

$$\tau(t)\frac{dT_{sp}(t)}{dt} + T_{sp}(t) = \gamma(t)\mathrm{T}_{ep} + \left(1 - \gamma(t)\right)T_{ec}(t), \tag{31}$$

where $\tau(t) = f\big(T_{sp}(t)\big) \in \mathbb{R}$ is a function that determines the dynamics of the system, $k_c \in \mathbb{R}^+$ and $m_c \in \mathbb{R}^+$ are water flow parameters. The valve flow function was determined as $F_c(t) = \frac{1}{\pi}\mathrm{atan}\big(k_v(V_c(t) - \overline{V}_c)\big) + \frac{1}{2} \in \mathbb{R}$ for the hot water and $F_p(t) = \frac{1}{\pi}\mathrm{atan}\big(k_v(V_p(t) - \overline{V}_p)\big) + \frac{1}{2} \in \mathbb{R}$ for the cold water, where $u_1(t) = V_c(t) \in \mathbb{R}$ is the input current of the motor-driven valve controlling the hot water inlet, $u_2(t) = V_p(t) \in \mathbb{R}$ is the input current of the motor-driven valve, which controls the cold water inlet, $\overline{V}_c$ and $\overline{V}_p$ are the average values of the two input currents of the valve, and $k_v \in \mathbb{R}$ is the valve parameter. The measurable state signal $d(t) = T_{ec}(t)$ is an oscillating signal generated by an on-off controller modeled with a sinusoidal function with an amplitude of $2\,^\circ\mathrm{C}$ and a period of $360\,\mathrm{s}$ added to the signal $\overline{T}_{ec}$. A white noise distribution $\mathcal{N}(0, 0.1)$ was added to the system output. The theoretical model parameters were determined as $\tau(T_{sp}(t)) = 55 - 0.5T_{sp}(t)$, $T_{ep} = 0\,^\circ\mathrm{C}$, $k_c = 0.65$, $\mathrm{m} = 1.93$, $k_v = 0.23$, $\overline{V}_c = \overline{V}_p = 10$, $\overline{T}_{ec} = 90\,^\circ\mathrm{C}$ [10]. The continuous-time theoretical PHE model (31) was simulated using Euler integration with a sampling time of $t_s = 4\,\mathrm{s}$.

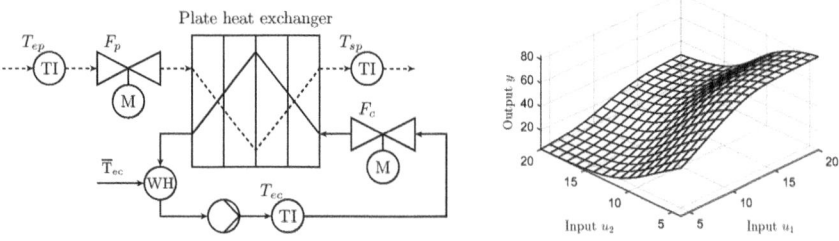

Fig. 3. Schematic diagram of the PHE pilot plant (left). The solid line shows the primary hot water loop, while the dashed line represents the secondary cold water loop. The static characteristic of the system (right).

Most of the parameters of the eDoE method were fixed in the experiment, and we investigated the effects of the design region variables $\underline{\Delta u}$ and the dc-gain merging thresholds $\underline{\kappa}_K$ on the quality and number of rules of the identified model. In this experiments $\underline{\Delta u}$ and $\underline{\kappa}_K$ are equal for both inputs. The remaining parameters were selected as follows: $u_{\min} = 4$, $u_{\max} = 20$, $\kappa_r = 3$, $\alpha_P = 10^5$, $\kappa_V = 3$, $\kappa_f = 0.05$, and $k_h = 300$. Each experiment was repeated 20 times and the mean and variance of the results were calculated. The OE-LLM structure of the evolving neuro-fuzzy model was designed as a first-order system with the filtered regressor and the clustering input as $\varphi_f^T(k) = [u_{1f}(k-1), u_{2f}(k-1),$ $-y_f(k-1), d_f(k-1), 1]$ and $\underline{z}^T(k) = [u_1(k), u_2(k), y(k)]$. The quality of the models was measured by the mean squared error (MSE) between the system output and the simulated output of the model as

$$\mathrm{MSE} = \frac{1}{N}\sum_{k=1}^{N}\big(y(k) - \hat{y}(k)\big)^2, \tag{32}$$

where N is the number of validation samples. The antecedent Gaussian clusters and the validation of the evolving model after 30 steps are shown in Fig. 4. The results of the eDoE experiment are gathered in Table 1.

6 Discussion

Initially, design parameters that produce larger clusters achieve higher simulation accuracy, but parameters with finer search regions (smaller $\underline{\Delta u}$) and merging criteria (smaller $\underline{\kappa}_K$) perform better in longer experiments. A small merging parameter leads to a model with low MSE but a comparatively larger number of clusters. The validation of the D-optimal design with parameters $\underline{\Delta u} = [1, 1]$ and $\underline{\kappa}_K = [0.75, 0.75]$ at 20 identification steps is significantly worse than that of the other models, probably due to the fact that an important part of the system has not yet been identified and some distant rule is reactivated. This area is identified later and the accuracy is high at 50 identification steps. A common problem in both DoE [14] and evolving systems [9] is the curse of dimensionality. For high-dimensional data, the number of samples for identification and the computational complexity grow exponentially with the dimensions of the problem, making it difficult to develop accurate models. The proposed sequential space-filling approach mitigates this somewhat by stimulating the regions with the least available information in the data to ensure that they are relevant and

Table 1. A comparison of the evolving neuro-fuzzy design of experiment with D-optimal and A-optimal design for different design ranges $\underline{\Delta u}$ and dc-gain merging parameters after 20, 30 and 50 space-filling steps.

D-optimal design

Num. of steps		20		30		50	
$\underline{\Delta u}$	$\underline{\kappa}_K$	MSE	c	MSE	c	MSE	c
[1, 1]	[0.5, 0.5]	5.6 ± 10	19 ± 0.91	2.7 ± 2	27 ± 3	1.6 ± 0.55	40 ± 6.9
[1, 1]	[0.75, 0.75]	$12 \pm 7.6 \cdot 10^2$	19 ± 1.3	5.1 ± 89	25 ± 1.6	2.3 ± 4.2	34 ± 7.8
[1, 1]	[1, 1]	6.4 ± 23	18 ± 2	4.4 ± 26	23 ± 8.6	3.6 ± 9.1	27 ± 14
[1.5, 1.5]	[0.5, 0.5]	4.5 ± 9	17 ± 1.3	2.7 ± 3	24 ± 3.5	$\mathbf{1.6 \pm 0.1}$	34 ± 7.4
[1.5, 1.5]	[0.75, 0.75]	4.7 ± 4.9	16 ± 3.1	2.6 ± 1.9	20 ± 3.6	2.4 ± 1.3	25 ± 11
[2, 2]	[0.5, 0.5]	3.7 ± 0.5	17 ± 2.4	$\mathbf{2.6 \pm 0.26}$	22 ± 4.5	2.5 ± 0.82	31 ± 8.9
[2, 2]	[0.75, 0.75]	$\mathbf{3.6 \pm 1.4}$	15 ± 3.5	3 ± 0.58	19 ± 5.5	4 ± 17	23 ± 8.9

A-optimal design

Num. of steps		20		30		50	
$\underline{\Delta u}$	$\underline{\kappa}_K$	MSE	c	MSE	c	MSE	c
[1, 1]	[0.5, 0.5]	2.6 ± 0.85	19 ± 0.34	1.6 ± 0.21	27 ± 2.9	1.1 ± 0.18	39 ± 17
[1, 1]	[0.75, 0.75]	4 ± 32	19 ± 1.7	3.1 ± 29	25 ± 6.2	1.4 ± 0.21	32 ± 13
[1, 1]	[1, 1]	3 ± 1.5	18 ± 1.9	1.8 ± 0.31	23 ± 5.4	4.3 ± 21	24 ± 14
[1.5, 1.5]	[0.5, 0.5]	$\mathbf{2 \pm 0.44}$	18 ± 1.1	$\mathbf{1.4 \pm 0.19}$	25 ± 4.6	$\mathbf{1.1 \pm 0.071}$	36 ± 9.8
[1.5, 1.5]	[0.75, 0.75]	2.3 ± 0.35	16 ± 1.3	1.8 ± 0.32	20 ± 3.7	2.6 ± 3.2	25 ± 5.4
[2, 2]	[0.5, 0.5]	2.6 ± 0.59	17 ± 3.1	2.2 ± 0.45	24 ± 3.7	2.2 ± 0.77	33 ± 7
[2, 2]	[0.75, 0.75]	2.8 ± 1.1	16 ± 2.9	$4.9 \pm 1.2e+02$	19 ± 6.2	6.6 ± 33	24 ± 9.8

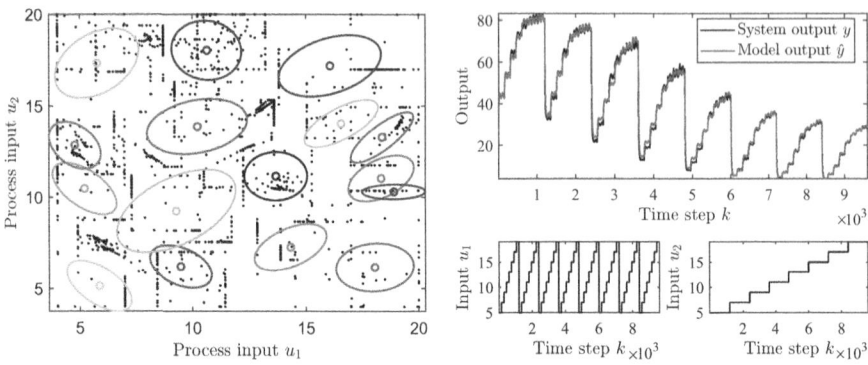

Fig. 4. Input samples (black) and clusters represented by centers and $1 - \sigma$ contours after 30 space-filling steps with A-optimal DoE, $\Delta u = 1.5$, and $\kappa_K = 1$ (left). Validation with simulation of the identified model (right). A staircase signal distributed in a grid pattern in the input space is used as input signal for validation.

useful for identification. The Bhattacharyya distance works best for clusters of similar size because it measures the area of overlap and clusters with small areas are not detected. Clusters of similar size are encouraged by constraining the design area with Δu. The most computationally intensive part of the algorithm is the space-filling maximin optimization, but it is computed only once after a rule is identified.

7   Conclusion

This paper presents an evolving design of experiments method that combines sequential space-filling design based on the Maximin criterion with optimal design based on the Fisher information matrix. The proposed hybrid design was demonstrated using a theoretical model of a plate heat exchanger. The proposed hybrid DoE results in a model that has high simulation accuracy with a small number of clusters and can be used analytically for other applications, such as soft sensor calibration, system monitoring, fault detection, predictive control, decision support system, etc. In future work, online space-filling DoE will be improved by selecting regions based on a measure of nonlinearity. The input computation will be extended to a larger time horizon and the prediction will be used to impose constraints on the output during the optimal DoE computation based on the fuzzy model structure.

References

1. Andonovski, G., Blažič, S., Angelov, P., Škrjanc, I.: Robust evolving cloud-based controller in normalized data space for heat-exchanger plant. In: IEEE International Conference on Fuzzy Systems, vol. 2015-November (2015). https://doi.org/10.1109/FUZZ-IEEE.2015.7337992

2. Bect, J., Bachoc, F., Ginsbourger, D.: A supermartingale approach to Gaussian process based sequential design of experiments. **25**, 2883–2919 (2019). https://doi.org/10.3150/18-BEJ1074

3. Fedorov, V.: Optimal experimental design. Wiley Interdisc. Rev. Comput. Stat. **2**, 581–589 (2010). https://doi.org/10.1002/WICS.100

4. Gill, P.E., Murray, W., Wright, M.H.: Numerical linear algebra and optimization. Numer. Linear Algebra Optim. (1991). https://doi.org/10.1137/1.9781611976571

5. Hametner, C., Stadlbauer, M., Deregnaucourt, M., Jakubek, S., Winsel, T.: Optimal experiment design based on local model networks and multilayer perceptron networks. Eng. Appl. Artif. Intell. **26**, 251–261 (2013). https://doi.org/10.1016/J.ENGAPPAI.2012.05.016

6. Jang, J.S.R., Sun, C.T.: Neuro-fuzzy modeling and control. Proc. IEEE **83**, 378–406 (1995). https://doi.org/10.1109/5.364486

7. Kasabov, N.: Evolving fuzzy neural networks for supervised/unsupervised online knowledge-based learning. IEEE Trans. Syst. Man Cybern. Part B: Cybern. **31**, 902–918 (2001). https://doi.org/10.1109/3477.969494

8. Kennedy, J., Eberhart, R.: Particle swarm optimization. In: Proceedings of ICNN 1995 - International Conference on Neural Networks, vol. 4, pp. 1942–1948 (1995). https://doi.org/10.1109/ICNN.1995.488968

9. Lughofer, E., Sayed-Mouchaweh, M.: Autonomous data stream clustering implementing split-and-merge concepts - towards a plug-and-play approach. Inf. Sci. **304**, 54–79 (2015). https://doi.org/10.1016/J.INS.2015.01.010

10. Ožbot, M., Lughofer, E., Škrjanc, I.: Evolving neuro-fuzzy systems based design of experiments in process identification. IEEE Trans. Fuzzy Syst. 1–11 (2022). https://doi.org/10.1109/TFUZZ.2022.3216992

11. Pronzato, L., Müller, W.G.: Design of computer experiments: space filling and beyond. Stat. Comput. **22** (2012). https://doi.org/10.1007/s11222-011-9242-3

12. Smucker, B., Krzywinski, M., Altman, N.: Optimal experimental design. Nat. Methods **15**, 559–560 (2018). https://doi.org/10.1038/S41592-018-0083-2

13. Stojanovic, V., Nedic, N., Prsic, D., Dubonjic, L.: Optimal experiment design for identification of ARX models with constrained output in non-gaussian noise. Appl. Math. Model. **40**, 6676–6689 (2016). https://doi.org/10.1016/J.APM.2016.02.014

14. Voigt, T., Kohlhase, M., Nelles, O.: Incremental doe and modeling methodology with gaussian process regression: an industrially applicable approach to incorporate expert knowledge. Mathematics **9**, 2479 (2021). https://doi.org/10.3390/MATH9192479

15. Waarde, H.J.V.: Beyond persistent excitation: online experiment design for data-driven modeling and control. IEEE Control Syst. Lett. **6**, 319–324 (2022). https://doi.org/10.1109/LCSYS.2021.3073860

16. Willems, J.C., Rapisarda, P., Markovsky, I., Moor, B.L.D.: A note on persistency of excitation. Syst. Control Lett. **54**, 325–329 (2005). https://doi.org/10.1016/J.SYSCONLE.2004.09.003

17. Zanchettin, C., Minku, L.L., Ludermir, T.B.: Design of experiments in neuro-fuzzy systems. Int. J. Comput. Intell. Appl. **9**, 137–152 (2010). https://doi.org/10.1142/S1469026810002823

18. Škrjanc, I.: Evolving fuzzy-model-based design of experiments with supervised hierarchical clustering. IEEE Trans. Fuzzy Syst. **23**, 861–871 (2015). https://doi.org/10.1109/TFUZZ.2014.2329711

19. Škrjanc, I.: Cluster-volume-based merging approach for incrementally evolving fuzzy Gaussian clustering-eGAUSS+. IEEE Trans. Fuzzy Syst. **28**, 2222–2231 (2020). https://doi.org/10.1109/TFUZZ.2019.2931874

20. Škrjanc, I.: An evolving concept in the identification of an interval fuzzy model of Wiener-Hammerstein nonlinear dynamic systems. Inf. Sci. **581**, 73–87 (2021). https://doi.org/10.1016/J.INS.2021.09.004
21. Škrjanc, I., Iglesias, J., Sanchis, A., Leite, D., Lughofer, E., Gomide, F.: Evolving fuzzy and neuro-fuzzy approaches in clustering, regression, identification, and classification: a survey. Inf. Sci. **490**, 344–368 (2019). https://doi.org/10.1016/J.INS.2019.03.060
22. Škrjanc, I., Matko, D.: Predictive functional control based on fuzzy model for heat-exchanger pilot plant. IEEE Trans. Fuzzy Syst. **8**, 705–712 (2000). https://doi.org/10.1109/91.890329

A Methodology for Limit Cycle Detection in Simulation Models

Francesco Bertolotti$^{(\boxtimes)}$ ⓘ and Luca Mari ⓘ

LIUC - Università Cattaneo, 21053 Castellanza, VA, Italy
fbertolotti@liuc.it

Abstract. The exploration of model behavior is often a necessary step to validate and generate information from them, and it permits modelers and users to identify critical parameters. This is even more crucial when simulation techniques, such as agent-based models, are employed. The paper contributes to the discipline by proposing a novel exploration methodology to detect parameter configurations that generate limit cycles. It differs from pre-existing methodologies since it automatically explores the parameter space with a strategy that, once detected, investigates all the neighboring space until the limit cycle region ends.

Keywords: Model exploration · Limit cycles · Simulation · Agent-based model

1 Introduction

The use of mathematical models to study the behavior of dynamic systems has a long history, particularly in some disciplines such as economics, biology and engineering [5]. Two notable examples are Adam Smith's model of economic growth [41], where it was posited that the sustained growth of the economy could result from the accumulation of capital and the division of labor, and Isaac Newton's model of universal gravitation [33], which described the motion of celestial bodies in terms of their masses and distances. At the beginning of the 20th century, one important development was the generalization of the mathematical meta-structure of models, particularly those based on the physical theories related to the tradition of Newtonian mechanics. This meta-structure typically involved systems of differential equations that described the evolution over time of state variables such as position, velocity, and acceleration. By developing more general mathematical frameworks for these models, researchers were able to apply them to a wider range of contexts.

The practice of mathematical modeling involves the creation of representations of real-world systems, that can be employed to understand and predict their behavior [16]. However, no model can perfectly capture the complexity and variability of a real-world system and no perfect knowledge can be grasp in advance regarding how the model works [10,46]. It is then crucial to study the behavior of models as the parameters that govern their behavior are varied. It is

ⓒ The Author(s), under exclusive license to Springer Nature Switzerland AG 2024
M. Mujica Mota and P. Scala (Eds.): EUROSIM 2023, CCIS 2033, pp. 317–331, 2024.
https://doi.org/10.1007/978-3-031-68438-8_23

then a way of assessing the validity of the model, reliability of the results, and to generates insight regarding the real-world model, which is the main goal even for abstract models [28,32]. Moreover, this parameter variation could be used to detect emergent behaviours, which are typically hard to find, as well as explain how they happen [9].

In recent years, the development of large-scale models and computer-implemented simulation models has made the exploration of model behavior as parameters vary even more important [36]. Since these models can involve thousands or even millions of parameters, exploring the behavior of the model across this high-dimensional parameter space is a daunting task [22]. Fortunately, many techniques have been developed to do this task, including sensitivity analysis, optimization, and uncertainty quantification [11,19,40]. These techniques can help researchers and practitioners to identify the most important parameters, estimate their values more accurately, and assess the reliability of the model's predictions [45,49].

When two time series results of represent on a 2-dimensional state space, one of the emerging features that can appear during a model exploration are limit cycles a feature of dynamical systems that have long captivated the attention of scientists and mathematicians alike [26,35,47]. These behavioral patterns arise when a system is driven to oscillate between two or more states over time, without settling into a stable equilibrium. They are an indicator of qualitative changes in the periodic behavior of dynamical systems when a distinguished parameter is varied. In some cases, the presence of these bifurcations may signal the emergence of chaos, leading to the sudden disappearance of periodic behavior [29,37]. So, the emergence of oscillatory behavior provides essential information about the interactions within the underlying system. Despite their apparent simplicity, limit cycles can be hard to predict or understand. As a result, the study of limit cycles remains an active area of research across many different fields, from physics and chemistry to biology and economics [13,30,38]. Extensive research has been conducted in the field of bifurcations, examining a diverse range of dynamical systems that encompass the Hopf bifurcation. Typically, investigations of these phenomena have employed either maps or invariant measures of mathematical variables as the analytical tool.

In this paper, we propose a novel exploration methodology that permits the detection of parameter configurations that generate limit cycles in simulation models. This follows an existing need for new analytical tools for exploring the dynamic behavior of agent-based models (ABMs) [15]. While this methodology can potentially be employed for different simulation models that require the investigation of a parameter space to find non-punctual equilibria and are analytically intractable [14], traditional techniques already exist for equation-based models to investigate the presence of limit cycles. To the best of our knowledge, nothing similar exists for agent-based models and other non-formalized simulation models. In many cases, it could be relevant to discover more about the dynamics of an agent-based system, even when, as is often the case, the global behavior is too complex to be formally depicted by a set of differential equations.

Moreover, in comparison to pre-existing techniques such as Lyapunov analysis [27], Poincaré maps [35], or the Bendixson's Criterion [3], the main advantage of this methodology is its ability to enable multi-dimensional parameter space exploration. This eliminates the need for a person to run the model, evaluate the simulation output, and adjust the parameters to get a better result. This is especially relevant for complex non-equation-based models with unknown behavior.

The paper is divided as follows. In the next section, we review and discuss the main research regarding simulation model exploration, providing a background for this work. In the method section, we present both the detection algorithm (which is the main focus of this research) and the case study models on which we tested it. Then, we present the results of applying the algorithm to the case study models, and draw some conclusions.

2   Background

The idea of limit cycles originated from the well-known works of Poincaré [35] and was subsequently popularized by the 23 mathematical problems presented by David Hilbert at the Second International Congress of Mathematicians [23]. System Theory has deeply investigated the topic, especially finding the set of differential equations and the parameter settings that could make them happen [29]. Later, the need for detecting limit cycle-like behavior emerged in various application fields, even when the unit of analysis was not aggregate (and then represented with an equation-based model) but individual (so developed with an agent-based modeling), from the detection of fashion cycles [1] to understanding how environmental parameters affect prey-predator systems [13]. Moreover, evolutionary game theory (which can be considered, from a structural perspective, a specific configuration of agent-based modeling) has a long tradition of detecting limit cycles [17,21].

While the methodology proposed in this work is original, it is far from new to study agent-based models by means of parameter space explorations [44], as an alternative to manually altering the parameters during long simulations to detect dynamical shifts [50]. Typically, these techniques consider the model from a black-box perspective, so what is "inside" the box (i.e., the model) is mostly ignored, and only the inputs U and the outputs Y are observed. Using this simplification, it is possible to observe two classes of simulation model exploration techniques. The first class is composed of methodologies that adjust the input and observe the subsequent effect on the output. The most famous of these models is called sensitivity analysis [39], which consists of evaluating the dependency of the model output on the model input (usually defining sensitivity as $\frac{\Delta y}{\Delta u}$) and investigating the role of each model input in the determination of the output to understand the principal factors [24,31,39]. Sensitivity analysis implicitly implies computing a measure (the sensitivity), but one could also be interested in exploring the effect on the output of specific combinations or studying the conditional variations between variables. In this case, the model can be simulated by sampling the target parameters from defined ranges of values and observing the outputs, and the goal would be to find a function f that connects U to Y (or two of their subsets). The simplest strategy for exploring a parameter space is called "sample grid sampling", which consists of exploring every combination of points of the parameter space [8, 25]. While this methodology permits not taking any preliminary stance regarding the potential outcome, and so it is especially suitable when there is no knowledge nor expectation towards the result of the simulation model, the computational time needed to complete it grows with the number of possible combinations, making it often unfeasible [15]. So, several techniques for enhancing sampling in numerical experiments under limited computational resources have been proposed, such as Sobol sequences or Latin hypercube sampling. The Sobol sequence is the sequence that minimizes the discrepancy for a set of parameters, where no fitting functions are used to decompose the output variance [42]. It only takes into account the average impact of parameters across the entire parameter space and does not examine various patterns within this space [11]. Sensitivity analysis can be used to address uncertainty regarding a single parameter, but it is not adequate to address uncertainty related to multiple parameters. To sample variations in multiple parameters without considering every possible permutation, one can use Latin hypercube sampling, which consists of taking one point in each dimension of a hypercube whose number of dimensions is equal to the number of parameters involved in the sampling [12, 20]. Other exploration techniques of this type include the quantification of tipping points where transitions between behavioral regimes occur due to external forcing, typical of closed biological systems [2].

The second class of exploration techniques regards the ones that fix a desired output and adjust the input to find the right combination to get the target behavior, calibrating the behavior of the model on a target. This is especially relevant for models of real-world systems, whose target is to be used for understanding or predicting their behavior [48].

For agent-based models of real-world systems with available data, the calibration of the behavior could be made by means of dynamic data assimilation, for example, using the ensemble Kalman filter (EnKF) to address the typical nonlinearity of ABM and their computational elevated cost [48]. Moreover, different studies highlight how to use genetic algorithms in agent-based modeling, either evaluating the possible methodologies [43] or applying to fit real-world behavior [7]. A comparison of economic agent-based model calibration methods also finds that Bayesian estimation consistently outperforms different other calibration methods in many contexts [34]. Moreover, there is a way to reduce the computational costs of both behavior and parameter explorations by learning a surrogate and approximate sub-model of the original model with a lower number of training points [51].

Finally, a methodology with an analog finality to the one presented in this paper was already tested on a prey-predator model [13], using Monte Carlo singular spectrum analysis to find statistically significant oscillatory patterns in the behavior of a prey-predator agent-based model. Nevertheless, the methodology is very computationally expensive and tailor-made for a specific setting whose analytical equations are known. In this paper, we are proposing a faster technique that can be generically applied to each analytical intractable simulation model with two outputs.

3   Methods

The code used to implement and test the methodology depicted in this section is written in Python 3.9. For the sake of transparency and replicability, all the code used and the experiments employed can be found at the following link: https://github.com/francescobertolotti/limitcycledetection/.

3.1   Detection Algorithm

The exploration methodology proposed in this paper presents many steps, depicted in Fig. 1.

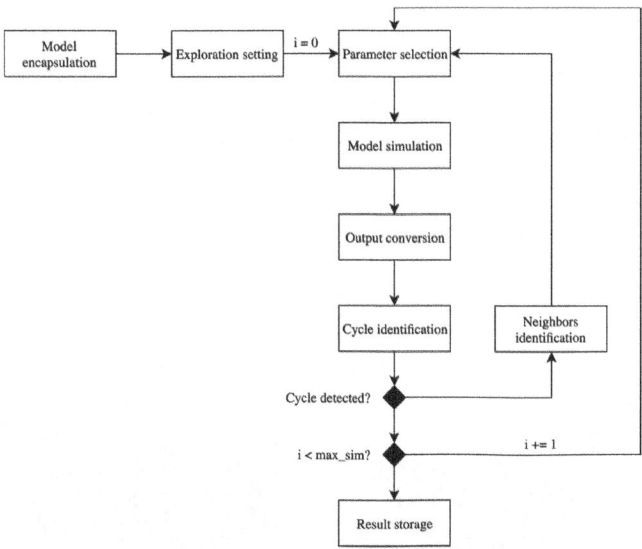

Fig. 1. The process of model exploration for limit cycles

Description. Firstly, the target simulation model is encapsulated into a function, so that the underlying structure is ignored and it can be treated as a black box. This encapsulation permit to get a given output from a specific combination of parameters.

Next, the encapsulated model reads a combination of parameters as input to generate an output. These parameters can be generated in any fashion. In the case of numeric parameters distributed on a single order of magnitude, sampling from a uniform distribution is sufficient. However, if the range of available values of a parameter exceeds a single order of magnitude, the values of the parameter could be alternatively sampled from a loguniform distribution (the discussion about which sampling method is better exceeds the scope of this work). In this phase, it is crucial to clearly pinpoint a division of the parameter space so that the overall number of cases can be computed, and it is even possible to identify the size of the parameter space addressed at each exploration. Additionally, this specification later permits the recognition of the neighbors of a point in the parameter space that generates a limit cycle once the model is run. At the end of this step, a sampling strategy is defined, from which a set of parameters can be randomly generated at each time and fed to the model.

Subsequently, a loop starts where the number of iterations is a parameter of the exploration meta-model. In the loop, a set of parameters is randomly generated (using the previously defined strategy) and provided as input to the simulation model. From this simulation, a two-dimensional time-series output is generated. The output is then converted into a boolean matrix of size m (Fig. 2 presents an example of the transformation process), where m is a parameter of the exploration algorithm discussed at the end of this section. The compression algorithm consists of two phases. First, it takes the original two-dimensional trajectory and makes it denser so that for each consecutive point, a fixed number of intermediate points is created. This operation is required so that even if the trajectory makes a long jump between two consecutive points, the path is completely fulfilled. Secondly, the graphical representation of the line in two dimensions and the matrix are superimposed in such a manner that the upper and lower boundaries of the line coincide with the extremes of the matrix. At this point, each point of the line is reported inside the matrix, marking each cell of the matrix where the line is passing with a 1, and a 0 otherwise. Figure 2 depicts this output conversion process.

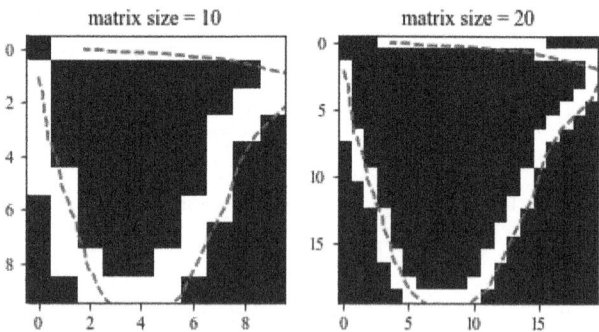

Fig. 2. The Figure presents the comparison between the two-dimensional behaviors of a single run of the Lotka-Volterra case study model (depicted by the red dotted line) and its conversion two matrices with different sizes. When the size of the matrix is equal to 10, it is possible to observe how a long trajectory is actually confused with a limit cycle. However, this does not happen in the case of a higher definition. (Color figure online)

After transforming the two-dimensional trajectory of the model's behavior into a matrix, an algorithm is used to evaluate the size of the cycle. The cycle dimension, denoted as C, can be defined as the number of cells in the largest area of cells with a value of 0 around which there is a continuous path of cells with a value of 1. The size of the cycle is then compared to the overall area of the matrix, obtaining a value c between 0 and 1.

At this point, whether or not the cycle is detected depends on two elements. The first element is the sensitivity of the detection, denoted by s, which is the threshold above which c is considered sufficiently high to indicate the presence of a limit cycle. In the case studies presented later in this paper, two different values of s are proposed. Additionally, if the model is stochastic, more than one simulation may be needed to accurately assess the value of c, in order to avoid missing configurations that return limit cycles only a fraction of the time. In this case, the trigger for further exploration of the surroundings of the detected point is not $c > s$, but rather $E[c] > s$, where $E[c]$ is the expected value of c. To avoid being misled by noise resulting from stochasticity and equifinality, which is the case where different rules and parameter settings lead to the same model output [18], it is necessary to repeat the experiment a statistically valid number of times for each parameter combination to ensure the robustness of the results.

So, if the condition is not satisfied, the cycle loop is exited and another parameter is randomly selected. Otherwise, the neighbors of the current point in the parameter space are identified based on the division of the parameter space defined in the second step of the process. Once the neighbors are identified, they are stored in a list and the exploration process is repeated for each neighbor, following the same cycle as before. This process continues as long as there are points in the list of neighbors that have not been explored yet. If a limit cycle is detected in any of the neighbors of the previously explored point, new points are added to the exploration list. This ensures that once a single point in the parameter space that generates a limit cycle in the encapsulated model is detected, the entire n-dimensional area around it is explored, and this exploration continues until no new points with limit cycles are found.

Finally, whenever the number of explorations i gets equal to the maximum number of simulations $max_s im$, the exploration stops and the results are stored in a dataframe. In this sense, this exploration meta-model can be seen as a black-box, which read in input an encapsulated model and ranges of a set of allowable values for each parameter, along with the distribution from which to sample them, and returns a table which columns are the parameters and the correspondent c (or $E[c]$, if the model is stochastic). Figure 3 presents this interpretation.

Limitations. The proposed methodology has some limitations and requirements for its application. Firstly, it can address both deterministic and stochastic models, since most simulation models have some degree of stochasticity, especially those that are adherent to reality. Nevertheless, the whole process is considerably less expensive in terms of computation required to explore the space when used with deterministic models. Whenever the model studied is deterministic or

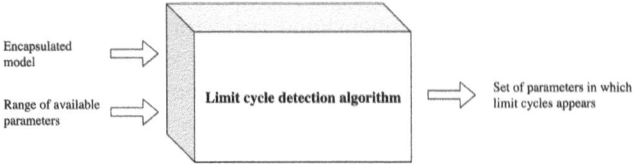

Fig. 3. A representation of the black box interpretation of the limit cycle exploration methodology

stochastic, the high sensitivity to initial conditions typically affects the result of a single run [4,6]. Even more problematic in terms of computational cost is addressing models with high dimensionality, due to the so-called "curse of dimensionality." This can be addressed by using a coarser subdivision of the parameter space.

Secondly, the methodology only considers two numeric outputs, allowing them to be represented in a two-dimensional state diagram. This limitation permits the graphical depiction of the behavior and identification of the hole.

Moreover, appropriate setting of the cycle detection threshold and the matrix granularity is required. A value that is too large or too small could compromise the achievement of the final result. Regarding granularity, a value that is too large could make trajectories resemble cycles even when they have not closed, while a value that is too small could create gaps between points, breaking the cycle and preventing the algorithm from identifying it later. On the other hand, a threshold for identification that is too large does not allow any limit cycle to be identified (trivially, $c < 1$ in any case since the area of the circle bounded by the curve is necessarily lower than that of the square because there will be matrix cells that delimit the perimeter of the figure), while a threshold that is too low captures configurations of parameters that create gaps in the state space between various lines (typically, at least pseudo-chaotic behavior), rendering the automatic recognition procedure unusable (which is, in fact, the true strength of this methodology).

3.2   Case Study Models

These are two models whose characteristics make them good candidates for applying the new methodology proposed in this paper. First, their behavior could not be derived analytically. Second, they are well-known in the scientific literature, so there is no need to discuss the results of the exploration, as the purpose of the case study exploration is solely to show that the procedure presented in the previous section works. Third, they are structurally different, as the scheduling strategy and the manner in which agents interact with each other vary. Fourth, they can be deeply simplified, so that a primitive version could be implemented for the mere scope to test the algorithm on it.

Prey-Predator Agent-Based Model. In the model, there are two types of entities positioned on a toroidal surface: preys and predators. Each agent is defined by a two-dimensional spatial location. Both agents move at every time step, and the size and direction of the movement are generated from a random uniform distribution $U \sim \text{Unif}(-1, 1)$. Finally, both agents can asexually reproduce, so that each prey and predator have a given probability, at every time step, to generate another prey or another predator agent. There are two differences between predator and prey agents. Firstly, predator agents can prey on prey agents. More specifically, whenever a predator has preys at a distance lower than a given threshold, it chooses one of them and eats it. While doing so, it acquires a fixed amount of energy. Furthermore, predators have energy levels, which decrease at a constant rate at each time step. When the level gets to 0, predators die. Whenever a predator breeds, half of its energy is given to the bred agent, so that the overall amount of energy in the system does not increase with a reproduction event. The pseudo-code in Algorithm 1 describes the scheduling of the model (Table 1).

Algorithm 1. Main Loop of the Simulation

```
 1: for t in [1...t_max] do
 2:     shuffle PREDATORSLIST
 3:     shuffle PREYSLIST
 4:     for PREY in PREYSLIST do
 5:         PREY move randomly
 6:     for PREDATOR in PREDATORSLIST do
 7:         PREDATOR move randomly
 8:         if any? PREY in PREYSLIST at distance < 2 then
 9:             energy of PREDATOR += energy per hunt
10:             set PREYTARGET one of preys in PREYSLIST at distance < 2
11:             die PREYTARGET
12:         energy of PREDATOR -= energy consumption
13:         if energy of PREDATOR <= 0 then
14:             die PREDATOR
    new preys = int(length(PREYSLIST) × growth rate prey)
15:     for i in [i...newpreys] do
16:         create NEWPREY
17:         append NEWPREY to PREYSLIST
    new predators = int(length(PREDATORSLIST) × growth rate predators)
18:     for i in [i...newpreys] do
19:         set PREDATORFATHER one of predators in PREDATORSLIST
20:         create NEWPREDATORS
21:         energy of NEWPREDATORS = energy of PREDATORFATHER × 0.5
22:         energy of PREDATORFATHER = energy of PREDATORFATHER × 0.5
23:         append NEWPREDATORS to PREDATORSLIST
```

Table 1. Parameters used in the exploration of the prey-predator model

Parameter	Min value	Max Value	Interval between values
$e_p y$	0	10	1
pd_g	0	0.1	0.01
py_c	0.2	0.8	0.1
py_g	0	0.2	0.01

Prisoner's Dilemma on a Network. The code defines an Agent-Based Model where agents play the Prisoner's Dilemma, a classic game theory model of strategic interaction. In the Prisoner's Dilemma, two individuals are arrested for a crime, and the police offer each of them a plea deal to testify against the other. If both individuals stay silent, they will be convicted for a lesser crime, but if one confesses and the other stays silent, the confessor will go free, and the silent one will be punished harshly. If both confess, they will receive a moderate sentence. The prisoner's dilemma is valid with different settings (as long as $T > R > P > S$). Table 2 depicts the payoff configuration employed in this implementation, where the specific values are input parameters of the model.

Table 2. Payoff Matrix for the Prisoner's Dilemma

	Cooperate	Defect
Cooperate	(R, R)	(S, T)
Defect	(T, S)	(P, P)

The model consists of n agents connected to each other by means of a complete network, which is a network in which every node is connected to every other node. At every round, two random agents are picked to play the prisoner's dilemma against each other. As in the classic prisoner's dilemma configuration, agents are endowed with a strategy that can be cooperative or defective. An agent can start the simulation being cooperative or defective with a probability defined at the beginning of the game, so that the expected share of agents starting with a cooperative strategy is given by the input parameter c. The strategy of any agent can change over time, since each agent has a memory of previous rounds. During each round, two agents are randomly selected to play the game, and their payoffs are updated based on their chosen strategy (as in Table 2). If both agents cooperate, they receive a reward (R); if one cooperates and the other defects, the defector receives the highest reward (T), and the cooperator receives the lowest reward (S); if both defect, they receive a punishment (P). After each round, the agents adjust their strategies based on their average payoffs. Specifically, if an agent's payoff is below the average payoff of all agents, the agent

Table 3. Parameters used in the exploration of the Prisoner Dilemma model

Parameter	Min value	Max Value	Interval between values
m	1	10	1
R	0	3	1
S	0	3	1
T	-3	0	1
P	3	5	1

has a probability of switching to the other strategy. The model continues for a given number of rounds T (a parameter of the model). Algorithm 2 presents the pseudo-code of this implementation of the prisoner's dilemma (Table 3).

Algorithm 2. Main Loop of the Simulation

1: **for** t in $[1...t_{max}]$ **do**
2:     set PLAYER1 one of agents in AGENTSLIST
3:     set PLAYER2 one of agents in AGENTSLIST
4:     payoff of PLAYER1 = f(strategy of PLAYER1, strategy of PLAYER2, payoff matrix)
5:     payoff of PLAYER2 = f(strategy of PLAYER1, strategy of PLAYER2, payoff matrix)
6:     append payoff of PLAYER1 to payofflist of PLAYER1
7:     append payoff of PLAYER2 to payofflist of PLAYER2
8:     **for** AGENT in AGENTSLIST **do**
9:         meanpayoff of AGENT = mean(payofflist of AGENT)
      mean payoffs = mean(meanpayoff of AGENTS in AGENTSLIST)
10:     **if** meanpayoff of PLAYER1 < mean payoffs **then**
11:         set strategy of PLAYER1 one of possible strategies chosen randomly
12:     **if** meanpayoff of PLAYER2 < mean payoffs **then**
13:         set strategy of PLAYER2 one of possible strategies chosen randomly
14:     remove last element to payofflist of PLAYER1
15:     remove last element to payofflist of PLAYER2

---

4   Case Studies Results

This section presents the results from the application of the proposed methodology to the case study simulation models. The results in this section are not discussed, as it is not within the scope of the paper to study the behavior of the models or derive insights from them. Instead, we aim to demonstrate the applicability of the methodology and highlight any differences encountered during the exploration.

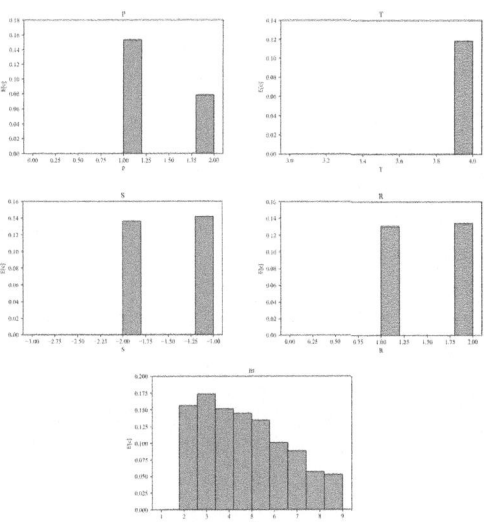

Fig. 4. Results of the limit cycles exploration for the prey predator model

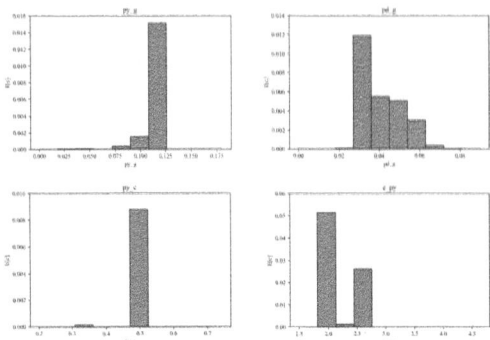

Fig. 5. Results of the limit cycles exploration for the prisoner dilemma model

The results generated from the proposed methodology are presented in tabular format, collecting the results from a 2000 simulation limit cycle exploration. Figure 4 and Fig. 5 illustrate the aggregation of the results, presenting the parameter space points where limit cycles appear. The methodology was applied to two different models to demonstrate its applicability, and the results are structurally similar. Specifically, both models exhibit the emergence of limit cycles, and parameters differently affect this phenomenon. The distribution of the results, such as for the variable m in the Prisoner's Dilemma model, indicates that the appearance of limit cycles is less sensitive to the presence of a given parameter. However, for some parameters, such as P in the Prisoner's Dilemma model or py_c in the Prey-Predator model, limit cycles are only detectable when the parameter has a specific value. This insight sheds light on the elements that generate an orbit in the state space, and is affected by the granularity of the parameter space.

5   Conclusions

After briefly reviewing the existing literature on simulation model exploration, this paper presented a method to detect limit cycles in non-formalized simulation models. The proposed methodology was described in detail, and its application to two case study models was presented. This methodology has great potential for enhancing the exploration of simulation models, particularly agent-based or complex equation-based models without formalization.

However, the proposed methodology has three main limitations. Firstly, it can only consider two output variables per experiment. Secondly, it can be computationally expensive, especially for high-dimensional models that require exploration with high granularity. Thirdly, it does not guarantee finding a result unless the entire parameter space is explored. Nonetheless, it represents an advance in the state-of-the-art, particularly for low-dimensional models.

Further developments could involve testing the methodology on simulation models with a higher number of parameters, to assess its feasibility and performance. Additionally, a new version of the algorithm could be implemented to compute the exploration direction following a gradient descent.

References

1. Apriasz, R., Krueger, T., Marcjasz, G., Sznajd-Weron, K.: The hunt opinion model-an agent based approach to recurring fashion cycles. PLoS ONE (2016). https://doi.org/10.1371/journal.pone.0166323
2. Ashwin, P., Wieczorek, S., Vitolo, R., Cox, P.: Tipping points in open systems: bifurcation, noise-induced and rate-dependent examples in the climate system. Philos. Trans. Roy. Soc.: Math. Phys. Eng. Sci. (2012). https://doi.org/10.1098/rsta.2011.0306
3. Bendixson, I.: Sur les courbes définies par des équations différentielles. Acta Mathematica **24**(1), 1 (1901). https://doi.org/10.1007/BF02403068
4. Berceanu, C., Patrascu, M.: Initial conditions sensitivity analysis of a two-species butterfly-effect agent-based model. In: Baumeister, D., Rothe, J. (eds.) EUMAS 2022. LNCS, pp. 60–78. Springer, Cham (2022). https://doi.org/10.1007/978-3-031-20614-6_4
5. von Bertalanffy, L.: General system theory: Foundations, development, applications. G. Braziller (1968)
6. Bertolotti, F., Locoro, A., Mari, L.: Sensitivity to initial conditions in agent-based models. In: Bassiliades, N., Chalkiadakis, G., de Jonge, D. (eds.) EUMAS/AT - 2020. LNCS (LNAI), vol. 12520, pp. 501–508. Springer, Cham (2020). https://doi.org/10.1007/978-3-030-66412-1_32
7. Bertolotti, F., Roman, S.: Risk sensitive scheduling strategies of production studios on the US movie market: an agent-based simulation. Intelligenza Artificiale **16**, 81–92 (2022). https://doi.org/10.3233/IA-210123
8. Bertolotti, F., Roman, S.: The evolution of risk sensitivity in a sustainability game: an agent-based model (2022)
9. Bodine, E.N., Panoff, R.M., Voit, E.O., Weisstein, A.E.: Agent-based modeling and simulation in mathematics and biology education. Bull. Math. Biol. (2020). https://doi.org/10.1007/s11538-020-00778-z
10. Box, G.E.P.: Science and statistics. J. Am. Stat. Assoc. **71**(356), 791–799 (1976)
11. ten Broeke, G., van Voorn, G., Ligtenberg, A.: Which sensitivity analysis method should i use for my agent-based model? JASSS (2016). https://doi.org/10.18564/jasss.2857
12. Collins, A.J., Seiler, M.J., Gangel, M., Croll, M.: Applying Latin hypercube sampling to agent-based models: understanding foreclosure contagion effects. Int. J. Hous. Markets. Anal. **6**(4), 422–437 (2013). https://doi.org/10.1108/IJHMA-Jul-2012-0027
13. Colon, C., Claessen, D., Ghil, M.: Bifurcation analysis of an agent-based model for predator-prey interactions. Ecol. Model. (2015). https://doi.org/10.1016/j.ecolmodel.2015.09.004
14. Cranmer, K., Brehmer, J., Louppe, G.: The frontier of simulation-based inference. Proc. Natl. Acad. Sci. U.S.A. (2020). https://doi.org/10.1073/pnas.1912789117
15. Daly, A.J., De Visscher, L., Baetens, J.M., De Baets, B.: Quo vadis, agent-based modelling tools? Environmental Modelling & Software p. 105514 (2022)
16. Epstein, J.M.: Why model? J. Artif. Soc. Soc. Simul. **11**(4), 12 (2008)

17. Ficici, S.G., Melnik, O., Pollack, J.B.: A game-theoretic and dynamical-systems analysis of selection methods in coevolution. IEEE Trans. Evol. Comput. **9**(6), 580–602 (2005). https://doi.org/10.1109/TEVC.2005.856203

18. Gräbner, C.: How to relate models to reality? An epistemological framework for the validation and verification of computational models. JASSS (2018). https://doi.org/10.18564/jasss.3772

19. Hales, D., Rouchier, J., Edmonds, B.: Model-to-model analysis. JASSS (2003)

20. Hamis, S., Stratiev, S., Powathil, G.G.: Uncertainty and sensitivity analyses methods for agent-based mathematical models: an introductory review, chap. Chapter 1, pp. 1–37 (2020). https://doi.org/10.1142/9789811223495_0001, https://www.worldscientific.com/doi/abs/10.1142/9789811223495_0001

21. Hauert, C., Wakano, J.Y., Doebeli, M.: Ecological public goods games: cooperation and bifurcation. Theor. Popul. Biol. **73**(2), 257–263 (2008). https://doi.org/10.1016/j.tpb.2007.11.007

22. Heppenstall, A.J., Evans, A.J., Birkin, M.H.: Genetic algorithm optimisation of an agent-based model for simulating a retail market. Environ. Plann. B. Plann. Des. (2007). https://doi.org/10.1068/b32068

23. Hilbert, D.: Sur les problèmes futurs des Mathématiques. In: 1900 International Congress of Mathematicians. Paris (1900)

24. Iooss, B., Saltelli, A.: Introduction to sensitivity analysis. In: Ghanem, R., Higdon, D., Owhadi, H. (eds.) Handbook of Uncertainty Quantification, pp. 1103–1122. Springer, Cham (2017). https://doi.org/10.1007/978-3-319-12385-1_31

25. Klabunde, A.: Computational economic modeling of migration. In: The Oxford Handbook of Computational Economics and Finance (2018). https://doi.org/10.1093/oxfordhb/9780199844371.013.41

26. Lotka, A.J., et al.: Elements of physical biology (1925)

27. Lyapunov, A.M.: The general problem of the stability of motion. Int. J. Control **55**(3), 531–534 (1992). https://doi.org/10.1080/00207179208934253

28. Matsiuk, V., Galan, O., Prokhorchenko, A., Tverdomed, V.: An agent-based simulation for optimizing the parameters of a railway transport system. In: ICTERI, pp. 121–128 (2021)

29. May, R.M.: Simple mathematical models with very complicated dynamics. Nature (1976). https://doi.org/10.1038/261459a0

30. Mittal, S., Mukhopadhyay, A., Chakraborty, S.: Evolutionary dynamics of the delayed replicator-mutator equation: Limit cycle and cooperation. Phys. Rev. E **101**(4), 42410 (2020)

31. Morris, M.D.: Factorial sampling plans for preliminary computational experiments. Technometrics (1991). https://doi.org/10.1080/00401706.1991.10484804

32. Murase, Y., Jo, H.H., Török, J., Kertész, J., Kaski, K.: Deep learning exploration of agent-based social network model parameters. Front. Big Data **4**, 739081 (2021)

33. Newton, I.: Philosophiae naturalis principia mathematica, vol. 1. G. Brookman (1833)

34. Platt, D.: A comparison of economic agent-based model calibration methods. J. Econ. Dyn. Control (2020). https://doi.org/10.1016/j.jedc.2020.103859

35. Poincaré, H.: Mémoire sur les courbes définies par une équation différentielle (I). Journal de Mathématiques Pures et Appliquées **7**, 375–422 (1881), http://eudml.org/doc/235914

36. Raimbault, J., Cottineau, C., Le Texier, M., Le Néchet, F., Reuillon, R.: Space matters: extending sensitivity analysis to initial spatial conditions in geosimulation models. JASSS (2019). https://doi.org/10.18564/jasss.4136

37. Robbio, F.I., Alonso, D.M., Moiola, J.L.: Detection of limit cycle bifurcations using harmonic balance methods. Int. J. Bifurcat. Chaos Appl. Sci. Eng. (2004). https://doi.org/10.1142/S0218127404011491

38. Roman, S., Bullock, S., Brede, M.: Coupled societies are more robust against collapse: a hypothetical look at Easter island. Ecol. Econ. (2017). https://doi.org/10.1016/j.ecolecon.2016.11.003

39. Saltelli, A., et al.: Global sensitivity analysis. The Primer (2008). https://doi.org/10.1002/9780470725184

40. Schouten, M., Verwaart, T., Heijman, W.: Comparing two sensitivity analysis approaches for two scenarios with a spatially explicit rural agent-based model. Environ. Model. Softw. (2014). https://doi.org/10.1016/j.envsoft.2014.01.003

41. Smith, A.: The wealth of nations. na (1776)

42. Sobol, I.M.: Global sensitivity indices for nonlinear mathematical models and their Monte Carlo estimates. Math. Comput. Simul. (2001). https://doi.org/10.1016/S0378-4754(00)00270-6

43. Stonedahl, F.J.: Proposal - Genetic Algorithms for the Exploration of Parameter Spaces in Agent-Based Models. ProQuest Dissertations and Theses (2011)

44. Terano, T.: Exploring the vast parameter space of multi-agent based simulation. In: Antunes, L., Takadama, K. (eds.) MABS 2006. LNCS (LNAI), vol. 4442, pp. 1–14. Springer, Heidelberg (2007). https://doi.org/10.1007/978-3-540-76539-4_1

45. Troost, C., et al.: How to keep it adequate: a protocol for ensuring validity in agent-based simulation. Environ. Model. Softw. **159**, 105559 (2023)

46. Vermeer, W.H., Smith, J.D., Wilensky, U., Brown, C.H.: High-fidelity agent-based modeling to support prevention decision-making: an open science approach. Prev. Sci. 1–12 (2021)

47. Volterra, V.: Variazioni e fluttuazioni del numero d'individui in specie animali conviventi, vol. 2. Societá anonima tipografica" Leonardo da Vinci" (1926)

48. Ward, J.A., Evans, A.J., Malleson, N.S.: Dynamic calibration of agent-based models using data assimilation. Roy. Soc. Open Sci. (2016). https://doi.org/10.1098/rsos.150703

49. Wikstrom, K., Nelson, H.T.: Spatial validation of agent-based models. Sustainability **14**(24), 16623 (2022)

50. Woods, J., Perilli, A., Barkmann, W.: Stability and predictability of a virtual plankton ecosystem created with an individual-based model. Prog. Oceanogr. (2005). https://doi.org/10.1016/j.pocean.2005.04.004

51. Zhang, Y., Li, Z., Zhang, Y.: Validation and calibration of an agent-based model: a surrogate approach. Discret. Dyn. Nat. Soc. (2020). https://doi.org/10.1155/2020/6946370

Learning Explanatory Coherence Models from Agent-Based Simulation Experiments

Levent Yilmaz[(✉)] [iD]

Auburn University, Auburn, AL 36849, USA
yilmale@auburn.edu

Abstract. In Agent-based models of complex adaptive systems, macro-level behavior is not engineered but results from local interactions among agents. Due to the consequences of complex, distributed interactions among decentralized agents, the causal chain of cross-cutting processes that give rise to behavioral regularities is difficult to explain. To provide a context for the explainability of agent-based models, a systematic review of philosophical and cognitive models of causal explanation is provided. For illustration purposes, the theory of explanatory coherence is used as a computational framework for learning explanatory cognitive maps of increasingly refined and broadened model features. The framework offers a perspective that signifies strategies for learning coherence-driven explanatory models with implications for simulation model development environments.

Keywords: Modeling methodology · Explainable model · Agent-based simulation · Cognitive coherence · Causal explanation

1 Introduction

Computational models are often complex monolithic entities that are difficult to comprehend and explain [6]. Agent-Based Models (ABMs) that rely on autonomous agency, adaptation, and context-sensitive interactions are particularly challenging to understand [7,25]. This challenge is akin to the explainability crisis in Artificial Intelligence that calls for interpretable systems that can communicate their reasoning [10]. Similar concerns exist in the broader modeling and simulation domain, partly due to the significance of reproducibility and transparency in the use of models [17,23], as well as the need for establishing trust in simulations [13,24].

Because cause-effect reasoning is a critical objective in science, the concept of explanation has long been a central focus of model-based science [1,12]. The scientific inquiry aims to discover an explanation Y for a phenomenon in terms of the underlying generative mechanisms. Similarly, engineers search for a design mechanism, X, which can produce a desirable property, Y, for an envisioned system. In both cases, modelers are driven by an intrinsic motivation to understand or generate targeted aspects of systems. To achieve this objective, models have

© The Author(s), under exclusive license to Springer Nature Switzerland AG 2024
M. Mujica Mota and P. Scala (Eds.): EUROSIM 2023, CCIS 2033, pp. 332–346, 2024.
https://doi.org/10.1007/978-3-031-68438-8_24

become instrumental in scientific thinking and communication [9,12]. The theory and methodology of modeling can benefit from insights offered by philosophy and cognitive science of science to advance its role further. Such insights can help discern abstractions conducive to effective explanations for (1) conveying causal relations to learners and (2) facilitating abductive model building [6] to generate plausible alternative explanations.

Besides, explanations can be used to study models independent of a target system. One can explore a model's behavior across its entire range to explain when and which tentative features cohere to generate behavioral regularities. Explanations are central to sensemaking by facilitating understanding of events and forming causal mental models. They help model users determine what would happen in counterfactual reasoning or perform prospective anticipatory thinking by projecting into future states. In practice, providing explanatory support via causal models improves understanding and helps formulate better questions and adapt a model to better align with the objectives of a simulation study.

This paper reviews and analyzes the extant literature on the philosophy and psychology of explanation. The overview lends itself to a methodical strategy grounded on the theory of explanatory coherence and reflective equilibrium for generating explanations of ABMs and outlining plausible strategies for learning explanatory models through simulation experiments. The requirements and strategies for explanatory modeling are delineated to advance simulation infrastructures toward supporting the development of self-aware and explanatory models.

2   Background

Both science and engineering involve model-based discovery and explanation activities. Therefore, issues concerning explanation have long been central to the philosophy of science to elucidate normative criteria for causal reasoning. Normative models often use idealized strategies such as *logic-based deduction, inductive generalization*, and *abductive reasoning*. On the other hand, because explanation involves cognitive effort and relates to mental models, cognitive models of explanation are worth considering to develop pragmatic model-based explanations. In this section, we examine philosophical, cognitive, and theoretical foundations of model-based explanations, in which the explanations refer to the properties and behavior of the idealized abstract model in describing the observed regularities.

Explanations involve causal claims to address why or how something occurred. Such explanations can either be process-centric representations that focus on the organization of entities and activities or rely on theory and data. One of the earliest models of explanation is the Deductive-Nomological (DN) model [11], which views an explanation as a deductive argument. Probabilistic and inductive methods extend the DN model for statistical explanation to support explanation under uncertainty. Deductive-Statistical explanations involve deriving regularities based on statistical laws, whereas Inductive-Statistical explanations attach a likelihood to outcomes based on probabilistic interpretation. The

Statistical Relevance (SR) model [15] leverages conditional dependence relations among events. Specifically, given features, A, B, and C, the feature C will be statistically relevant to attribute B if $P(B|A \cap C) \neq P(B|A)$; that is if the probability of event B conditional on A and C is not the same as the probability of B conditional A.

Inspired by the DN and AR models, the *simulacrum* strategy for explanation [2] requires finding a model that fits into the framework of the theory and serves as an analog of the system. Similarly, the theory of explanatory coherence [18] provides a computational framework to establish relations of local coherence between plausible explanations. The inference to the best explanation measures how well the conjectured explanations cohere with observations and other accepted principles and explanations.

The abductive reasoning process starts with a trigger. In ABM, this may be macro behavior that is either surprising and thus requires explanation, or expected and needs justification. According to [14], this observed event needs to be nontrivial. Following the observation of the event, one or more explanations are generated. This process is often a creative act, but context, prior domain knowledge, and heuristics help narrow relevant explanations. Generated explanations are then evaluated to identify possible, more likely, explanations supporting the observed event. The *manipulative abduction* strategy underlines the role that the context and its creative manipulation play in revealing new explanatory hypotheses [12].

Consistent with the AR model, explanatory coherence [20] views scientific explanation from a cognitive perspective and categorizes explanation into three major processes and four critical technical methods. The processes are (1) selecting an explanation based on available information, (2) creating new hypotheses for an explanation, and (3) evaluating plausible competing explanations to make an inference to the best explanation. The methods commonly used as part of these processes are then categorized into deductive (e.g., rule or logic-based strategies), schematic models based on explanation patterns and analogies, probabilistic (e.g., Bayesian models), and connectionist network models of explanation.

The selection process is not always based on utility and expected value calculations when there are multiple explanations. For instance, in the theory of explanatory coherence [18], explanations are evaluated based on coherence judgments via constraint satisfaction mechanisms. The selected tentative explanation is expected to be reevaluated for revision and refinement based on further evidence. The use of coherence as a strategy for explanation is also evident in the unificationist account [8] of explanation. A theory that unifies a broad range of phenomena provides a compelling account of explanatory relevance. Under the unificationist view, understanding something fits it into a broader pattern. The wider the pattern, the more the explanatory power. Pattern-oriented explanations are also analogical due to their ability to explain distinct phenomena that can be subsumed by the same set of schematic explanation patterns.

The Causal Mechanical (CM) model of explanation [16] is a process-centric theory of explanation. Because CM emphasizes spatio-temporal processes in formulating explanations, it has gained traction in the context of discovering and explaining biological mechanisms [3]. The causal mechanistic view considers entities, their activities, and their organization as the central elements of an explanation. The activities within and between entities produce, underlie, and maintain the regularities observed in the system [5]. In accord with the CM model, an explanation is characterized as information that is relevant to manipulation and control, implying the synergy between explanation and exploration [22]. The manipulation of causes facilitates *counterfactual reasoning*, which determines the variation in the outcomes had the factors referenced in the explanations been different.

3   Explanatory Coherence Maps

Agent-based models are often used to specify complex adaptive systems comprised of a large number of spatially connected entities that interact with each other and the context, resulting in macro-level behavior that is difficult to attribute to specific features of the model. Designing ABMs that explain the regularities observed in a system is a *target-directed explanation* activity. Scientists conjecture a plausible explanation by finding a model capable of generating the targeted regularity. Following the convergence to a plausible model, *model-directed explanation* shifts the focus to exploring the consequences of the model's assumptions and explaining how the consequences manifest as a function of the features of the model.

3.1   Theory of Explanatory Coherence

In evaluating alternative explanations, coherent justification is needed to account for the explanation. According to the *Reflective Equilibrium* theory [4] in the presence of conflicts among explanations, adjustment is needed until explanations are in equilibrium. In the most general sense, Reflective Equilibrium can be construed as an attractor state in a complex adaptive system.

The attractor state emerges at the end of a deliberation process by which explanations and goals about an area of inquiry are revised. If we can view equilibrium as a stable state that resolves conflicts, the equilibrium state serves as a coherence account of justification. An optimal equilibrium can be attained when revision is unnecessary because explanations have the highest degree of acceptability [4,19]. The principles and judgments arrived at the equilibrium can account for the context and the situation examined. The *theory of explanatory coherence* provides a set of principles that lends itself to a computational strategy in the form of constraint satisfaction to compute the state of reflective equilibrium [19].

The constraint satisfaction strategy is similar to viewing a state in an N-dimensional space. The activation levels of explanatory hypotheses (i.e., nodes in a coherence graph) are analogous to the acceptability of the respective propositions. Each node receives input from and is reinforced by every other node that is explanatorily connected. The inputs are moderated by the weights of the links from which the input arrives. The activation value is updated as a function of the weighted sum of the inputs it receives. The process continues until the activation values of all units settle. Formally, if we define the activation level of each node j as a_j, where a_j ranges from -1 (rejected) to $+1$ (accepted), the update function can be defined as follows:

$$a_j(t+1) = \begin{cases} a_j(t)(1 - \theta) + net_j(M - a_j(t)), & \text{if } net_j > 0 \\ a_j(t)(1 - \theta) + net_j(a_j(t) - m), & \text{otherwise} \end{cases}$$

In this formulation, the variable θ is a decay parameter that decrees each unit's activation level at every cycle. In the absence of input from other units, the activation level of the unit gradually decays, with m being the minimum activation, M denoting the maximum activation, and net_j representing the net input to a unit, as defined by the following equation: $\sum_i w_{ij} a_i(t)$. These computations can be carried out for every node until the activation levels of elements stabilize and the network reaches an equilibrium via self-organization. Nodes with positive activation levels at the equilibrium state can be distinguished as maximally coherent propositional explanations.

As an illustrative example, consider the Prey-Predator dynamics, which is used in modeling the behavior of biological systems with multiple interacting species. These species play the roles of prey and predator, and their populations change as a result of competitive as well as cooperative interactions.

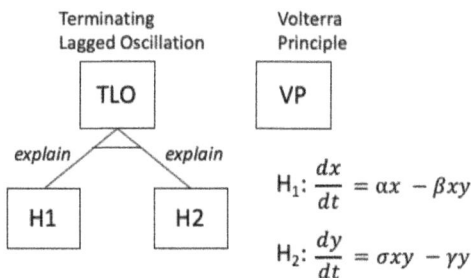

Fig. 1. Explanatory relations. Lotka-Volterra equations are used to explain population dynamics in the presence of two species. The rate of change of population sizes is described in terms of parameters that describe the interaction between species.

Figure 1 depicts two significant properties of interest, lagged oscillation and the Volterra principle, along with two hypothesized equation-based mechanisms that are conjectured to explain these properties. H1 (prey dynamics) and H2 (predator dynamics) together explain the *lagged oscillation* behavior while falling short of exhibiting behavior consistent with the Volterra principle. Such equation-based models are highly idealized and easy to explain abstracted phenomena; however, they become intractable as the complexity increases due to factors associated with the context and the specific activities of the individual members of the population. Equation-based models provide averages across populations while ignoring details that involve theories of community structure, measures of environmental diversity, food chain and stability, and diverse individual behavior.

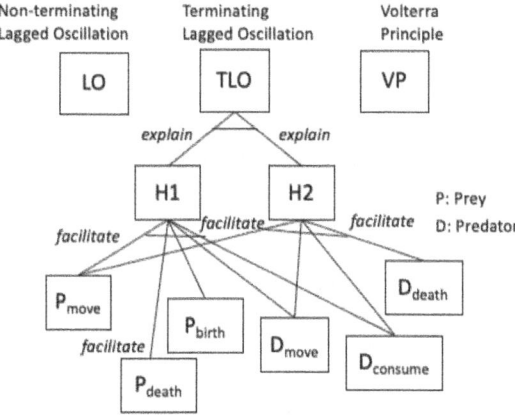

Fig. 2. Refinement of explanatory hypotheses. Idealized equation models are replaced with their respective ABM specifications,

3.2   Abductive Reasoning with Explanatory Cognitive Maps

To improve the accuracy and realism of explanations, increasingly detailed and refined hypotheses are introduced to explain such generic hypotheses and to account for the impact of specific factors such as community structure and individual activities. ABMs mitigate the concerns associated with equation-based models, by describing idealized equation-based models in terms of causal mechanisms that involve entities, their activities, and interactions. On the other hand, as ABMs introduce programmatic simplifications and utilities that are not necessarily connected to the underlying theory, determining whether the results are due to the essentials of the model or its externalities becomes a challenge. For instance, consider the following rules [21] that aim to realize H1 and H2 with two species, prey and predator:

- **Movement rule**: Move in a random direction.
- **Consume rule**: Check if there is prey at the current location. If there is prey, randomly choose one prey and consume it to acquire energy.
- **Reproduction rule**: If the agent has enough energy, pick a random number between 0 and 1. If the number is less than or equal to the predator reproduction rate, hatch a new predator agent at the current location.

– **Removal rule**: If the agent has an energy level of 0, then remove it from simulation.

Notice that these rules roughly represent the constraints of the equation-based model in the original Lotka-Volterra model. Representing an equation-based model in terms of an ABM requires making explicit assumptions about individual behavior that was either implicit or undefined in the equation-based model. Therefore, there can be multiple ABM realizations that aim to generate targeted behaviors such as stabilized oscillations in population sizes. Determining which one of many alternatives, competing models is a valid representation of the system for the intended purposes of the study emerges as a significant challenge.

For instance, the equation-based model does not make any assumptions about the prey or predator's movement. The movement rule in the ABM can vary from completely random moves to more realistic representations that include group behavior and risk-averse strategies to avoid considered risky regions based on experiential learning. An ABM developer needs to make explicit decisions about such representational issues. To complete the model specification and test its ability to generate stabilized oscillation as an emergent behavior of population dynamics, prey rules can similarly be defined as follows:

– **Movement rule**: Move in a random direction.
– **Reproduction rule**: Sample a random number. If the number is less than or equal to the prey-reproduction rate, then reproduce.
– **Removal rule**: If the agent is captured by a predator, it is removed from simulation.

Figure 2 depicts which rules are used as refined explanatory mechanisms for each high-level idealized equation model. The birth, death, and move rules are used to characterize the growth rate in H1. Predators' *move* and *consume* behaviors are also associated with H1 to account for the rate at which preys and predators interact, possibly resulting in the death of preys. Similarly, the move, consume, and death rules of predators, along with the *move* behavior of preys are used to develop a causal mechanistic explanation of H2. Replacing idealized equation models with their respective ABM specifications, we observe *terminating lagged* oscillation. That is, the model generates lagged oscillation for a period, but the population of either the prey or the predator reaches 0 or its maximum, resulting in the termination of the simulation. Therefore, the hypothesized behavioral rules are not robust realizations of the original hypotheses.

Because neither the availability of food nor its density is explicitly modeled, the prey population grows exponentially in the original hypothesized equation-based model without interacting with the predators. The size of the spatial context and the density and distribution of either the prey or predator population influence the rate at which they interact. Yet, in reality, the growth of populations is counterbalanced by the degree of availability of food resources. To facilitate such counterbalancing behavior, the reproduction rule of the prey is specified as contingent on the energy acquired from the food resource in the environment. To this end, an environment component (i.e., grass) is introduced as a food resource for the prey (i.e., sheep). The density of the food resource is controlled by the `grass-growth-rate`. The prey agent consuming the grass receives energy that stimulates the reproduction process. However, uncontrolled increase in the prey population is now suppressed by diminishing levels of food, imposing negative feedback that balances the positive feedback caused by reproduction.

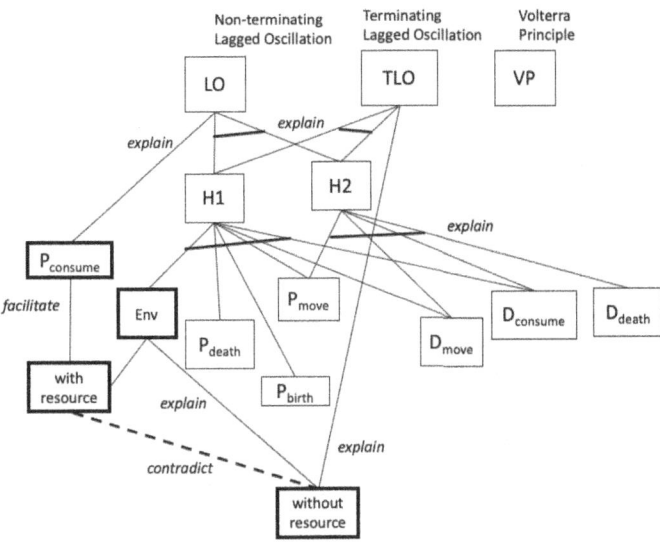

Fig. 3. Revising Explanatory Hypotheses. An environment feature with two alternative realizations, with and without resource, is included to further refine hypothesis **H1**

The prey's consumption rule is facilitated if the environment has resources to help explain the desired lagged oscillation behavior. Because the environment features with and without resource elements are alternatives, they are not compatible and hence contradict and suppress each other. Therefore, they cannot be used as simultaneous explanatory mechanisms. While the environment with the resource feature supports lagged oscillation, the alternative feature explains terminating lagged oscillation, as observed in the previous model version.

The cognitive explanatory map shown in Fig. 3 suggests which hypotheses and associated behavioral mechanisms can be selected to support a specific targeted behavior. The explanatory power of hypothesized causal explanations increases as they succeed in (1) broadening their support to explain an increasing number of targeted behaviors and (2) deepening their explanation in terms of lower-level fundamental mechanisms that explain how and why high-level explanations work. An important finding of the Lotka-Volterra prey-predator dynamics model is the Volterra principle, which is an empirical behavior observed in fishery statistics.

According to this principle, as characterized by H1 and H2, when prey and predator dynamics are negatively coupled, injecting biocide (e.g., toxic element) into the environment increases the abundance of the prey and decreases the abundance of the predator. To broaden the applicability of H1 and H2, the model is modified with a new causal mechanism that introduces *biocide* into the context. Moreover, the consumption behavior of both the prey and predator are refined by two alternative explanatory causal mechanisms; one that allows consumption of biocide and the other that considers only the food resource. Figure 4 depicts how H1 and H2, along with the resource consumption and environmental biocide features support explaining the Volterra principle.

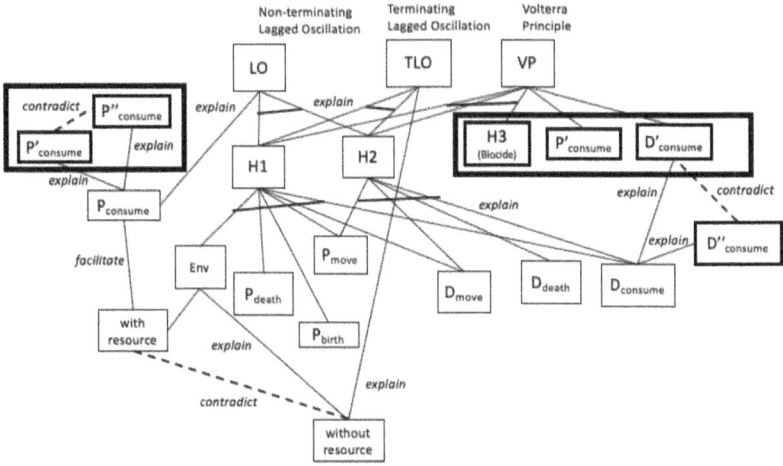

Fig. 4. Specific configurations of H1 and H2, coupled with refined resource consumption and environmental biocide features support explaining the Volterra principle.

The discussion above is intended to highlight the three critical elements of target-directed explanation. First, we need to be able to select a set of coherent explanatory features from among alternatives. However, if existing explanatory features are insufficient to generate the target behavior, alternative features are created and included in the set of plausible explanations for consideration. Second, explanations are evaluated to make inference to the best explanation. That is, alternative explanatory features are ranked in terms of their degree of relevance and success in generating the desired or expected behavior. Finally, the scope and resolution of explanatory features are expanded further to instill confidence in the explanatory power of identified features. Specifically, the explanations are broadened to target additional behavior while also refined into increasingly detailed and high-resolution causal mechanistic features.

4 Explanatory Modeling Process

The illustrative example presented above suggests that the generation of explanatory models involves the interplay of multiple activities and is a highly dynamic process that involves specific inferential processes. The example and its underlying coherence-driven strategy motivate the following guidelines.

- *Explanation is an iterative, incremental process*: The provision of a model-based explanation stimulates further inquiry to deepen and broaden the scope of the plausible explanations. Initial explanations often provide a template to continue the search process, allowing model builders to iteratively refine the model's causal mechanisms by adding details to increase its level of resolution. During the process, the focus of inquiry can shift due to the evaluation of alternative explanatory mechanisms.
- *Explanation is a symbiotic adaptive process:* As the process of searching for an explanation unfolds, a symbiotic search process takes place between the hypothesis (e.g., model structure) and experiment spaces. Following the search within the architectural space of models, the experimental conditions are created so that they provide new information that would otherwise be unavailable. Such new information is then used to influence the search process within the hypothesis space [12].
- *Emergence prompts explanation:* Agent-based models of complex systems reveal emergent properties as a result of interactions among a diverse set of agents in a spatial context. Emergent behavior can be unexpected and surprising, as it may be an indicator of new knowledge that cannot simply be inferred as a linear function of agent attributes. The recognition of such emergent behavior prompts the need for an explanation to provide an account of the causal mechanisms responsible for the observed behavior.
- *Explanation requires understanding via self-awareness*: Having an introspective capability to assess the consequences as a function of the premises of one's behavior is critical for reflection. Explanatory reasoning requires reflecting on the underlying causes of observed behavior, and such reflection enables the explanation of behavior in terms of beliefs, objectives, and intentions that drive observable actions. A model with self-awareness capabilities can compare its simulated behavior to objectives and evolve an understanding of its features and how and when they contribute to the desired behavior.
- *Explanation involves an exploratory learning process*: The generation of an explanation involves the use of three vital inferential processes: deduction, induction, and abduction. Deduction occurs in the form of simulation of the model to derive the consequences of its underlying assumptions. Initially, multiple competing explanations can be represented by a distinct model. Distinct models in the ensemble represent plausible explanations examined under simulated experiments to reveal their behavior. The exploration phase requires the selection of experiments and the model's variation to improve model users' ability to distinguish between alternative models concerning their explanatory

power. Simulation-generated data are then generalized via inductive mechanisms to support learning rules that separate successful results from those that fail to satisfy the objectives.

5   Implications for Simulation Environments

Explanatory modeling involves exploratory learning to support the generation and evaluation of explanations. Exploration requires variations of model structure and representations. The ability to generate numerous models with different architectures across multiple resolutions and perspectives must be managed transparently. Such management should be extended to computational experiments so that alternative scenarios and model variants can be coupled adaptively in the context of an evolving analysis.

To generate explanatory models, the exploration results need to be generalized and presented using visual, causal, and conceptual representations meaningful to model users. Furthermore, the set of possible models needs to be explicitly represented to be queried, sampled, and analyzed to support incremental and adaptive modeling. Self-adaptive models impose additional constraints for representing and selecting model features.

5.1   Variation of Causal Assumptions in an Evolving Analysis

Models are developed with specific assumptions in mind. The rules that define agents' actions are often theoretical and pragmatic elements. Pragmatic constructs include simplifying assumptions that are not related to the underlying theory of the system. For instance, the movements of prey and predator agents in an agent-based model of ecological dynamics can be either random or intentional to mimic the flocking behavior of collectives. Supporting seamless variation of model structures and representations can be in implicit model design strategies or based on explicit modeling constructs intended to facilitate systematic variability management. Implicit design strategies include well-known design patterns such as the *Strategy* and the *Factory Method*, which are used in model behavior variation when the specific behavior cannot be anticipated in advance.

Alternatively, explicit modeling constructs and paradigms can be considered to design models with variability in mind. For instance, both the feature and aspect-oriented modeling paradigms promote relatively simple features and cross-cutting aspects that can be assembled into aggregate models subject to composition constraints. Modeling environments can make such constraints and feature models explicit by using declarative specifications, which can allow the use of powerful tools that can manipulate and adapt declarative specifications transparently, and hence enable manipulative abductive reasoning [12]. A modeling environment needs to support an explicit variability model to support such variation and customization. Such a model can include at least a mechanism for specifying the behavior of individual features, defining acceptable and well-formed feature combinations in terms of syntactic and semantic composition constraints, and providing the means for evolving these specifications as a result of the information gathered through simulation experiments.

5.2   Controlling for Confounding Factors

A model development environment needs to facilitate interventions and counterfactual reasoning to facilitate explanations that move beyond the provision of associations. Experiments are critical instruments in search of explanations, as they enable intervening with the conditions under which a model is simulated to observe their effects. As such, interventions allow discerning effects of carefully manipulated causes. Whereas interventions focus on prospective reasoning, the counterfactual mode of experimentation involves revisiting the assumptions underlying the model by asking questions such as "What is the probability of observing the outcome $Y = y$ had we used the causal mechanism M1, given that we observed the outcome $Y = y'$ and designed the model with the mechanism M2". While intervention-based exploration requires searching the experiment space, counterfactual analysis requires searching within the model design and experiment spaces.

To experiment on a model is to place it under the control of an experiment manager to prune and revise the space of possible causal mechanisms. The focus on evaluating alternative causal explanations suggests the design of targeted experiments to reveal the role and significance of elements of putative mechanisms. One experiment strategy is to determine whether an entity, activity, property or organizational feature is causally relevant to another. This strategy helps model developers determine the contextual conditions that trigger a phenomenon or understand if a specific element is sufficient for what happens in the next stage of the mechanism. Another strategy is to discern whether an entity, activity or organizational characteristic is relevant to the overall behavior generated by the whole mechanism. These experiments facilitate determining causal relations across the levels of hierarchically organized models. Once sufficient confidence is achieved, various questions can be answered by intervening with the model and the experimental conditions.

Targeted experiments are mechanism-aware in that they allow direct interference, stimulation, and activation of components in a model. Interference experiments should provide facilities to inhibit, diminish, or disable the components of the underlying causal structure. In *stimulation experiments*, one intervenes to excite or amplify the model's behavior and detect the impact of that change. Interference and stimulation experiments are bottom-up due to their intervention in the lower-level mechanistic components. On the other hand, activation experiments are top-down experiments that influence the conditions to activate higher-level behavior and detect the impact of such activation on the lower-level components.

Whereas intervention experiments are intended to determine the role of relatively stable mechanisms comprising specific components, the exploratory analysis starts by asking specific mechanistic questions to identify these components in the first place. There are various kinds of such experiments [5].

- In *by-what-activity* experiments, the goal is to discover which kind of activity connects the causes to effects. Such experiments require reasoning about the alignment of invariants with the preconditions of activities. One needs to identify the preconditions of plausible activities and compare the situations in which that precondition is met with cases in which it is not. These comparisons can be used to discriminate among competing hypothesized activity features.
- Activities are produced by components that collaborate to produce observed behavior. In *by-what-component* experiments, artificial conditions are created to disable components in a controlled manner to discern whether the intervention prevents the phenomenon. Alternatively, different combinations and organizational representations of components are conducive to generating the desired behavior. Activity-based and component-oriented experiments can be combined to prune the space of possible explanatory causal mechanisms.

5.3   Stepwise Refinement with Explanatory Modeling

As experiments are conducted to intervene with the context and mechanisms of the target system, the results are expected to gradually converge to increasingly credible mechanisms that successfully produce and maintain the desired behavior. As a result, trust in the model's behavior evolves due to its ability to account for a broadening set of accumulating evidence. Successful models continue to be refined to include increasingly detailed characterization of the high-level mechanisms. This learning process should be facilitated through appropriate tools that can evaluate the experiments' results and provide necessary criticism to guide the search process further.

Evaluation of the efficacy of the components and activities requires the explicit separation of a learning element from the model so that the *learning component* can use the feedback received from the critic and determine how the model should be modified to perform better concerning the goals of the experiment. The *critic* informs the learning element via reward (or penalty) how well the model is performing for the performance standard. The learning process requires an additional *generative component*, which facilitates model generation in accord with the abductive reasoning strategies outlined earlier. Unless a model generator explores alternative representations, the model can quickly converge to a representation that can perform well under specific scenarios but may fail to behave robustly as the range of scenarios is broadened.

6   Conclusion

The application domains of successful models broaden while refinements to a model deepen the level of resolution to improve accuracy and fidelity as learning takes place. However, explaining the cause-effect relations in such models is a critical challenge, primarily due to agents' autonomous, decentralized decision-making that adapts and interacts with each other. Explanation strategies are grounded in philosophy, and cognitive science of science are reviewed. Due to the need for aligning explanations with the cognitive requirements of the target audience, a strategy based on cognitive coherence is adopted.

The coherence-driven strategy demonstrates how explanatory models can evolve with a simulation model based on the results of experiments. Specifically, the theory of explanatory coherence and reflective equilibrium can show how an inquiry in population dynamics can be broadened and deepened to revise beliefs across multiple levels and scales about causal premises of expected behavioral regularities.

Following the demonstration of the theory of explanatory coherence, the principles of explanatory modeling are delineated to characterize the highly dynamic process that results in the formation and growth of explanatory models. The underlying inferential processes are explained within a framework that interfaces with exploration and learning. Based on the principles and the evaluation criteria, we conclude with specific guidelines for developing modeling and simulation environments that support explanatory modeling.

References

1. Bokulich, A.: Models and explanation. In: Magnani, L., Bertolotti, T. (eds.) Springer Handbook of Model-Based Science. SH, pp. 103–118. Springer, Cham (2017). https://doi.org/10.1007/978-3-319-30526-4_4
2. Cartwright, N., McMullin, E.: How the laws of physics lie (1984)
3. Craver, C.F., Darden, L.: In Search of Mechanisms: Discoveries Across the Life Sciences. University of Chicago Press, Chicago (2013)
4. Daniels, N.: Justice and Justification: Reflective Equilibrium in Theory and Practice, vol. 22. Cambridge University Press, Cambridge (1996)
5. Darden, L.: Strategies for discovering mechanisms: schema instantiation, modular subassembly, forward/backward chaining. Philos. Sci. **69**(S3), S354–S365 (2002)
6. Davis, P.K., O'Mahony, A., Gulden, T.R., Osoba, O.A., Sieck, K.: Priority Challenges for Social and Behavioral Research and its Modeling. RAND Corporation Santa Monica, CA (2018)
7. Egli, L., Weise, H., Radchuk, V., Seppelt, R., Grimm, V.: Exploring resilience with agent-based models: state of the art, knowledge gaps and recommendations for coping with multidimensionality. Ecol. Complexity (2018). https://doi.org/10.1016/J.ECOCOM.2018.06.008, https://www.sciencedirect.com/science/article/pii/S1476945X18301089
8. Friedman, M.: Explanation and scientific understanding. J. Philos. **71**(1), 5–19 (1974)
9. Gelfert, A.: Assessing the credibility of conceptual models. In: Beisbart, C., Saam, N.J. (eds.) Computer Simulation Validation. SFMA, pp. 249–269. Springer, Cham (2019). https://doi.org/10.1007/978-3-319-70766-2_10
10. Gunning, D.: Explainable artificial intelligence (XAI). Defense Advanced Research Projects Agency (DARPA) (2017). https://www.darpa.mil/attachments/XAIProgramUpdate.pdf
11. Hempel, C.G., Oppenheim, P.: Studies in the logic of explanation. Philos. Sci. **15**(2), 135–175 (1948)

12. Magnani, L.: Abductive Cognition: The Epistemological and Eco-cognitive Dimensions of Hypothetical Reasoning, vol. 3. Springer, Heidelberg (2009)

13. Onggo, S., Yilmaz, L., Klugl, F., Terana, T., Macal, M, C.: Credible agent-based simulation – an illusion or only a step away? In: Proceedings of the Winter Simulation Conference, pp. in–press. ACM (2019)

14. Peirce, C.S.: The Essential Peirce: Selected Philosophical Writings, vol. 2. Indiana University Press, Bloomington (1992)

15. Salmon, W.C.: Statistical Explanation and Statistical Relevance, vol. 69. University of Pittsburgh Press, Pittsburgh (1971)

16. Salmon, W.C.: Scientific Explanation and the Causal Structure of the World. Princeton University Press, Princeton (1984)

17. Teran-Somohano, A., Dayıbas, O., Yilmaz, L., Smith, A.: Toward a model-driven engineering framework for reproducible simulation experiment lifecycle management. In: Proceedings of the Winter Simulation Conference 2014, pp. 2726–2737. IEEE (2014)

18. Thagard, P.: Explanatory coherence. Behav. Brain Sci. **12**(3), 435–467 (1989)

19. Thagard, P.: Coherence in Thought and Action. MIT press, Cambridge (2002)

20. Thagard, P.: The Cognitive Science of Science: Explanation, Discovery, and Conceptual Change. MIT press, Cambridge (2012)

21. Weisberg, M.: Simulation and Similarity: Using Models to Understand the World. Oxford University Press, Oxford (2012)

22. Woodward, J.: Making Things Happen: A Theory of Causal Explanation. Oxford University Press, Oxford (2005)

23. Yilmaz, L.: Reproducibility in M&S research: issues, strategies and implications for model development environments. J. Exp. Theor. Artif. Intell. **24**(4), 457–474 (2012)

24. Yilmaz, L., Liu, B.: Model credibility revisited: concepts and considerations for appropriate trust. J. Simul. 1–14 (2020)

25. Yilmaz, L., Ören, T.: Agent-directed simulation. Agent-Directed Simul. Syst. Eng. 111–143 (2009)

Deep Neural Input-Filtered Gaussian Process Model

Tadej Krivec[1,2] and Juš Kocijan[1,3(✉)]

[1] Institut Jožef Stefan, Jamova cesta 39, Ljubljana, Slovenia
{tadej.krivec,jus.kocijan}@ijs.si
[2] Jožef Stefan International Postgraduate School,
Jamova cesta 39, Ljubljana, Slovenia
[3] University of Nova Gorica, Vipavska 13, Nova Gorica, Slovenia

Abstract. This paper presents a new model for addressing the error-in-variables problem in Gaussian process autoregressive models by combining a Gaussian process with a deep neural network. The Gaussian process autoregressive model is a simple and effective method for modeling dynamical systems due to its nonlinear, nonparametric, and Bayesian nature. The analytic solution for the marginal log-likelihood is obtained by considering the training input-output relationship to be static, with dynamics modeled through the inclusion of lagged observations in the input regressor. The limitation of this autoregressive method is that both the outputs and inputs are affected by noise. The simulation is obtained iteratively by propagating the Gaussian distribution through a nonlinear function. This results in a costly estimation of the simulated response with Monte Carlo integration. We propose an alternative approach in which a pre-filtering step is performed using a deep neural network to approximate the intractable recurrent filtering of the latent states. The proposed model improves the autoregressive approach while reducing the computational time of the simulation. The proposed model is validated on two case studies: a synthetic example and a real-world problem.

Keywords: Gaussian process models · Deep neural networks ·
Error-in-variables · Dynamical systems · Simulation

1 Introduction

1.1 Literature Review

Gaussian process (GP) models are suitable for problems where the model uncertainty is desired, such as safety-critical applications, robust control, and fault detection. This is because they provide a well-calibrated predictive distribution at a model output, which can be used to systematically quantify the model's accuracy. GP models use an infinite number of basis functions [17,31] which makes them a part of a class of rich probabilistic models.

While the marginal and conditional distributions can be computed in closed-form, the GP training algorithm has a cubic computational complexity and a

© The Author(s), under exclusive license to Springer Nature Switzerland AG 2024
M. Mujica Mota and P. Scala (Eds.): EUROSIM 2023, CCIS 2033, pp. 347–361, 2024.
https://doi.org/10.1007/978-3-031-68438-8_25

quadratic complexity in memory. Over the last few decades, researchers have developed various techniques for reducing the computational complexity of GP models while preserving their elegant properties. An overview of these techniques can be found in [21].

Modeling dynamical systems can be achieved by augmenting the inputs with a set of lagged output observations, resulting in a nonlinear autoregressive model with exogenous inputs (NARX). When a GP is used to model the functional relationship between the input regressor and the output, the model is known as a GP-NARX model. The elegance of the NARX model lies in its ability to simplify the training process, making it more scalable for large datasets and easier to experiment with. Furthermore, the NARX approach has been shown to outcompete more specialized models that struggle with handling large data sets or require approximated inference methods [9].

The downside of the NARX model is the introduction of noise in the inputs. An approach that can filter the states recurrently is the GP state-space model [7,10,11] or GP nonlinear output-error model [18]. When the dynamics are linear, the problem can be efficiently solved using Kalman filtering [15]. However, for nonlinear dynamics, one has to resort to approximations or numerical inference. This results in specialized solutions that are hard to reuse, and can also fail miserably where the approximate inference is not adequate for the problem at hand [9].

A compromise between the simplicity of an autoregressive model and the complexity of the recurrent filter can be achieved by utilizing a model with uncertain inputs. This approach, which has been supported by various literature [3,5,24,25] offers a good balance, but can still be relatively complex. An alternative approach that has a simpler implementation is to pre-process the noisy observations in the input regressor using a Butterworth filter as proposed in [12]. However, the last paper does not address the simulation, which can result in instability even when the prediction is accurate.

1.2　Limitations of the Current Approach

The uncertain inputs pose a significant challenge in GP models, as the propagation of a distribution through the nonlinear function cannot be obtained in closed-form. This issue is especially relevant in simulation, where many applications, such as predicting the future state of a system for multiple steps ahead, require inputs that are currently unknown and need to be determined iteratively. In GP-NARX models, the Gaussian distribution is sequentially propagated through the nonlinear model, making the simulation response analytically intractable. To overcome this limitation, researchers have proposed various approximations, such as those found in [4,6,13,26].

Monte Carlo (MC) integration is a widely used method for estimating the simulated latent response in GP-NARX models. However, it comes with a major drawback: MC sampling of GP-NARX trajectories requires storing consecutive MC samples in memory. This results in an increase of computational complexity, especially for GP-NARX models with long prediction horizons [2]. This can be computationally infeasible for some systems and applications.

1.3 Proposed Approach

In this paper, a new approach to address the error-in-variables problem and the problem of computational complexity in simulation is proposed by introducing a deep neural network (DNN) pre-filtering step that predicts the estimated latent states. The estimated noise-free states are then used in the input to the GP model. The integration of GPs with DNNs has already been considered in the existing literature [32]. However, this approach has not yet been applied to address the error-in-variables problem in autoregressive GP models. The proposed method aims to take advantage of the ability of DNNs to learn a flexible latent state filter and the ability of GP models to model the dynamics robustly with uncertainty quantification.

2   GP-NARX

Let us assume that the true dynamics are governed by

$$f_t = f\left(f_{t-n_a}, \ldots, f_{t-1}, x_{t-n_b}, \ldots, x_{t-1}\right), \tag{1}$$

where f is a nonlinear function, f_t denotes an evaluation of the nonlinear function at the respective timestep, x_t represents exogenous variables at the respective timestep and $n_a, n_b > 0$ denote the number of lags for the delayed outputs and delayed exogenous variables, respectively. We represent the model in the matrix form as

$$\mathbf{f}_{1:t} = f(\mathbf{Z}_{1:t}), \tag{2}$$

where $f_{1:t}$ denotes the vector of latent function values. The input matrix $\mathbf{Z}_{1:t} \in \mathcal{R}^{t \times (n_a + n_b)}$ is represented with a NARX model, where t denotes the number of observations. The i-th row of the matrix $\mathbf{Z}_{1:t}$ is defined by

$$\mathbf{z}_i^T = [f_{i-n_a}, \ldots, f_{i-1}, x_{i-n_b}, \ldots, x_{i-1}]. \tag{3}$$

Instead of the true output vector, we observe a noisy one, i.e. $\mathbf{y}_{1:t} = \mathbf{f}_{1:t} + \epsilon$, where $\epsilon \sim \mathcal{N}(\epsilon | \mathbf{0}, \mathbf{I}\sigma_n^2)$. In the autoregressive model, the dynamical system is then modeled by

$$\mathbf{y}_{1:t} = f(\mathbf{Z}_{1:t}) + \epsilon, \tag{4}$$

where the i-th row of the input matrix $\mathbf{Z}_{1:t}$ is corrupted by noisy observations, i.e.

$$\mathbf{z}_i^T = [y_{i-n_a}, \ldots, y_{i-1}, x_{i-n_b}, \ldots, x_{i-1}]. \tag{5}$$

The goal is to infer the latent function f from Eq. (4). We model the latent function by a GP.

GP is a collection of random variables, any finite number of which have a joint Gaussian distribution [31]. It is completely specified by its mean $m(\cdot)$ and covariance function $k(\cdot, \cdot)$

$$m(\mathbf{z}) = \mathbb{E}[f(\mathbf{z})], \tag{6a}$$

$$k(\mathbf{z}, \mathbf{z}') = \mathbb{E}[(f(\mathbf{z}) - m(\mathbf{z}))(f(\mathbf{z}') - m(\mathbf{z}')^T)]. \tag{6b}$$

A GP will be denoted by

$$f(\cdot) \sim \mathcal{GP}\left(f(\cdot)\,|m(\cdot), k(\cdot, \cdot)\right), \tag{7}$$

where the $m(\cdot)$ and $k(\cdot, \cdot)$ are parametrized by hyperparameters $\boldsymbol{\theta}$.

The prior over functions, represented by a GP, is transformed to the posterior using the Bayes theorem. The posterior is defined by

$$\mathcal{GP}\left(f(\cdot)\,|\mathbf{y}_{1:t}, \sigma_n^2, m(\cdot), k(\cdot, \cdot)\right) = \frac{p\left(\mathbf{y}_{1:t}|\mathbf{f}_{1:t}, \sigma_n^2\right)\mathcal{GP}\left(f(\cdot)\,|m(\cdot), k(\cdot, \cdot)\right)}{p\left(\mathbf{y}_{1:t}|\sigma_n^2, m(\cdot), k(\cdot, \cdot)\right)}, \tag{8}$$

where the likelihood is Gaussian, i.e. $p(\mathbf{y}_{1:t}|\mathbf{f}_{1:t}) \sim \mathcal{N}(\mathbf{y}_{1:t}|\mathbf{f}_{1:t}, \sigma_n^2)$. Hereafter, the conditioning on the choice of $m(\cdot)$ and $k(\cdot, \cdot)$ will be replaced with the parameters $\boldsymbol{\theta}$ that parametrize the aforementioned functions or completely omitted, depending on the context, for notational brevity. Additionally, the mean function will be without the loss of generality considered as $m(\cdot) = \mathbf{0}$. The posterior GP is specified by

$$\begin{aligned}\mathbb{E}\left[p(f(\cdot)\,|\mathbf{y}_{1:t})\right] &= k\left(\cdot, \mathbf{Z}_{1:t}\right)\left(\mathbf{K}_{f_{1:t}, f_{1:t}} + \mathbf{I}\sigma_n^2\right)^{-1}\mathbf{y}_{1:t}, \\ \mathbb{V}\left[p(f(\cdot)\,|\mathbf{y}_{1:t})\right] &= k\left(\cdot, \cdot\right) - k\left(\cdot, \mathbf{Z}_{1:t}\right)\left(\mathbf{K}_{f_{1:t}, f_{1:t}} + \mathbf{I}\sigma_n^2\right)^{-1}k\left(\mathbf{Z}_{1:t}, \cdot\right),\end{aligned} \tag{9}$$

where $[\mathbf{K}_{f_{1:t}, f_{1:t}}]_{ij} = k(\mathbf{z}_i, \mathbf{z}_j)$.

2.1 Kernel Learning

The hyperparameters of the GP-NARX model can be obtained through the maximization of the marginal log-likelihood (MLL) defined by

$$\begin{aligned}\log p\left(\mathbf{y}_{1:t}|\boldsymbol{\theta}, \sigma_n^2\right) = &-\frac{1}{2}\mathbf{y}_{1:t}^T\left(\mathbf{K}_{f_{1:t}, f_{1:t}} + \mathbf{I}\sigma_n^2\right)^{-1}\mathbf{y}_{1:t} \\ &-\frac{1}{2}\log\left|\mathbf{K}_{f_{1:t}, f_{1:t}} + \mathbf{I}\sigma_n^2\right| - \frac{n}{2}\log 2\pi.\end{aligned} \tag{10}$$

2.2 Prediction

The prediction at $t+1$ is obtained by marginalizing out the latent values that do not belong to timestep $t+1$ out of the posterior, i.e.

$$p(f_{t+1}|\mathbf{y}_{1:t}) = \int \mathcal{GP}\left(f(\cdot)\,|\mathbf{y}_{1:t}\right)d\mathbf{f}_{\backslash\{f_{t+1}\}} = \frac{\int p\left(\mathbf{y}_{1:t}|\mathbf{f}_{1:t}\right)p\left(\mathbf{f}_{1:t}, f_{t+1}\right)d\mathbf{f}_{1:t}}{p\left(\mathbf{y}_{1:t}\right)}, \tag{11}$$

where $\mathbf{f}_{\backslash\{f_{t+1}\}}$ represents all latent values except f_{t+1}. The mean and the variance of the predictive posterior are specified by

$$\mathbb{E}[f_{t+1}|\mathbf{y}_{1:t}] = \mathbf{K}_{f_{t+1},f_{1:t}} \left(\mathbf{K}_{f_{1:t},f_{1:t}} + \mathbf{I}\sigma_n^2\right)^{-1} \mathbf{y}_{1:t}, \tag{12a}$$

$$\mathbb{V}[f_{t+1}|\mathbf{y}_{1:t}] = \mathbf{K}_{f_{t+1},f_{t+1}} - \mathbf{K}_{f_{t+1},f_{1:t}} \left(\mathbf{K}_{f_{1:t},f_{1:t}} + \mathbf{I}\sigma_n^2\right)^{-1} \mathbf{K}_{f_{1:t},f_{t+1}}. \tag{12b}$$

A block diagram representing the GP model prediction is shown in Fig. 2a.

2.3 Simulation

Marginalizing the GP posterior over all the latent quantities that are not of interest, we arrive at the posterior prediction at timesteps from $t+1$ up to $t+n$

$$\begin{aligned} p\left(\mathbf{f}_{t+1:t+n}|\mathbf{y}_{1:t}\right) &= \int \mathcal{GP}\left(f\left(\cdot\right)|\mathbf{y}_{1:t}\right) d\mathbf{f}_{\backslash\{f_{t+1:t+n}\}} \\ &= \frac{\int p\left(\mathbf{y}_{1:t}|\mathbf{f}_{1:t}\right) p\left(\mathbf{f}_{1:t}, \mathbf{f}_{t+1:t+n}\right) d\mathbf{f}_{1:t}}{p\left(\mathbf{y}_{1:t}\right)}. \end{aligned} \tag{13}$$

It is important to note that the estimation of the latent function values, up to the timestep $t+n$, can only be obtained in closed-form if the inputs at that timestep are known. However, in simulation, the inputs are not known in advance and have to be determined iteratively. This is achieved by using the past predicted latent values to replace the output data in the input $\mathbf{z}$ in a recursive manner. The recursive relationship is illustrated in Fig. 2b with a backward loop.

2.4 Computational Complexity

The computational complexity of the GP-NARX model can be quite high. A single evaluation of the MLL requires $\mathcal{O}(t^3)$ operations, where t is the number of observed data points. Making predictions with the GP-NARX model has a computational complexity of $\mathcal{O}(t^2)$, and MC simulation of the GP-NARX model has a complexity of $\mathcal{O}(n \cdot (n + t)^2)$, where n is the number of predicted steps into the future.

Sparse approximations have been proposed as a way to reduce the computational complexity of GP-NARX models for large training data sets. These methods aim to reduce the number of data points t to a smaller number m of pseudo-inputs, which is a user-defined parameter. An overview of these sparse approximations can be found in [27]. However, despite using sparse approximations, the cubic computational complexity with respect to the number of predicted steps into the future still persists.

For long prediction horizons, the simulation can become impractical, and further approximations are needed. A popular approximation considers the latent values that do not belong to the training data independent given the observations [19]. This is a rough approximation, but it is quite fast. It can be obtained in $\mathcal{O}(n \cdot t^2)$ (Fig. 1).

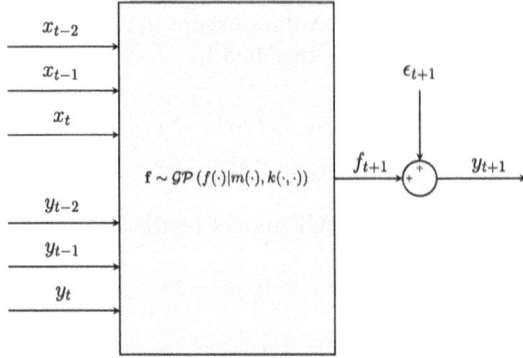

(a) GP-NARX model in prediction.

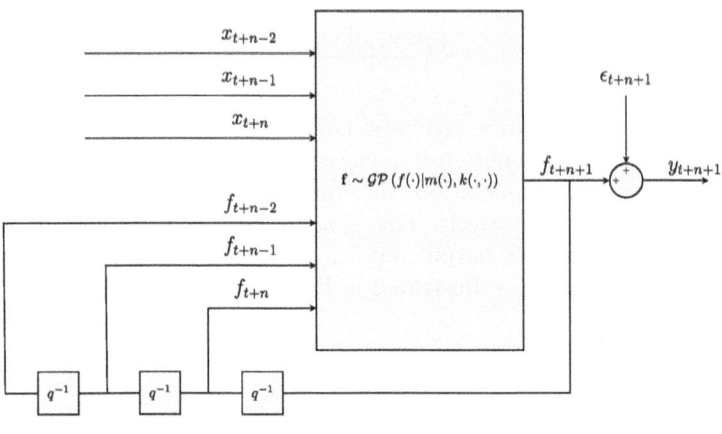

(b) GP-NARX model in simulation.

Fig. 1. Block diagram of a GP-NARX model for $n_a = n_b = 3$. The noisy observations in the inputs of the model training and the model prediction result in an error-in-variables problem. The backward loop in simulation results in intractable simulation that has to be addressed with computationally demanding numerical integration.

3   Deep Neural Input-Filtered GP

The proposed method called Deep Neural Input-Filtered GP (DNIF-GP) aims to learn the latent states directly from the exogenous variables using a DNN pre-filtering step. In this paper, we use a bi-directional Long Short-Term Memory (B-LSTM) network [14, 28] for this purpose. We choose the LSTM model because its deterministic nature is well-suited for modeling nonlinear sequential problems without explicitly incorporating output lags in the input regressor. The pre-filter is, therefore, a nonlinear finite impulse response (NFIR) model and costly autoregression is avoided.

The forward pass of a single LSTM cell can be found in [14]. The DNN used in DNIF-GP is trained separately from the GP model, and the length of the time series used for training the DNN does not have to match the number of lags in the inputs of the GP model. To aid in the learning process, a longer series of n_c timesteps is typically selected for training the DNN. In this context, the length of a time series (i.e. timesteps) refers to the length of a single learning example. Figure 2 depicts an example of the DNN that uses 5 consecutive timesteps for learning, but the GP part of the model ingests the last $n_a = 3$ estimated latent states. This approach aims to provide a noise-free version of the inputs to the GP.

3.1   Kernel Learning

Let us denote the mapping by a neural network with g. The neural network part of the model is defined by

$$\mathbf{y}_{t-n_c-1:t-1} = g\left(x_{t-n_c-1}, \ldots, x_{t-1}\right) + \epsilon, \tag{14}$$

which specifies the input/output pairs of time series examples of the length of n_c. The estimated latent states $\hat{\mathbf{f}}$ are obtained by predicting through the neural network, i.e.

$$\hat{\mathbf{f}}_{t-n_c-1:t-1} = g\left(x_{t-n_c-1}, \ldots, x_{t-1}\right), \tag{15}$$

The GP part of the model is then defined by

$$y_t = f(\mathbf{z}_t) + \epsilon, \tag{16}$$

where the input uses the estimated states instead of noisy observations, i.e.

$$\mathbf{z}_t^T = [\hat{f}_{t-n_a}, \ldots, \hat{f}_{t-1}, x_{t-n_b}, \ldots, x_{t-1}]. \tag{17}$$

The GP is then trained on neural network's estimated states in the input. However, it retains the noisy observations at the model output.

3.2   Prediction and Simulation

The prediction can be obtained by predicting through the neural network, constructing the inputs $\mathbf{z}_z$, and then through the GP, i.e.

$$\hat{\mathbf{f}}_{t-n_c:t} = g\left(x_{t-n_c}, \ldots, x_t\right), \tag{18a}$$

$$\mathbf{z}_{t+1}^T = [\hat{f}_{t-n_a+1}, \ldots, \hat{f}_t, x_{t-n_b+1}, \ldots, x_t], \tag{18b}$$

$$f_{t+1} = f(\mathbf{z}_{t+1}). \tag{18c}$$

Note that the dynamical input $\mathbf{z}_t$ is predicted purely from the exogenous variables. For that reason, the prediction can also be viewed as a simulation.

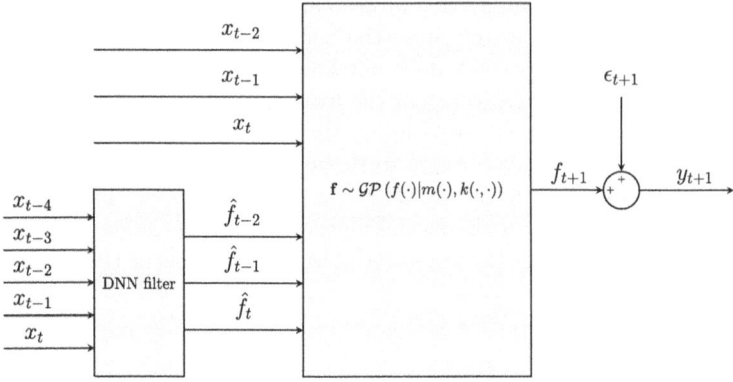

(a) DNIF-NARX model in prediction. NFIR model structure avoids autoregression.

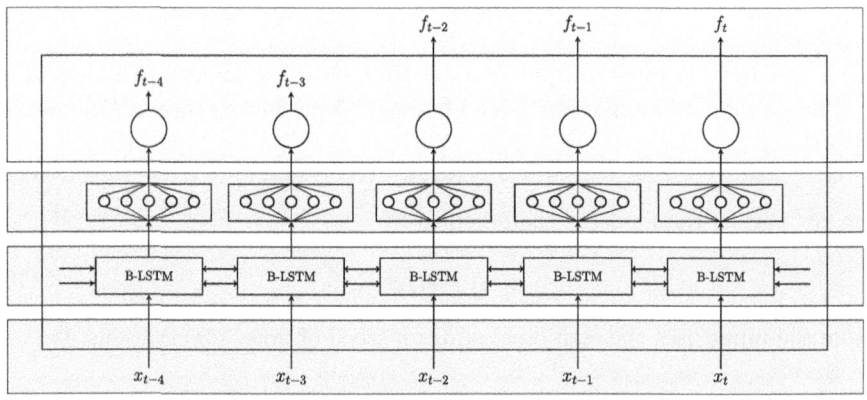

(b) Deep neural network that filters the latent states.

Fig. 2. Block diagram of a DNIF-GP model for $n_a = n_b = 3$ and $n_c = 5$. The latent values in the inputs of the GP model training and prediction are replaced by the estimated states from the neural network. Red color represents the input layer, blue the hidden layers, and green the output layer. (Color figure online)

3.3   Computational Complexity

The proposed DNIF-GP method uses a DNN as a pre-filtering step to estimate the latent states from the exogenous variables. The computational complexity of a single LSTM layer is assumed to be $\mathcal{O}(h \cdot (h + d))$, where h represents the dimensionality of the hidden states and d is the dimensionality of the exogenous variables. Since we have to predict for n_c timesteps to construct a single input for the GP part, the computational complexity is multiplied by a factor n_c.

The rest of the neural architecture used in this paper is simple and its computational complexity is assumed to be negligible compared to the recurrent layer. The computational complexity of the DNIF-GP in training is therefore $\mathcal{O}(h \cdot (h+d) \cdot n_c) + \mathcal{O}(t^3)$; in prediction is $\mathcal{O}(h \cdot (h+d) \cdot n_c) + \mathcal{O}(t^2)$; and in simulation is $\mathcal{O}(h \cdot (h+d) \cdot n_c \cdot n) + \mathcal{O}(n \cdot t^2)$, where t denotes the number of observed data in training and n denotes the number of predicted steps into the future. For sparse approximations, the number of observed data t can be replaced by the number of pseudo-inputs m. It is important to note that the computational complexity of the GP part of the model dominates the computational complexity of the neural network during the training and prediction stages, resulting in only a marginal increase in complexity compared to the GP-NARX model. However, the proposed method significantly reduces the computational complexity of the simulation compared to the numerical estimation in the GP-NARX case.

4   Case Studies

In this section, the validation of the newly proposed approach is presented.

4.1   Methodology, Hardware and Software Specifications

We considered two examples to validate the proposed approach. Each individual experiment presented in this section was repeated 30 times with different random seed to generate various noise realizations. The computer on which the experiments were conducted had the following specifications: Intel(R) Core(TM) i7-8700K CPU 3.70 GHz, 2×16 GB DIMM DDR4 3200 MHz. The models were implemented on top of Tensorflow 2.4.0. [1] and GPflow 2.4.0 [23] software libraries.

4.2   Synthetic Example

Here, we will consider a synthetic problem. This will be elegant as the ground truth latent states will be available for comparison.

Case Study 1. Let us consider an illustrative problem [2], where the true process is governed by the difference equation

$$
\begin{aligned}
f_{t+1}^1 &= f_t^1 \exp(1 - 0.4 f_t^1 - \frac{\left(2 + 1.2 x_t^1\right) f_t^2}{1 + \left(f_t^1\right)^2}), x_t^1 = \cos\left(2\pi t\right), \\
f_{t+1}^2 &= f_t^2 \exp(1 - 0.5 x_t^1 - \frac{\left(1.5 - x_t^2\right) f_t^2}{f_t^1}), x_t^2 = \sin\left(2\pi t\right).
\end{aligned}
\tag{19}
$$

We observe noisy values, i.e. $y_t^1 \sim \mathcal{N}\left(y_t^1 | f_t^1, \sigma_n^2\right)$ and $y_t^2 \sim \mathcal{N}\left(y_t^2 | f_t^2, \sigma_n^2\right)$. We generated the training dataset of 10000 samples and the test dataset of 1000 samples. The training data were generated from the initial condition $[f_0^1, f_0^2] = [0.3, 0.8]$ and the test data from initial condition $[f_0^1, f_0^2] = [0.268, 0.4]$.

Model. The architecture of the proposed DNN is shown in Fig. 2b. The network consists of two hidden layers: the first layer is a B-LSTM layer, with a timestep of

$n_c = 100$, and a hidden state dimensionality of 5. The activation function for this layer is the hyperbolic tangent function, and the recurrent activation function is the sigmoid function. The second layer is a dense layer with a hyperbolic tangent activation function. The output layer is a linear transformation that corresponds to the output dimensionality of the synthetic problem. The parameters of the network were trained using the Adam optimizer to minimize the mean squared error [16]. The covariance function used in the GP component of the model is a combination of a linear kernel and a squared exponential kernel with Automatic Relevance Determination (ARD) property [8]. The lags used in the GP model are $n_a = n_b = 5$. A variational approximation method was used to reduce the computational complexity of the GP, using $m = 100$ pseudo-inputs [29]. The GP model was trained using MLL maximization with the Limited-memory Broyden-Fletcher-Goldfarb-Shanno (LBFGS) optimizer [20].

4.3   Silverbox

The Silverbox dataset is a real-world system, where data are collected from an electric circuit. The data set was used to validate the proposed approach on a real-world example.

Case Study 2. The benchmark represents a second-order linear time-invariant system with a third-degree polynomial static nonlinearity in the feedback loop. The system is described by the equation

$$m\frac{d^2y(t)}{dt^2} - d\frac{dy(t)}{dt} + ay(t) + by(t)^3 = u(t). \tag{20}$$

The training data were selected from samples 40,586 to 127,410, and the test data were taken from the first 40,495 samples [22]. The system was excited with a white Gaussian noise sequence filtered by a ninth-order discrete-time Butterworth filter with a cut-off frequency of 200 Hz for the test data. Additional information about the benchmark can be found in [30].

Model. The Silverbox model was defined similarly to the synthetic case, with the exception of the number of timesteps selected for training, which were set to $n_c = 300$. This change was found to improve the performance of the model during training. A variational inference approximation was used, as described in [29], with $m = 100$ pseudo-inputs and trained using the LBFGS optimizer. The covariance function used was the squared exponential kernel with ARD property.

4.4   Results and Disussion

Figure 3, Fig. 4, and Fig. 5 compare the performance of the GP-NARX models to the DNIF-GP model for different levels of added noise. In the synthetic case, the standardized mean squared error (SMSE) [17] between the ground truth latent states and the estimated latent states was computed, as the ground truth is known. In the Silverbox dataset, the data were already corrupted with a small amount of noise, and additional artificial noise was added. Since the ground truth was not available, the SMSE was computed from the noisy observations. The simulation was estimated with numerical approximation, where the latent values that do not belong to the training data were considered independent given the observations, as described in [19].

In the synthetic example experiment, presented in Fig. 3, it can be seen that the DNIF-GP model significantly outperforms the GP-NARX model for all noise levels. An interesting observation from the results is that the combination of a DNN and a GP model performs much better than using only a B-LSTM neural network to process the noisy observations in the first step, as shown in the right subfigure. This suggests that the GP can explain some of the residual variance left by the imperfectly trained neural network, assuming the neural network does not greatly distort the space in which the GP projects from inputs to outputs.

In the Silverbox experiment presented in Fig. 4 and Fig. 5, two scenarios are considered, as the benchmark includes an extrapolated part. Figure 4 shows the results on the part that is not extrapolated, and Fig. 4 shows the results on the original data set with the extrapolated part. In both cases, the DNIF-GP model outperforms the GP-NARX model when the noise level is relatively high. However, for the data set that includes the extrapolated part, the results are slightly worse with DNIF-GP than with the GP-NARX model for low noise levels. This could be because the DNN having worse extrapolating properties compared to a GP model, which assumes some prior knowledge about the smoothness of the latent function, which could aid in extrapolation. However, the DNIF-GP model still performs better than the stand-alone neural network, although the improvement is not as significant as in the synthetic case.

Figure 6 shows the comparison of the wall times for obtaining the ground truth simulation of the GP-NARX model and the simulation of the DNIF-GP model on the synthetic example. Since the latent states in the DNIF-GP model are estimated directly from the control inputs or exogenous variables, the simulation can be obtained statically, without the need for recursion. This allows for efficient and faster computation. The experiment was not repeated for the Silverbox case study, since the ground truth simulation response could not be obtained in the GP-NARX model due to the high computational complexity. Figure 6

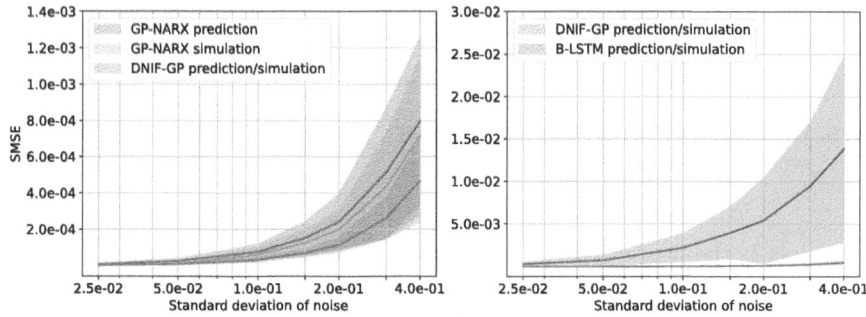

Fig. 3. SMSE with increasing the noise levels for the synthetic problem.

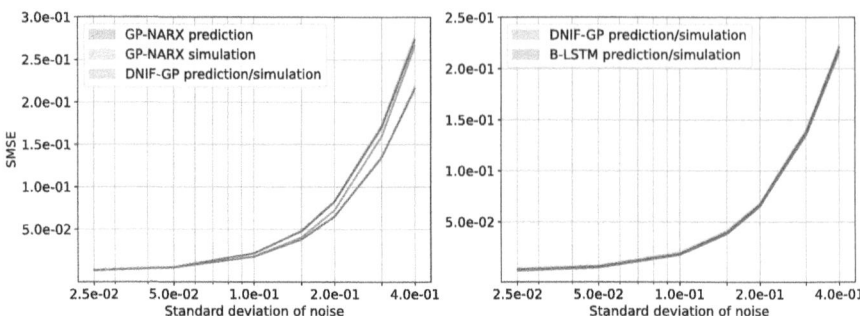

Fig. 4. SMSE with increasing the noise level for the Silverbox problem where the extrapolated region is discarded (from the 30000th timestep forward).

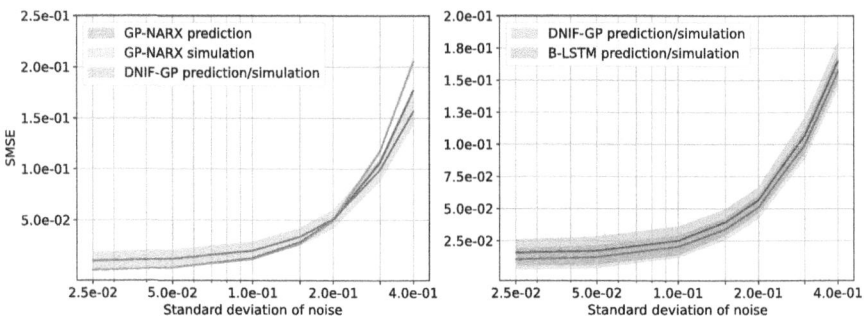

Fig. 5. SMSE with increasing the noise level for the Silverbox problem where the extrapolated region is retained.

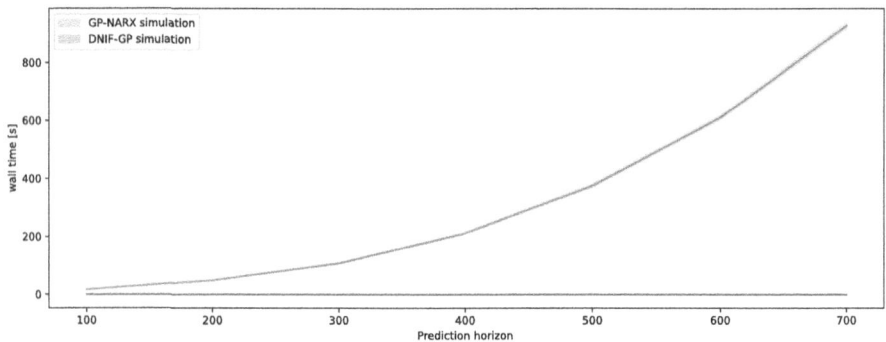

Fig. 6. Wall time comparison between GP-NARX simulation and DNIF-GP simulation for the synthetic problem.

shows that as the prediction horizon increases, the wall time also increases with cubic dependency. For larger prediction horizons, the ground truth simulation becomes computationally infeasible.

5   Conclusion

GP-NARX model belongs to a family of rich nonlinear, probabilistic, and non-parametric dynamical models that are well known for their simplicity in training. This allows for flexible and robust models that can quantify uncertainty systematically. Prior knowledge about the system can be incorporated into the model through the design of the covariance function. However, despite these benefits, GP-NARX models are also affected by certain issues. One such issue is error-in-variables problem, which arises with the introduction of noisy observation to the input regressor. Additionally, simulating the GP-NARX model is intractable and numerical approximations computationally demanding.

In this paper, we proposed a new model, named DNIF-GP, that addresses the limitations of the GP-NARX model by introducing a DNN pre-filtering step that estimates the latent states directly from the exogenous variables. This allows the latent states to be filtered and eliminates the need for recursion in simulation which can be obtained computationally efficiently and in closed-form. In summary, the advantages of the proposed model are the sequentially filtered latent states and computationally efficient and robust simulation.

However, one potential drawback of the proposed approach is that the DNN filter may introduce significant distortion to the data. If this distortion is not relatively smooth, but rather a step-like transformation, the GP model could have significant problems as GPs are known to not perform well with such data. The space of functions that can be modeled by a GP is assumed to be relatively smooth. The DNIF-GP, therefore, heavily relies on the neural network's ability to correctly filter the latent states. This problem was already highlighted in the Silverbox experiment, where extrapolation was considered, showing that when noise levels are low, the DNIF-GP model results were slightly worse than the GP-NARX model because of the neural network's worse extrapolation properties.

For future work, the method could be further validated on more data sets. Another interesting idea would be to consider training the DNN jointly with a GP in an end-to-end fashion. Other filters could be considered for the pre-filtering step.

Acknowledgements. The authors acknowledge the research core funding No. P2-0001, and Ph.D. grant funding for Tadej Krivec which were all financially supported by the Slovenian Research Agency.

References

1. 1 Abadi, M., Agarwal, A., Barham, P., et al.: TensorFlow: Large-scale machine learning on heterogeneous systems (2015). https://www.tensorflow.org/. software available from tensorflow.org
2. Beckers, T., Hirche, S.: Prediction with approximated Gaussian process dynamical models. IEEE Trans. Autom. Control 1–1 (2021). https://doi.org/10.1109/TAC.2021.3131988

3. Bijl, H., Schön, T.B., van Wingerden, J.W., Verhaegen, M.: System identification through online sparse Gaussian process regression with input noise. IFAC J. Syst. Control **2**, 1–11 (2017)
4. Candela, J.Q., Girard, A., Larsen, J., Rasmussen, C.E.: Propagation of uncertainty in Bayesian kernel models-application to multiple-step ahead forecasting. In: 2003 IEEE International Conference on Acoustics, Speech, and Signal Processing, 2003. Proceedings.(ICASSP'03), vol. 2, pp. II–701. IEEE (2003)
5. Damianou, A.C., Titsias, M.K., Lawrence, N.D.: Variational inference for uncertainty on the inputs of gaussian process models. arXiv preprint arXiv:1409.2287 (2014)
6. Deisenroth, M.P., Fox, D., Rasmussen, C.E.: Gaussian processes for data-efficient learning in robotics and control. IEEE Trans. Pattern Anal. Mach. Intell. **37**(2), 408–423 (2013)
7. Deisenroth, M.P., Turner, R.D., Huber, M.F., Hanebeck, U.D., Rasmussen, C.E.: Robust filtering and smoothing with Gaussian processes. IEEE Trans. Autom. Control **57**(7), 1865–1871 (2011)
8. Duvenaud, D.: Automatic model construction with Gaussian processes. Ph.D. thesis, University of Cambridge (2014)
9. Frigola, R.: Bayesian time series learning with Gaussian processes. Ph.D. thesis, University of Cambridge (2015)
10. Frigola, R., Chen, Y., Rasmussen, C.E.: Variational Gaussian process state-space models. In: Advances in Neural Information Processing Systems, vol. 27 (2014)
11. Frigola, R., Lindsten, F., Schön, T.B., Rasmussen, C.E.: Bayesian inference and learning in Gaussian process state-space models with particle MCMC. In: Advances in Neural Information Processing Systems, vol. 26 (2013)
12. Frigola, R., Rasmussen, C.E.: Integrated pre-processing for Bayesian nonlinear system identification with Gaussian processes. In: 52nd IEEE Conference on Decision and Control, pp. 5371–5376. IEEE (2013)
13. Girard, A., Rasmussen, C., Candela, J.Q., Murray-Smith, R.: Gaussian process priors with uncertain inputs application to multiple-step ahead time series forecasting. In: Advances in Neural Information Processing Systems, vol. 15 (2002)
14. Hochreiter, S., Schmidhuber, J.: Long short-term memory. Neural Comput. **9**(8), 1735–1780 (1997)
15. Kalman, R.E.: A new approach to linear filtering and prediction problems (1960)
16. Kingma, D.P., Ba, J.: Adam: a method for stochastic optimization. arXiv preprint arXiv:1412.6980 (2014)
17. Kocijan, J.: Modelling and Control of Dynamic Systems Using Gaussian Process Models. Springer, Cham (2016)
18. Kocijan, J., Petelin, D.: Output-error model training for Gaussian process models. In: Dobnikar, A., Lotrič, U., Šter, B. (eds.) ICANNGA 2011. LNCS, pp. 312–321. Springer, Heidelberg (2011). https://doi.org/10.1007/978-3-642-20267-4_33
19. Krivec, T., Papa, G., Kocijan, J.: Simulation of variational Gaussian process NARX models with GPGPU. ISA Trans. **109**, 141–151 (2021)
20. Liu, D.C., Nocedal, J.: On the limited memory BFGS method for large scale optimization. Math. Program. **45**(1), 503–528 (1989)
21. Liu, H., Ong, Y.S., Shen, X., Cai, J.: When Gaussian process meets big data: a review of scalable GPs. IEEE Trans. Neural Netw. Learn. Syst. **31**(11), 4405–4423 (2020)
22. Ljung, L., Zhang, Q., Lindskog, P., Juditski, A.: Estimation of grey box and black box models for non-linear circuit data. IFAC Proc. Vol. **37**(13), 399–404 (2004)

23. Matthews, A.G.D.G., et al.: GPflow: a Gaussian process library using TensorFlow. J. Mach. Learn. Res. **18**(40), 1–6 (2017), http://jmlr.org/papers/v18/16-537.html
24. Mattos, C.L.C., Damianou, A., Barreto, G.A., Lawrence, N.D.: Latent autoregressive Gaussian processes models for robust system identification. IFAC-PapersOnLine **49**(7), 1121–1126 (2016)
25. Mchutchon, A., Rasmussen, C.: Gaussian process training with input noise. In: Shawe-Taylor, J., Zemel, R., Bartlett, P., Pereira, F., Weinberger, K. (eds.) Advances in Neural Information Processing Systems. vol. 24. Curran Associates, Inc. (2011)
26. Quinonero-Candela, J., Girard, A., Rasmussen, C.E.: Prediction at an uncertain input for Gaussian processes and relevance vector machines-application to multiple-step ahead time-series forecasting (2003)
27. Quinonero-Candela, J., Rasmussen, C.E.: A unifying view of sparse approximate Gaussian process regression. J. Mach. Learn. Res. **6**, 1939–1959 (2005)
28. Schuster, M., Paliwal, K.K.: Bidirectional recurrent neural networks. IEEE Trans. Signal Process. **45**(11), 2673–2681 (1997)
29. Titsias, M.: Variational learning of inducing variables in sparse Gaussian processes. In: Artificial Intelligence and Statistics, pp. 567–574. PMLR (2009)
30. Wigren, T., Schoukens, J.: Three free data sets for development and benchmarking in nonlinear system identification. In: 2013 European Control Conference (ECC), pp. 2933–2938. IEEE (2013)
31. Williams, C.K., Rasmussen, C.E.: Gaussian processes for machine learning, vol. 2. MIT press, Cambridge (2006)
32. Wilson, A.G., Hu, Z., Salakhutdinov, R., Xing, E.P.: Deep kernel learning. In: Artificial Intelligence and Statistics, pp. 370–378. PMLR (2016)

Author Index

© The Editor(s) (if applicable) and The Author(s), under exclusive license
to Springer Nature Switzerland AG 2024
M. Mujica Mota and P. Scala (Eds.): EUROSIM 2023, CCIS 2033, pp. 363–365, 2024.
https://doi.org/10.1007/978-3-031-68438-8

SPRINGER NATURE

GPSR Compliance

The European Union's (EU) General Product Safety Regulation (GPSR) is a set of rules that requires consumer products to be safe and our obligations to ensure this.

If you have any concerns about our products, you can contact us on ProductSafety@springernature.com

In case Publisher is established outside the EU, the EU authorized representative is:

Springer Nature Customer Service Center GmbH
Europaplatz 3
69115 Heidelberg, Germany

The manufacturer's authorised representative in the EU is Springer
Nature Customer Service Centre GmbH, Europaplatz 3, 69115 Heidelberg,
Germany. If you have any concerns regarding our products, please
contact ProductSafety@springernature.com

Printed and bound by CPI Group (UK) Ltd, Croydon, CR0 4YY

24/04/2026

02096358-0011